METHODS IN MOLECULAR BIOLOGY™

For other titles published in this series, go to
www.springer.com/series/7651

Meiosis

Volume 1, Molecular and Genetic Methods

Edited by

Scott Keeney

Howard Hughes Medical Institute
Memorial Sloan-Kettering Cancer Center, New York, NY, USA

Humana Press

Editor
Scott Keeney
Howard Hughes Medical Institute
Memorial Sloan-Kettering
Cancer Center
Molecular Biology Program
1275 York Ave
New York, NY 10065-6007
USA
s-keeney@ski.mskcc.org

ISSN 1064-3745 e-ISSN 1940-6029
ISBN 978-1-934115-66-4 e-ISBN 978-1-59745-527-5
DOI 10.1007/978-1-59745-527-5
Springer Dordrecht Heidelberg London New York

Library of Congress Control Number: 2009926989

Printed on acid-free paper

Springer is part of Springer Science+Business Media (www.springer.com)

Preface

Each generation in a sexually reproducing organism such as a fly or a mouse passes through the bottleneck of meiosis, which is the specialized cell division that gives rise to haploid reproductive cells (sperm, eggs, spores, etc.). The principal function of meiosis is to reduce the genome complement by half, which is accomplished through sequential execution of one round of DNA replication followed by two rounds of chromosome segregation. Within the extended prophase between DNA replication and the first meiotic division in most organisms, homologous maternal and paternal chromosomes pair with one another and undergo homologous recombination, which establishes physical connections that link the homologous chromosomes until the time they are separated at anaphase I. Recombination also serves to increase genetic diversity from one generation to the next by breaking up linkage groups.

The unique chromosome dynamics of meiosis have fascinated scientists for well over a century, but in recent years there has been an explosion of new information about how meiotic chromosomes pair, recombine, and are segregated. Progress has been driven by advances in three main areas: (1) genetic identification of meiosis-defective mutants and cloning of the genes involved; (2) development of direct physical assays for DNA intermediates and products of recombination; and (3) increasingly sophisticated cytological methods that describe chromosome behaviors and the spatial and temporal patterns by which specific proteins associate with meiotic chromosomes. Often, the biggest insights have been obtained at the intersection between these historically separate approaches. New assays are being developed and classical methods are being applied in new ways, all in a diverse range of organisms from single-celled fungi, to plants, to animals both big and small.

These two volumes provide detailed protocols for genetic, molecular, and cytological methods for studying meiotic chromosome dynamics, in particular homologous recombination, higher-order chromosome structures, and chromosome segregation. Broad coverage is provided of many of the experimental organisms in which meiosis is often studied (e.g., the yeasts *Saccharomyces cerevisiae* and *Schizosaccharomyces pombe*, the nematode *Caenorhabditis elegans*, the fruit fly *Drosophila melanogaster*, the plant *Arabidopsis thaliana*, and the house mouse *Mus musculus*). Coverage is also provided of methods applicable to study of meiosis in humans, as well as in other organisms which often offer distinct experimental advantages or unique mechanistic or evolutionary insights.

These books are aimed at scientists in (at least) three main categories: (1) Students of meiosis who want to "cross over" and apply basic techniques from other disciplines to the biological problems in which they are most interested. For example, cytologists who want to connect microscopic observations to the underlying DNA events will find guides to molecular methods for studying recombination. Likewise, geneticists who want to connect mutant phenotypes to the details of chromosome dynamics will find cytological methods that allow them to do so. (2) Students of meiosis in one organism who wish to examine similar processes and conserved proteins in another organism, or who wish to gain a better understanding of both the possibilities and the limitations of methods for studying meiosis in other organisms. Ideally, the hope is that this book will play at least a small role in fostering crosstalk between investigators working in different experimental systems. (3) Students of basic chromosome biology in mitotically dividing cells who want to extend studies into meiosis, and, more generally, geneticists studying any biological process who find in hand a mutation that unexpectedly affects fertility and who need a handy primer on how to study this phenotype further. Indeed, this latter group has not been uncommon in the era of reverse genetic gene targeting in mouse.

The first volume, *Meiosis: Volume 1, Molecular and Genetic Methods*, is divided into two parts. The chapters in Part I of the first volume are devoted to genetic analyses, including methods for culturing and manipulating commonly used model organisms and methods for detecting and quantifying meiotic recombination or other aspects of chromosome dynamics. Part II of the first volume describes techniques for the direct study of meiotic recombination events through physical analysis of DNA or of protein-DNA interactions. Numerous approaches are described in budding and fission yeasts and in mouse and human.

The second volume, *Meiosis: Volume 2, Cytological Methods*, is subdivided for convenience by the general type of organism: fungi in Part I, plants and small animals (mostly invertebrates) in Part II, and mammals in Part III. Although there is some redundancy in certain aspects of the cytological methods, there are also many instances of species-specific differences—or even gender-specific differences within a species—that make it important to provide separate detailed protocols for different organisms. Cytology is a visual science, and the use of color and animation is often critical to the appropriate display of experimental results. As a consequence, Springer has graciously agreed to provide a companion CD for the second volume, on which can be found color versions of many of the figures that are reproduced in grayscale in the printed volume. The CD also contains a number of movies that illustrate results of real-time imaging of chromosome dynamics in yeasts, that show animations of three-dimensional reconstructions of meiotic nuclei, or that demonstrate particular experimental manipulations. A computer macro can also be found that provides analytical tools for evaluating the spatial distribution of cytological protein complexes on chromosomes. It is hoped that the contents of this CD will be a useful resource for readers of this volume.

I thank the many colleagues in the meiosis community for advice and suggestions on content, and the authors of chapters in these volumes for their hard work and willingness to share their expertise.

New York, NY
June 2008

Scott Keeney

Contents

Contributors

MOHAMMED F. F. ABDULLAH • *Mara Institute of Technology, Shah Alam, Malaysia*

THORSTEN ALLERS • *Laboratory of Biochemistry and Molecular Biology, Center for Cancer Research, National Cancer Institute, Bethesda, MD, USA*

NORMAN ARNHEIM • *Molecular and Computational Biology Program, University of Southern California, Los Angeles, CA, USA*

FRÉDÉRIC BAUDAT • *Institute of Human Genetics, Centre National de la Recherche Scientifique, Montpellier, France*

LUKE E. BERCHOWITZ • *Department of Biology and The Carolina Center for Genome Sciences, University of North Carolina, Chapel Hill, NC, USA*

VALÉRIE BORDE • *Institut Curie, Centre de Recherche, UMR7147-CNRS, Université Pierre et Marie Curie, Paris, France*

RHONA H. BORTS • *Department of Genetics, University of Leicester, Leicester, UK*

CYRIL BUHLER • *Center for Cancer Research, National Cancer Institute, Bethesda, MD, USA*

SEAN M. BURGESS • *Section of Molecular and Cellular Biology, College of Biological Sciences, University of California, Davis, CA, USA*

SABRINA M. CHEN • *Department of Pulmonary and Critical Care Medicine, The Permanente Medical Group, Sacramento, CA, USA*

GREGORY P. COPENHAVER • *Department of Biology and The Carolina Center for Genome Sciences, University of North Carolina at Chapel Hill, Chapel Hill, NC, USA*

VICTORIA E. COTTON • *Department of Genetics, University of Leicester, Leicester, United Kingdom*

BERNARD DE MASSY • *Institute of Human Genetics, Centre National de la Recherche Scientifique, Montpellier, France*

SUSAN L. ELROD • *Center for Excellence in Science and Mathematics Education, College of Science and Mathematics, California Polytechnic State University, San Luis Obispo, CA, USA*

TOMOYUKI FUKUDA • *Department of Life Sciences, Graduate School of Arts and Sciences, The University of Tokyo, and Shibata Distinguished Senior Scientist Laboratory, RIKEN Discovery Research Institute, Tokyo, Japan*

KENNETH J. HILLERS • *Biological Sciences, California Polytechnic State University, San Luis Obispo, CA, USA*

KOUJI HIROTA • *Department of Life Sciences, Graduate School of Arts and Sciences, The University of Tokyo, and Shibata Distinguished Senior Scientist Laboratory, RIKEN Discovery Research Institute, Tokyo, Japan*

EVA R. HOFFMANN • *MRC Genome Damage and Stability Centre, University of Sussex, Falmer, UK*

ELIZABETH A. HOUSWORTH • *Department of Mathematics and Department of Biology, Indiana University, Bloomington, IN, USA*

NEIL HUNTER • *Departments of Microbiology and Molecular & Cellular Biology, University of California, Davis, CA, USA*
RANDY W. HYPPA • *Division of Basic Sciences, Fred Hutchinson Cancer Research Center, Seattle, WA, USA*
ALEC J. JEFFREYS • *Department of Genetics, University of Leicester, Leicester, UK*
LEA JESSOP • *Laboratory of Biochemistry and Molecular Biology, Center for Cancer Research, National Cancer Institute, Bethesda, MD, USA*
LIISA KAUPPI • *Molecular Biology Program, Memorial Sloan-Kettering Cancer Center, New York, NY, USA*
SCOTT KEENEY • *Howard Hughes Medical Institute and Molecular Biology Program, Memorial Sloan-Kettering Cancer Center, New York, NY, USA*
FRANZ KLEIN • *Max F. Perutz Laboratories of the University of Vienna, Vienna, Austria*
KAZUTO KUGOU • *Department of Life Sciences, Graduate School of Arts and Sciences, The University of Tokyo, and Shibata Distinguished Senior Scientist Laboratory, RIKEN Discovery Research Institute, Tokyo, Japan*
JESSICA P. LAO • *Departments of Microbiology and Molecular & Cellular Biology, University of California, Davis, CA*
MICHAEL LICHTEN • *Laboratory of Biochemistry and Molecular Biology, Center for Cancer Research, National Cancer Institute, Bethesda, MD, USA*
DORIS LUI • *Section of Molecular and Cellular Biology, College of Biological Sciences, University of California, Davis, CA, USA*
MARCO ANTONIO MENDOZA • *Max F. Perutz Laboratories of the University of Vienna, Vienna, Austria*
CELIA A. MAY • *Department of Genetics, University of Leicester, Leicester, UK*
HAJIME MURAKAMI • *Molecular Biology Program, Memorial Sloan-Kettering Cancer Center, New York, NY, USA*
MATTHEW J. NEALE • *MRC Genome Damage and Stability Centre, University of Sussex, Falmer, UK*
ALAIN NICOLAS • *Institut Curie, Centre de Recherche, UMR7147-CNRS, Université Pierre et Marie Curie, Paris, France*
STEVE D. OH • *Departments of Microbiology and Molecular & Cellular Biology, University of California, Davis, CA, USA*
KUNIHIRO OHTA • *Department of Life Sciences, Graduate School of Arts and Sciences, The University of Tokyo, and Shibata Distinguished Senior Scientist Laboratory, RIKEN Discovery Research Institute, Tokyo, Japan*
JING PAN • *Molecular Biology Program, Memorial Sloan-Kettering Cancer Center, New York, NY, USA*
SILVIA PANIZZA • *Max F. Perutz Laboratories of the University of Vienna, Vienna, Austria*
JIAN QIN • *Fluidigm Corporation, South San Francisco, CA, USA*
KATJA SCHWARTZ • *Department of Genetics, Stanford University, Stanford, CA, USA*
ROBERT SHROFF • *Center for Cancer Research, National Cancer Institute, Bethesda, MD, USA*
ELIZABETH O. SHUSTER • *Academic Advising, General College and College of Arts and Sciences, University of North Carolina, Chapel Hill, NC, USA*

GERALD R. SMITH • *Division of Basic Sciences, Fred Hutchinson Cancer Research Center, Seattle, WA, USA*

FRANKLIN W. STAHL • *Institute of Molecular Biology, University of Oregon, Eugene, OR, USA*

JAICHANDAR SUBRAMANIAN • *National Institutes of Health, Bethesda, MD, USA*

ANNE M. VILLENEUVE • *Department of Developmental Biology, Stanford University School of Medicine, Stanford, CA, USA*

TAKATOMI YAMADA • *Department of Life Sciences, Graduate School of Arts and Sciences, The University of Tokyo, and Shibata Distinguished Senior Scientist Laboratory, RIKEN Discovery Research Institute, Tokyo, Japan*

Section I

Genetic Methods for Studying Meiotic Recombination and Chromosome Dynamics

Chapter 1

Interaction of Genetic and Environmental Factors in *Saccharomyces cerevisiae* Meiosis: The Devil is in the Details

Victoria E. Cotton, Eva R. Hoffmann, Mohammed F.F. Abdullah, and Rhona H. Borts

Abstract

One of the most important principles of scientific endeavour is that the results be reproducible from lab to lab. Although research groups rarely redo the published experiments of their colleagues, research plans almost always rely on the work of someone else. The assumption is that if the same experiment were repeated in another lab, results would be so similar that the same interpretation would be favoured. This notion allows one researcher to compare his/her own results to earlier work from other labs. An essential prerequisite for this is that the experiments are done in identical conditions and therefore the methodology must be clearly stated. While this may be scientific common sense, adherence is difficult because "standard" methods vary from one laboratory to another in subtle ways that are often not reported. More importantly, for many years the field of yeast meiotic recombination considered typical differences to be innocuous. This chapter will highlight the documented environmental and genetic variables that are known to influence meiotic recombination in *Saccharomyces cerevisiae*. Other potential methodological sources of variation in meiotic experiments are also discussed. A careful assessment of the effects of these variables, has led to insights into our understanding of the control of recombination and meiosis.

Key words: meiosis, yeast, recombination, temperature, nutrients.

1. Introduction

Understanding the interplay between recombination, chromosome dynamics, and chromosome segregation during meiosis relies on the integration of data obtained from genetic, cytological, and physical analyses. Genetic studies are based on tetrad dissection and "random spore" analysis of endproducts, while cytological and physical studies follow progression temporally from a relatively

Scott Keeney (ed.), *Meiosis, Volume 1, Molecular and Genetic Methods, vol. 557*

DOI 10.1007/978-1-59745-527-5_1 Springerprotocols.com

synchronous (e.g. SK1 and Y55) or relatively asynchronous (e.g. BR) population of *S. cerevisiae* cells undergoing meiosis. The genetic studies rely on the successful completion of meiosis, sporulation, and the viability of the resulting spore colonies. When these factors are optimal, tetrad analysis is a very powerful tool for analysing meiotic recombination models and other phenomena such as interference. For example, the first unambiguous observation of non-Mendelian segregation occurred in tetrads from budding yeast *(1)* and it was analysis of the fungal data that led to the double-strand break (DSB) repair model *(2–4)*. It was also genetic analysis that first challenged the proposition that the two constituent Holliday junctions that make up "double Holliday junctions" are resolved independently of each other to generate both crossovers and non-crossovers *(5)*. Problems arise when meiotic mutants are less viable than wild-type controls. Although the pattern of spore death itself can provide some information on the nature of the problem in mutants, the intimate relationship between crossing-over and chromosome segregation might lead experimentalists to underestimate the defect of these mutants because only survivors that presumably had relatively high levels of recombination are considered. To overcome viability issues, random spore analysis is often employed, but the same scientific concern remains: that only a sub-population of meiotic cells and their resulting gametes, where recombination occurred at "higher" frequencies, are viable and therefore included in the analysis.

In contrast, physical analysis allows an unbiased estimate of end-products. This is particularly useful in the analysis of mutants that fail to make viable progeny *(6–11)*. Moreover, physical analyses also allow intermediates to be identified and have thus provided definitive evidence to support many aspects of the double-strand break repair model *(2, 4, 9, 10, 12–17)*. Physical analyses have also led to modifications of this model *(18)*. The inherent asynchrony of meiotic cultures *(19)* and the transient nature of intermediates can make quantification of intermediates and endpoints technically challenging. In particular, when no endpoints such as crossovers are generated, accurate interpretation of intermediate formation is impossible. For example, a mutant with wild type-like appearance and disappearance of an intermediate but with decreased levels of endproducts (crossovers) could denote fewer intermediates with a prolonged half-life or the same number of intermediates that are processed, but not to a crossover. Finally, physical studies measure total recombination product. To classify molecules as crossovers requires a number of assumptions, for example, that only a single DSB influences the recombination event. One also has to assume and verify genetically that the two markers spanning the DSB are located sufficiently distant from the initiating lesion such that they are not converted. If either marker is converted, then the recombinant non-crossover event would erroneously be classified as a crossover. It can be unclear whether these assumptions are always tested. For example, quantification of crossover products of mutants that significantly increase the gene conversion tract length (e.g. *mms4* and *mus81*) *(20, 21)* may complicate physical assessment of crossover frequencies. Where possible, a combination of genetic and physical analysis is therefore needed.

Further complications arise from the integration of cytological studies that describe the morphological changes in chromosome structure, for example the assembly of the synaptonemal complex and chromosomal movements, with DNA events. For example, the highly synchronous SK1 strain used in physical assays is not as amenable to cytological assessment by nuclear spreading as BR or Y55 (*see* for example, **Chapter 1** of Vol. 2). Correlating the timing of molecular and cytological events can therefore be problematic.

Integrating data from the three different types of experimental analyses is also hampered by genetic and environmental heterogeneity in experimental design. For example, there is still significant controversy over some very basic aspects of recombination models such as whether DSB repair is inherently one or two sided, and what the source(s) of gene conversion gradients are *(22–28)*. Much of the controversy rests on conflicting genetic data obtained in different laboratories, the use of different genetic markers [e.g. palindromes *(22, 23)* versus single nucleotide polymorphisms *(24, 25)*], genetic heterogeneity of the various strain backgrounds employed, and the different environmental conditions (e.g. sporulation media and temperature). Thus, even within a single type of analysis, there are a tremendous number of sources of variation. Therefore, knowing the sources of experimental variation is vital to understanding what aspects of the data are influenced by the environment and which are inherent to the organism. This knowledge is essential for optimal refinement of recombination models. Here, we highlight some of the documented sources of variation such as strain background, temperature, and sporulation conditions that affect meiotic phenotypes.

2. Sources of Variability

2.1. Genetic Heterogeneity (Strain Background)

Genetic heterogeneity is one of the most important factors underlying penetrance of disease and other traits. It is therefore important to understand the role that strain background can play. In *S. cerevisiae*, there are a number of strain backgrounds that are commonly used for meiotic studies and they vary significantly in their meiotic phenotypes. The first physical analyses *(6, 7)* were performed in strains that were congenic with Y55, which sporulates quite synchronously although it is slow to enter meiosis *(29)*.

However, it was found shortly thereafter that SK1 sporulates more synchronously and the majority of subsequent physical analyses have been performed in that strain background *(8, 13, 30)*. Both Y55 and SK1 sporulate very efficiently (75–80% *(29)* and 95% *(31)*, respectively), while S288c sporulates poorly (12–15%) *(32)*, although this can be raised to as much as 60% depending on the conditions used *(33)* (*see* **Chapter 2** in this volume). Y55 and SK1 have excellent spore viability **(Fig. 1.1)** as does S288c. The "BR" strain background, which has been used extensively for genetic and cytological studies, generates about 60% tetrads *(34)* with good viability **(Fig. 1.1)**.

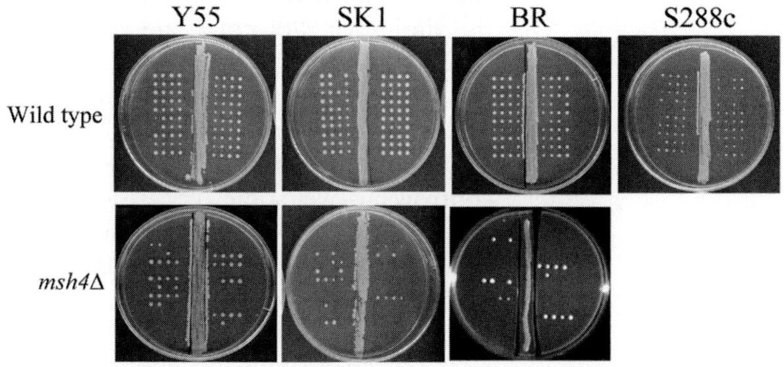

Fig. 1.1. The penetrance of *msh4* mutation in different strain backgrounds. The strain designations are given across the top while the genotype is indicated to the *left*. All strains were sporulated on the same complete KAc plate (Base C, **Table 1.2**, and McCusker/Borts/Supplement Mix) at 23°C. Twenty tetrads were dissected for each strain. The viability of the BR strain (20%) is significantly different from that of Y55 (44%, $p = 0.002$) but not SK1 (30%). The lack of significance is due solely to small numbers.

Historically, S288c has been used because of the availability of markers, the ease with which it can be transformed, and more recently because it is the fully sequenced strain. For example, the gene deletion library in budding yeast was created in S288c. However, prior to the era of straight-forward molecular manipulation of strains, genetic studies were performed in highly mixed genetic backgrounds. Generally, different strains were selected on the basis of marker availability and/or properties (e.g. sporulation). They were then crossed or backcrossed to obtain useful strains. Once transformation became possible, investigators started using congenic pairs of haploid parents derived from backcrossing to a common parent such as Y55 *(26, 35–38)* or S288c *(22, 39–41)*. Others chose to use different S288c congenic isolates *(42, 43)*. These were subsequently transformed to mutate genes or to introduce new markers. In these experiments, the haploid parents were derived from different backcrossed lineages that were used to create congenic diploids. Although all diploids that were subsequently compared were "isogenic" to each other, they were not identical to the pure diploids of the parental strains. For example the "S288c" line developed to analyse gene conversion at *ARG4* sporulates to about 50% *(43)* and has 99% spore viability while the pure S288c sporulates poorly (see above). The reasons for differences in sporulation are beginning to be elucidated and indicate that the genetic architecture of this trait is complex *(32, 44)*. However, a substantial fraction of the differences seem to be attributable to the control of entry into meiosis *(32, 44, 45)*. The complete sequencing of a number of the strains discussed here *(46)*, as well as other strains will reveal how much genetic variation exists and will allow further analysis of how this variation is influencing our understanding of the meiotic process.

Perhaps more important for the purpose of this review are differences in the meiotic phenotypes of the various strain backgrounds and mutants therein. These include response to temperature *(11, 47)*, genetic map distances *(48–54)* and frequency of gene conversion *(24, 41, 55)*. An understanding of the genetic basis of this variability could be extremely informative as the phenotype of a single gene deletion might be modified by the presence of other genetic differences. Indeed it should soon be technically feasible to perform quantitative trait loci mapping experiments similar to those that identified the basis for sporulation defects.

One phenotype that varies tremendously between different strains is the spore viability of meiotic recombination mutations (e.g. *msh4*, *dmc1*, *zip1*). Because the assessments of the different strains are also carried out under different environmental conditions, it can be difficult to discern the relative contribution of genetic heterogeneity. However, experiments where the different strains have been assessed in parallel [illustrated for spore viability of *msh4*Δ strains in **Fig. 1.1** and for other meiotic mutants in Fig. 7 of reference *(11)* and in reference *(56)*], demonstrate significant contribution of strain background to efficiency of sporulation as well as spore viability. Since genetic map distances and gene conversion frequencies are estimated from viable offspring, distinguishing whether viability affects estimates of map distances in mutants or the other way around, is impossible from genetic analysis alone.

Even amongst wild-type strains, however, there are differences in the map distances of commonly used genetic intervals. For example, the map distance from *HIS4* to *LEU2* varies ~2-fold, from 9–20 cM in SK1 *(51, 57–59)* to 19–25 cM in Y55 *(52, 53)*, and 24 cM in S288c *(54)*. The map differences of two intervals on chromosome VII, *MET13-TRP5* and *TRP5-CYH2* also vary (by about 1.4-fold) between Y55 and SK1. Among the possible explanations for these differences is that of strain background, differing auxotrophies, sporulation media and temperature. Until we understand the genetic/molecular nature of these strain differences, we will not be able to account for this variation. Furthermore, the genetic/molecular basis of this variation may explain the controversy over such things as the effect of *mlh1* and *mlh3* mutants on interference *(53, 60, 61)*. A less well documented, but nonetheless interesting, example of a potential effect of strain background on estimates of interference can be seen in the studies of Stahl and Hollingsworth *(57, 62)*, both of which explored the relationship between chromosome size and interference. In the Stahl *(62)* experiments, decreasing chromosome size increased interference as had been suggested previously *(63)*. In the Hollingsworth experiments *(57)*, no chromosome size effect was detected. Among the possible explanations for this difference is that of strain background, differences in sporulation media, and temperature.

2.2. External Environment

Estimating the effect of genetic heterogeneity is complicated by environmental influences on meiotic phenotypes. For example, the important contribution of genetic heterogeneity to the very different meiotic progression phenotypes of *dmc1* mutants in BR and SK1 was only resolved after sporulating both strains on the same medium in the same lab *(11, 57)*. Importantly, there are a number of well-documented cases where external environmental factors such as temperature *(11, 33, 47)*, potassium acetate (KAc) concentration *(11)*, and nutritional factors affect meiosis *(11)* and/or recombination *(11, 64)*.

2.2.1. Temperature

Y55 is unusual in its ability to sporulate at 37°C whereas SK1 sporulates well up to 33°C. Analysis of gene conversion of *his4-atc* indicates a 7-fold decrease in gene conversion events **(Table 1.1)** and crossovers (our unpublished data) between 23°C and 37°C. The decrease in conversion at *HIS4* can be attributed to decreases in double-strand breaks **(Fig. 1.2A)**. Furthermore, the effects of temperature on double-strand break formation are not uniform across the genome so there are locus and interval-specific responses to temperature **(Fig. 1.2B)**. Interestingly, conversion of the *his4-atc* allele also demonstrates temperature dependency in S288c-like strains. Isogenic strains carrying this allele displayed a small but significant difference in gene conversion frequency ($p<0.01$, Fisher exact test) when sporulated at 18°C or 30°C.

Other important aspects of meiotic recombination are also influenced by temperature. For example, crossover formation in all of the "ZMM"/"SIC"/*MSH4*-epistasis group mutants *(65)* is decreased from ~20% at 23°C to ~5% at 33°C. These observations have led to the hypothesis that there is a barrier to recombination, which is overcome less efficiently at 33°C in the absence of "ZMM" genes. At 33°C, the mutants also show decreased meiotic progression efficiency *(11)*. Therefore, temperature has been hypothesised to influence the "mode" of recombination and subsequent meiotic progression *(11)*. The temperature at which the

Table 1.1
Gene conversion at *his4-atc*

			% Gene conversion (Number of gene conversion events/Total number of tetrads)					
		Medium:	**Minimal**	**Supplemented**		**Complete**		
Strain	**Phenotype**	**Temperature:**	**23°C**	**18°C**	**30°C**	**23°C**	**30°C**	**37°C**
S288c[a] (PD84)	$Ade^- Ura^- Lys^+$		–	45[b] (302/677)	36[c] (37/102)	–	–	–
S288c (JS102)	$Ade^- Ura^- Lys^+$		–	52[b] (134/256)	–	–	–	–
Y55 (ERY103)	$Ade^+ Ura^- Lys^-$		22.6 (45/199)	–	–	14.0 (243/1731)	11.3 (44/390)	2.2 (8/372)
Y55[d] (VCD97)	$Ade^- Ura^- Lys^-$		48.2 (105/218)	–	–	26.7 (113/423)	–	–

[a]S288c strains were sporulated on medium A **(Table 1.2)** supplemented with 6 μgmL adenine *(41)*
[b]Data from Detloff et al. *(41)*
[c]Data from Alani et al. *(55)*
[d]Data from *(64)*. Minimal plates are KAc supplemented with 16 μg/mL uracil, 50 μg/mL lysine, and 41.67 μg/mL adenine as needed. Complete plates are Base C supplemented with the McCusker/Borts/Haber nutrient mix.

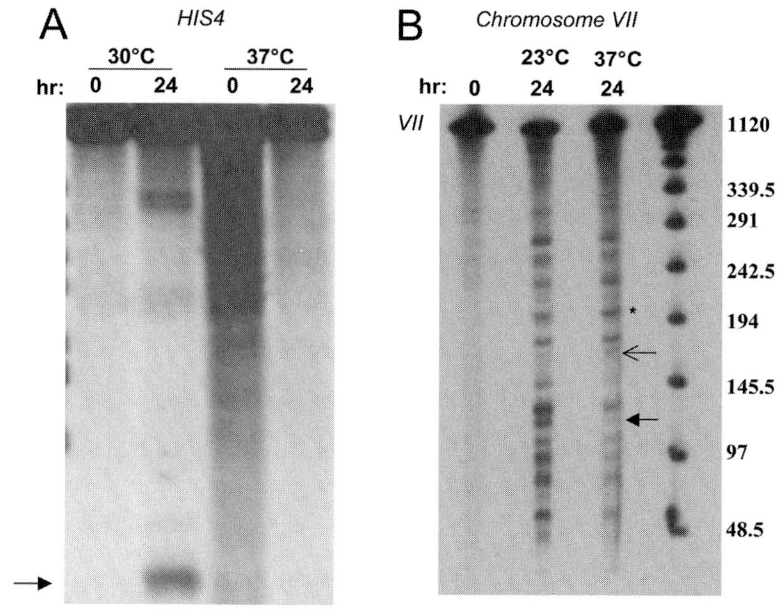

Fig. 1.2. Changes in double-strand break formation in response to temperature. (**A**) Double-strand breaks at *HIS4* (indicated by the *arrow*) are very temperature sensitive and mirror the decreased gene conversion frequency observed at 37°C. (**B**) Double-strand breaks on whole chromosomes can be detected using CHEF gel and Southern blot analysis. The accumulation of breaks is visualised in *rad50S* mutants after 0 or 24 h in sporulation medium at either 23 or 37°C. The size standard in the fourth lane is ~50 kb ladder. Double-strand breaks decrease the chromosome size causing the fragments to migrate below the intact chromosomes. Double-strand break frequencies are non-uniformally changed at 37°C. For instance, some hotspots (marked by the *closed arrowhead*) are decreased, while other hotspots marked by * remain the same, and some may even increase (*open arrowhead*).

various strain backgrounds switch from one mode to the other is also postulated to vary [Fig. 7 in reference *(11)*] and to occur as a binary switch on a *per* cell basis, thus accounting for the apparent differences in meiotic phenotypes. There are however, other explanations for these phenotypes including the speed of sporulation, genetic differences in checkpoint genes and differential temperature sensitivity of relevant proteins.

2.2.2. Nutrient Composition

It has been known since the late 1960s that nitrogen and carbon metabolism influence entry into meiosis [reviewed in *(66)*]. However, it has not previously been documented that the nutritional environment can affect recombination frequencies, although there have been hints in the literature. For example, the data presented concerning gene conversion at *his4-xho* (96 bp from the start codon) *(37, 38)* and those presented here for *his4 -atc* (G-C at the 3rd base pair of the start codon) *(64)* were significantly lower than reported previously for alleles in similar positions *(41, 55)* (**Table 1.1**). Although these experiments were carried out at a different temperature, they also varied in two additional parameters: phenotype for amino acid and/or purine auxotrophy and the sporulation media. As can be seen by the comparison of the Y55 strains, complete nutrient supplementation significantly lowers gene conversion frequency between 1.6 and 1.8-fold (ERY103, VCD97 respectively, **Table 1.1**). Thus, the apparent discrepancy between the S288c (JS102) and Y55 (VCD97) gene conversion data (52 vs. 26.7%) can be attributed to sporulation conditions. As will be discussed below, transcription factors play an important role in the initiation of recombination. We suggest that the effect of media on recombination is a result of modulating transcription factors in response to nutrient conditions.

2.3. Media

2.3.1. Solid Media

Two types of media are used, solid and liquid. Solid media are commonly used for ease in generating spores for tetrad dissection. Thus, most of the genetic data relating to gene conversion and crossing over come from slow, asynchronous meiosis. There are a number of basic types of solid media in use, those based on minimal amounts of supplements and those based on complete nutrient mixtures **(Table 1.2)**. Minimal sporulation medium consists of 1% potassium acetate (KAc) and 2% agar. More commonly, media rely on a synthetic mixture containing between 0.1 to 0.25% bacto-yeast extract, 1 to 2% KAc, and 0.05% D-glucose **(Table 1.2)**. This is then supplemented with nutrients for any auxotrophies the strain might have at levels between 20 to 100% of the "normal" levels of nutrients used **(Table 1.3)**. Similarly, complete sporulation medium is supplemented with 25–200% of the amount of "standard" complete mix **(Table 1.3)**. This leads to

Table 1.2
Components of solid media

Sporulation media

Minimal[a]	**Supplemented**[b]		
	A[a]	**B**[b]	**C**[c]
1% Potassium acetate	1% Potassium acetate	1.5% Potassium acetate	2% Potassium acetate
–	0.1% Bacto yeast extract	0.25% Bacto yeast extract	0.22% Bacto yeast extract
–	0.05% Dextrose	0.05% Dextrose	0.05% Dextrose
2% Bacto Agar	2% Bacto Agar	2% Bacto Agar	2.0% Bacto Agar

[a]From Sherman *(81)*
[b]From Kassir and Simchen *(67)*
[c]From Campbell, Fogel and Lusnak *(82)*

Table 1.3
Supplements for sporulation media

Nutrient	Amount (mg/L)							
	Lichten/ Goldman[a]	McCusker/Borts/ Haber[b]	Sherman[c]	Roeder/ Rockmill[a,d]	Alani/ Kleckner[a,c]	Simchen/ Kassir[e]	Fogel[f]	Hollingsworth[a]
Adenine	29.5	31.6	20.0	10.0	32.64	40.0	10.0	10.0
Arginine	29.5	31.6	20.0	50.0	32.64	20.0	50.0	10.0
Asparagine	147.5	158.0	100.0	80.0	–	–	–	–-
Histidine	29.5	31.6	20.0	20.0	32.64	20.0	20.0	10.0
Leucine	88.5	47.5	30.0	100.0	97.96	20.0	50.0	16.0
Lysine	44.2	47.5	30.0	50.0	48.96	20.0	50.0	16.0
Methionine	29.5	31.6	20.0	20.0	32.64	20.0	50.0	10.0
Phenylalanine	73.7	79.0	50.0	50.0	81.6	100.0	50.0	25.0
Threonine	294.9	316	200.0	100.0	326.52	350.0	600.0	16.0

Tryptophan	29.5	31.6	20.0	50.0	32.64	20.0	50.0	10.0
Tyrosine	44.2	47.5	30.0	50.0	48.8	40.0	50.0	16.0
Uracil	29.5	31.6	20.0	20.0	32.64	40.0	20.0	10.0
Isoleucine	–	–	30.0	50.0	–	–	50.0	16.0
Valine	–	–	150.0	140.0	–	–	–	16.0
Glutamate	–	158.0	100.0	–	–	–	–	–
Base composition[g]	C	C	A	–	A	B	C	A

[a]Personal communication.
[b]McCusker (http://www.duke.edu/web/microlabs/mccusker/) and Abdullah and Borts *(38)*.
[c]Used at ¼ × of total amount *(81)*.
[d]BIO 101 commercial nutrient mixture (www.qbiogene.com); base C, except with 0.2% Bacto yeast extract.
[e]*(67)*.
[f]Used at 2 × total amount *(82)*.
[g]*See* **Table 1.2**

extreme variability of sporulation conditions. Exactly how these factors influence parameters of meiosis such as sporulation frequency, gene conversion and crossing over is for the most part unclear.

2.3.2. Liquid Media

Liquid sporulation is used when a synchronous meiosis is required (*see* **Chapters 2, 9, 10, and 14** in this volume for examples of current protocols) such as when measuring intermediates and products of recombination. This is particularly useful for measuring recombination in mutant strains where spore viability is compromised. To obtain synchrony, it is necessary to pre-grow the cells under limiting nutrient conditions prior to transfer to sporulation medium. There are two types of common pre-growth media, YPA *(13, 67)* and buffered ones such SPS *(68)* and BYTA *(69)* **(Table 1.4)**. After washing, the cells are resuspended in sporulation medium which contains variously 0.3% *(8)*, 0.4% *(11)* (both with 0.02% raffinose), 1% *(68)* or 2% KAc. As above, liquid sporulation media may be supplemented with required nutrients. The concentration of KAc affects meiotic recombination and progression although the reason for this is not known *(11)*.

Table 1.4
Components of Liquid Pre-sporulation Media

YPA[a]	SPS[b]	BYTA[c]
2% Potassium acetate	1% Potassium acetate	1% Potassium acetate
2% Peptone	1% Peptone	–
1% Yeast extract	0.5% Yeast extract	1% Yeast extract
–	0.17% Nitrogen base	–
–	0.5% Ammonium sulphate	–
–	0.05 M Potassium biphthalate	50 mM Potassium phthalate
–	–	2% Bacto tryptone

[a]From Kassir and Simchen *(67)*
[b]From Resnick *(68)*
[c]From Blitzblau et al. *(69)*

2.4. Intracellular Environment

It has been known for some time that recombination at *HIS4* is dependent on the presence of three transcription factors, Rap1p, Bas1p/Bas2p and Gcn4p *(38, 70, 71)* and that these facilitate double-strand break formation *(72, 73)*. An indirect effect of auxotrophy on the frequency of recombination was first shown by Abdullah and Borts *(38)* and is supported by the data presented here (**Table 1.1**, ERY103 vs. VCD97). We have previously demonstrated that auxotrophy for lysine or adenine and prototrophy for leucine elevates gene conversion at *HIS4 (38)*. This effect is mediated by Gcn4p and we hypothesised that auxotrophies affect the intracellular levels of Gcn4p *(38)*. Gcn4p regulates between 500 and 1,000 genes involved in response to starvation for amino acids, purines and glucose as well as other stresses such as high salinity *(74)*. Thus, we suggested that the effects of auxotrophy are genome wide and can profoundly affect the genetic map *(38)*. In support of this, genome-wide analysis of DSB formation by Bas1/2p found 153 "hotspots" that appear to be regulated by Bas1/2p *(73)*. Of these, DSB activity is decreased at 68 hotspots by Bas1/2p, indicating that the effects of transcription factors on recombination are complex.

The effects of auxotrophies are important as they have led to the misattribution of a recombination phenotype (frequency of gene conversion) to deletion of the gene assessed, when the effect was actually due to the insertion of *LEU2* to make the gene deletion *(37, 38, 75)*. Another example comes from the use of the *HOM3-TRP2* interval to measure crossing over in a number of meiotic mutant strains. Recombination in this interval appears to be independent of the *MSH4* gene in strains that are *ade2 (49, 51)*. However, in the same genetic background when the strains are *ADE2*, crossing over is the expected two-fold lower in the *msh4* strain *(51)*. This could suggest that there is a link between Bas1p stimulated recombination and the dependence of crossing over on *MSH4*. Furthermore, the existence of an interval whose crossing over is entirely independent of Msh4p has profound implications for our understanding of crossover regulation.

An alternative, but not mutually exclusive interpretation, is that some of the enzymes involved in intermediary metabolism have direct roles in recombination. One such example is Fpr3, which maintains the recombination checkpoint independently of its function as a proline isomerase *(76)*. Both of these hypotheses might explain why genes involved in amino acid biosynthesis such as *ARG3*, *URA4*, and *TRP2*, have been identified in genome-wide screens designed to find chromosome segregation genes *(77, 78)*.

3. Conclusions

We are only beginning to unravel the complex interactions between genetic and environmental factors that determine the outcome of meiosis and there is a wealth of knowledge waiting to be teased out from these types of studies. However, the devil *is* in the details. As Louis Pasteur noted "chance favours only the prepared mind" *(79)*. Careful attention to apparently subtle differences in methodology, such as those discussed here, has led to at least two important observations. First, basic metabolism and temperature can influence recombination frequencies. If such influences are seen genome-wide, then the DSB landscape and the crossover distribution may be profoundly altered by environmental stimuli. This has implications for our understanding of meiosis not only in yeast but in other organisms and perhaps even man. For example, understanding the influence of temperature on meiosis in yeast may shed some light on why human spermatogenesis is so sensitive to temperature. Understanding these influences may also guide genetic and epidemiological studies of the genetic and environmental causes for human infertility. Second, despite substantial differences in DSB distribution at elevated temperature (**Fig. 1.2**), cells are resilient and manage not only to complete meiosis, but also to produce viable progeny. This may suggest that in wild-type cells at least, mechanisms that ensure recombination and segregation have evolved to be robust and efficient over a wide range of environmental conditions. This conclusion is supported by two recent observations. Using a series of *spo11* alleles, Martini et al. *(59)* demonstrated that cells experiencing only 20–30% of normal DSB frequencies, nevertheless had relatively high frequencies of crossovers and viability (>70%). In another study, Borde et al. *(80)* altered DSB hotspot sites by fusing Spo11 to the DNA-binding domain of Gal4p. This targeted Spo11p and in most cases DSBs to other genomic locations. However, as with the decreasing DSB levels, the gametes produced are viable. These two genetic observations support our proposition that cells are adapted to cope with a wide range of environmental conditions. One implication of this hypothesis is that mutants may be affected not only in specific aspects of recombination, but may also have a much narrower window in which their response to environmental fluctuations is compatible with meiotic progression and success.

Acknowledgments

We would like to thank all of our colleagues for their prompt responses to our many requests for the details of their sporulation protocols. We would also like to thank Rebecca Keelagher, Amit Dipak Amin and Robert Mason for technical assistance. R.H.B is a Royal Society/Wolfson Foundation Research Merit Award Holder. E.R.H. is a Royal Society Dorothy Hodgkin Fellow. The unpublished work was supported by the Wellcome Trust and the MRC.

References

1. Fogel, S. and Hurst, D. D. (1967) Meiotic gene conversion in yeast tetrads and the theory of recombination. *Genetics* **57,** 455–481.
2. Resnick, M. A. (1976) The repair of double-strand breaks in DNA: a model involving recombination. *J. Theor. Biol.* **59,** 97–106.
3. Orr-Weaver, T. L. and Szostak, J. W. (1985) Fungal recombination. *Microbiol. Rev.* **49,** 33–58.
4. Szostak, J. W., Orr-Weaver, T. L., Rothstein, R. J., and Stahl, F. W. (1983) The double-strand-break repair model for recombination. *Cell* **33,** 25–35.
5. Gilbertson, L. A. and Stahl, F. W. (1996) A test of the double-strand break model for meiotic recombination in *Saccharomyces cerevisiae. Genetics* **144,** 27–41.
6. Borts, R. H., Lichten, M., Hearn, M., Davidow, L. S., and Haber, J. E. (1984) Physical monitoring of meiotic recombination in *Saccharomyces cerevisiae. Cold Spring Harbor Symp. Quant. Biol.* **49,** 67–76.
7. Borts, R. H., Lichten, M., and Haber, J. E. (1986) Analysis of meiosis-defective mutations in yeast by physical monitoring of recombination. *Genetics* **113,** 551–567.
8. Alani, E., Padmore, R., and Kleckner, N. (1990) Analysis of wild-type and *rad50* mutants of yeast suggests an intimate relationship between meiotic chromosome synapsis and recombination. *Cell* **61,** 419–436.
9. Schwacha, A. and Kleckner, N. (1994) Identification of joint molecules that form frequently between homologs but rarely between sister chromatids during yeast meiosis. *Cell* **76,** 51–63.
10. Hunter, N. and Kleckner, N. (2001) The single-end invasion: an asymmetric intermediate at the double-strand break to double-Holliday junction transition of meiotic recombination. *Cell* **106,** 59–70.
11. Borner, G. V., Kleckner, N., and Hunter, N. (2004) Crossover/noncrossover differentiation, synaptonemal complex formation, and regulatory surveillance at the leptotene/zygotene transition of meiosis. *Cell* **117,** 29–45.
12. Sun, H., Treco, D., and Szostak, J. (1991) Extensive 3′-overhanging, single-stranded DNA associated with the meiosis-specific double-strand breaks at the *ARG4* recombination initiation site. *Cell* **64,** 1155–1161.
13. Cao, L., Alani, E., and Kleckner, N. (1990) A pathway for generation and processing of double-strand breaks during meiotic recombination in *S. cerevisiae. Cell* **61,** 1089–1101.
14. Schwacha, A. and Kleckner, N. (1995) Identification of double Holliday junctions as intermediates in meiotic recombination. *Cell* **83,** 738–791.
15. Storlazzi, A., Liuzhong, X., and Kleckner, N. (1995) Crossover and noncrossover recombination during meioisis:timing and pathway relationships. *Proc. Natl. Acad. Sci.* **92,** 8512–8516.
16. Goyon, C. and Lichten, M. (1993) Timing of molecular events in meiosis in *Saccharomyces cerevisiae*: stable heteroduplex is formed late in meiotic prophase. *Mol. Cell. Biol.* **13,** 373–382.
17. Allers, T. and Lichten, M. (2001) Intermediates of yeast meiotic recombination contain heteroduplex DNA. *Mol. Cell.* **8,** 225–231.
18. Allers, T. and Lichten, M. (2001) Differential timing and control of noncrossover and crossover recombination during meiosis. *Cell* **106,** 47–57.
19. Nachman, I., Regev, A., and Ramanathan, S. (2007) Dissecting timing variability in yeast meiosis. *Cell* **131,** 544–556.
20. de los Santos, T., Loidl, J., Larkin, B., and Hollingsworth, N. M. (2001) A role for *MMS4* in the processing of recombination intermediates during meiosis in *Saccharomyces cerevisiae. Genetics* **159,** 1511–1525.
21. de los Santos, T., Hunter, N., Lee, C., Larkin, B., Loidl, J., and Hollingsworth, N. M. (2003) The Mus81/Mms4 endonuclease acts independently of double-Holliday junction resolution to promote a distinct subset of crossovers during meiosis in budding yeast. *Genetics* **164,** 81–94.
22. Merker, J. D., Dominska, M., and Petes, T. D. (2003) Patterns of heteroduplex formation associated with the initiation of meiotic recombination in the yeast *Saccharomyces cerevisiae. Genetics* **165,** 47–63.
23. Jessop, L., Allers, T., and Lichten, M. (2005) Infrequent co-conversion of markers flanking a meiotic recombination initiation site in *Saccharomyces cerevisiae. Genetics* **169,** 1353–1367.
24. Hoffmann, E. R., Eriksson, E., Herbert, B. J., and Borts, R. H. (2005) *MLH1* and *MSH2* promote the symmetry of double-strand break repair events at the *HIS4* hotspot in *Saccharomyces cerevisiae. Genetics* **163,** 1292–1303.

25. Schultes, N. P. and Szostak, J. W. (1990) Decreasing gradients of gene conversion on both sides of the initiation site for meiotic recombination at the *ARG4* locus in yeast. *Genetics* **126,** 813–822.
26. Khazanehdari, K. and Borts, R. H. (2000) *EXO1* and *MSH4* differentially affect crossing-over and segregation. *Chromosoma* **109,** 94–102.
27. Borts, R. H., Chambers, S. R., and Abdullah, M. F. F. (2000) The many faces of mismatch repair in meiosis. *Mutat. Res.* **451,** 129–150.
28. Kirkpatrick, D. T., Dominska, M., and Petes, T. D. (1998) Conversion-type and restoration-type repair of DNA mismatches formed during meiotic recombination in *Saccharomyces cerevisiae. Genetics* **149,** 1693–1705.
29. Roth, R. and Halvorson, H. O. (1969) Sporulation of yeast harvested during logarithmic growth. *J. Bact.* **98,** 831–832.
30. Padmore, R., Cao, L., and Kleckner, N. (1991) Temporal analysis of reciprocal recombination and synaptonemal complex morphogenesis during meiosis in *S. cerevisiae. Cell* **66,** 1239–1256.
31. Kane, S. and Roth, R. (1974) Carbohydrate metabolism during ascospore development in yeast. *J. Bact.* **118,** 8–14.
32. Deutschbauer, A. M. and Davis, R. W. (2005) Quantitative trait loci mapped to single-nucleotide resolution in yeast. *Nat. Genet.* **37,** 1333–1340.
33. Codon, A. C., Gasent-Ramirez, J. M., and Benitez, T. (1995) Factors which affect the frequency of sporulation and tetrad formation in *Saccharomyces cerevisiae* baker's yeasts. *Appl. Environ. Microbiol.* **61,** 630–638.
34. Rockmill, B. personal communication.
35. Borts, R. H. and Haber, J. E. (1989) Length and distribution of meiotic gene conversion tracts and crossovers in *Saccharomyces cerevisiae. Genetics* **123,** 69–80.
36. Borts, R. H., Leung, W.-Y., Kramer, K., Kramer, B., Williamson, M. S., Fogel, S., and Haber, J. E. (1990) Mismatch repair-induced meiotic recombination requires the *PMS1* gene product. *Genetics* **124,** 573–584.
37. Hunter, N. and Borts, R. H. (1997) Mlh1p is unique among mismatch repair proteins in its ability to promote crossing-over during meiosis. *Genes & Dev.* **11,** 1573–1582.
38. Abdullah, M. F. F. and Borts, R. H. (2001) Meiotic recombination frequencies are affected by nutritional states in *Saccharomyces cerevisiae. Proc. Nat. Acad. Sci.* **98,** 14524–14529.
39. Symington, L. S. and Petes, T. D. (1988) Expansions and contractions of the genetic map relative to the physical map of yeast chromosome III. *Mol. Cell. Biol.* **8,** 595–604.
40. Detloff, P., White, M. A., and Petes, T. D. (1992) Analysis of a gene conversion gradient at the *HIS4* locus in *Saccharomyces cerevisaie. Genetics* **132,** 113–123.
41. Detloff, P., Sieber, J., and Petes, T. (1991) Repair of specific base pair mismatches during meiotic recombination in the yeast *Saccharomyces cerevisiae. Mol. Cell. Biol.* **11,** 737–745.
42. Nicolas, A., Treco, D., Schultes, N. P., and Szostak, J. W. (1989) An initiation site for meiotic gene conversion in the yeast *Saccharomyces cerevisiae. Nature* **338,** 35–39.
43. Treco, D., Thomas, B., and Arnheim, N. (1985) Recombination hot spot in the human beta-globin gene cluster: meiotic recombination of human DNA fragments in *Saccharomyces cerevisiae. Mol. Cell. Biol.* **5,** 2029–2038.
44. Ben-Ari, G., Zenvirth, D., Sherman, A., David, L., Klutstein, M., Lavi, U., Hillel, J., and Simchen, G. (2006) Four linked genes participate in controlling sporulation efficiency in budding yeast. *PLoS Genet.* **2,** 1815–1823.
45. Primig, M., Williams, R. M., Winzeler, E. A., Tevzadze, G. G., Conway, A. R., Hwang, S. Y., Davis, R. W., and Esposito, R. E. (2000) The core meiotic transcriptome in budding yeasts. *Nat. Genet.* **26,** 415–423.
46. Liti, G., Carter, D. M., Moses, A. M., Warringer, J., Parts, L., James, S. A., Davey, R. P., Roberts, I. N., Burt, A., Koufopanou, V., Tsai, I. J., Bergman, C. M., Bensasson, D., O'Kelly, M. J., van Oudenaarden, A., Barton, D. B., Bailes, E., Nguyen, A. N., Jones, M., Quail, M. A., Goodhead, I., Sims, S., Smith, F., Blomberg, A., Durbin, R., and Louis, E. J. (2009) *Nature* **458,** 337–341.
47. Hoffmann, E. R. and Borts, R. H. unpublished observations.
48. Nakagawa, T. and Ogawa, H. (1999) The *Saccharomyces cerevisiae MER3* gene, encoding a novel helicase-like protein, is required for crossover control in meiosis. *EMBO J* **18,** 5714–5723.
49. Tsubouchi, H. and Ogawa, H. (2000) Exo1 roles for repair of DNA double-strand breaks and meiotic crossing over in *Saccharomyces cerevisiae. Mol. Bio. Cell* **11,** 2221–2233.
50. Sym, M., Engebrecht, J. A., and Roeder, G. S. (1993) *ZIP1* is a synaptonemal complex protein required for meiotic chromosome synapsis. *Cell* **72,** 365–378.

51. Novak, J. E., Ross-Macdonald, P. B., and Roeder, G. S. (2001) The budding yeast Msh4 protein functions in chromosome synapsis and the regulation of crossover distribution. *Genetics* **158,** 1013–1025.
52. Hoffmann, E. R., Shcherbakova, P. V., Kunkel, T. A., and Borts, R. H. (2003) *MLH1* mutations differentially affect meiotic functions in *Saccharomyces cerevisiae. Genetics* **163,** 515–526.
53. Abdullah, M. F., Hoffmann, E. R., Cotton, V. E., and Borts, R. H. (2004) A role for the MutL homologue *MLH2* in controlling heteroduplex formation and in regulating between two different crossover pathways in budding yeast. *Cytogenet. Genome. Res.* **107,** 180–190.
54. Stone, J. E. and Petes, T. D. (2006) Analysis of the proteins involved in the in vivo repair of base-base mismatches and four-base loops formed during meiotic recombination in the yeast *Saccharomyces cerevisiae. Genetics* **173,** 1223–1239.
55. Alani, A., Reenan, R. A., and Kolodner, R. D. (1994) Interaction between mismatch repair and genetic recombination in *Saccharomyces cerevisiae. Genetics* **137,** 19–39.
56. Rockmill, B., Sym, M., Schertan, H., and Roeder, G. S. (1995) Roles for two RecA homologs in promoting meiotic chromosome synapsis. *Genes & Dev.* **9,** 2648–2695.
57. Turney, D., de Los Santos, T., and Hollingsworth, N. M. (2004) Does chromosome size affect map distance and genetic interference in budding yeast? *Genetics* **168,** 2421–2424.
58. Oh, S. D., Lao, J. P., Hwang, P. Y., Taylor, A. F., Smith, G. R., and Hunter, N. (2007) BLM ortholog, Sgs1, prevents aberrant crossing-over by suppressing formation of multichromatid joint molecules. *Cell* **130,** 259–272.
59. Martini, E., Diaz, R. L., Hunter, N., and Keeney, S. (2006) Crossover homeostasis in yeast meiosis. *Cell* **126,** 285–295.
60. Argueso, J. L., Kijas, A. W., Sarin, S., Heck, J., Waase, M., and Alani, E. (2003) Systematic mutagenesis of the *Saccharomyces cerevisiae MLH1* gene reveals distinct roles for Mlh1p in meiotic crossing over and in vegetative and meiotic mismatch repair. *Mol. Cell. Biol.* **23,** 873–886.
61. Argueso, J. L., Wanat, J., Gemici, Z., and Alani, E. (2004) Competing crossover pathways act during meiosis in *Saccharomyces cerevisiae. Genetics* **168,** 1805–1816.
62. Stahl, F. W., Foss, H. M., Young, L. S., Borts, R. H., Abdullah, M. F., and Copenhaver, G. P. (2004) Does crossover interference count in *Saccharomyces cerevisiae*? *Genetics* **168,** 35–48.
63. Kaback, D. B., Barber, D., Mahon, J., Lamb, J., and You, J. (1999) Chromosome size-dependent control of meiotic reciprocal recombination in *Saccharomyces cerevisiae*: the role of crossover interference. *Genetics* **152,** 1475–1486.
64. Cotton, V. (2007) *A structural and functional analysis of mismatch repair proteins in meiosis.* University of Leicester Ph. D., Leicester.
65. Hoffmann, E. R. and Borts, R. H. (2004) Meiotic recombination intermediates and mismatch repair proteins. *Cytogenet. Genome. Res.* **107,** 232–248.
66. Esposito, R. E. and Klapholz, S. (1981) Meiosis and ascospore development, in *The molecular biology of the yeast Saccharomyces,* vol. 1 (Strathern, J. N., Jones, E. W., and Broach, J. R., eds.), Cold Spring Harbor Laboratory, Cold Spring Harbor, pp. 211–287.
67. Kassir, Y. and Simchen, G. (1991) Monitoring meiosis and sporulation in *Saccharomyces cerevisiae. Meth. Enzymol.* **194,** 94–110.
68. Resnick, M. A., Stasiewicz, S., and Game, J. C. (1983) Meiotic DNA metabolism in wild-type and excision-deficient yeast following UV exposure. *Genetics* **104,** 583–601.
69. Blitzblau, H. G., Bell, G. W., Rodriguez, J., Bell, S. P., and Hochwagen, A. (2007) Mapping of meiotic single-stranded DNA reveals double-strand break hotspots near centromeres and telomeres. *Curr. Biol.* **17,** 2003–2012.
70. White, M. A., Detloff, P., Strand, M., and Petes, T. D. (1992) A promoter deletion reduces the rate of mitotic, but not meiotic, recombination at the *HIS4* locus in yeast. *Curr. Genet.* **21,** 109–116.
71. White, M. A., Dominska, M., and Petes, T. D. (1993) Transcription factors are required for the meiotic recombination hotspot at the *HIS4* locus in *Saccharomyces cerevisiae. Proc. Natl. Acad. Sci. USA* **90,** 6621–6625.
72. Fan, Q., Xu, F., and Petes, T. D. (1995) Meiosis-specific double-strand breaks at the *HIS4* recombination hotspot in the yeast *Saccharomyces cerevisiae*: control in *cis* and *trans. Mol. Cell. Biol.* **15,** 1679–1688.
73. Mieczkowski, P. A., Dominska, M., Buck, M. J., Gerton, J. L., Lieb, J. D., and Petes, T. D. (2006) Global analysis of the relationship between the binding of the Bas1p transcription factor and meiosis-specific double-strand DNA breaks in *Saccharomyces cerevisiae. Mol. Cell. Biol.* **26,** 1014–1027.

74. Hinnebusch, A. G. and Natarajan, K. (2002) Gcn4p, a master regulator of gene expression, is controlled at multiple levels by diverse signals of starvation and stress. *Eukaryot. Cell* **1,** 22–32.

75. Chua, P. R. and Roeder, G. S. (1998) *Zip2*, a meiosis-specific protein required for the initiation of chromosome synapsis. *Cell* **93,** 349–359.

76. Hochwagen, A., Tham, W. H., Brar, G. A., and Amon, A. (2005) The FK506 binding protein Fpr3 counteracts protein phosphatase 1 to maintain meiotic recombination checkpoint activity. *Cell* **122,** 861–873.

77. Marston, A. L., Tham, W. H., Shah, H., and Amon, A. (2004) A genome-wide screen identifies genes required for centromeric cohesion. *Science* **303,** 1367–1370.

78. Blunt and Hoffman, E. R. (2007) personal communication.

79. Kubinyi, H. (1999) Chance favors the prepared mind–from serendipity to rational drug design. *J. Recept. Signal. Transduct. Res.* **19,** 15–39.

80. Robine, N., Uematsu, N., Amiot, F., Gidrol, X., Barillot, E., Nicolas, A., and Borde, V. (2007) Genome-wide redistribution of meiotic double-strand breaks in *S. cerevisiae*. *Mol. Cell. Biol.* **27,** 1868–1880.

81. Sherman, F., Fink, G. R., and Hicks, J. B. (1986) *Methods in yeast genetics*, Cold Spring Harbor Laboratory, Plainview, NY.

82. Campbell, D. A., Fogel, S., and Lusnak, K. (1975) Mitotic chromosome loss in a disomic haploid of *S. cerevisiae*. *Genetics* **79,** 383–396.

Chapter 2

Optimizing Sporulation Conditions for Different *Saccharomyces cerevisiae* Strain Backgrounds

Susan L. Elrod, Sabrina M. Chen, Katja Schwartz, and Elizabeth O. Shuster

Abstract

Different sporulation and pre-sporulation regimens were compared for a number of commonly used laboratory strains of *S. cerevisiae* to define conditions that support high-efficiency sporulation.

Key words: yeast, meiosis, culture conditions.

1. Introduction

Numerous strain backgrounds of the budding yeast *S. cerevisiae* are in common use in the laboratory, and these differ genetically and phenotypically from one another, sometimes substantially (Saccharomyces Genome Resequencing Project, http://www.sanger.ac.uk/Teams/Team71/durbin/sgrp/). Differences have been noted between strain backgrounds with respect to meiosis and sporulation *(e.g., 1, 2, 3)* (for more detail, refer to **Chapter 1** in this volume). In most cases, the genetic basis for these differences is not understood, although a few quantitative trait loci have been identified that contribute to differences in the efficiency with which cells execute the meiotic program and form mature asci *(2)*.

Sporulation in *S. cerevisiae* is a response to starvation for nitrogen in the presence of a nonfermentable carbon source, such as acetate *(reviewed in 4)*. Numerous culture media have been described for inducing sporulation, and in many cases specific pre-sporulation growth regimens are applied to promote a more

Scott Keeney (ed.), *Meiosis, Volume 1, Molecular and Genetic Methods, vol. 557*

DOI 10.1007/978-1-59745-527-5_2 Springerprotocols.com

efficient response to the starvation conditions. The purpose of this study was to systematically compare effects of different pre-sporulation and sporulation conditions on different strain backgrounds. Results of this analysis should be useful both for studies of meiotic processes themselves and for obtaining optimal sporulation for genetic crosses or other purposes in many commonly used laboratory strains.

Strains used in this study are summarized in **Table 2.1** and include fast, efficient sporulating backgrounds such as SK1 and Y55 as well as relatively inefficiently sporulating backgrounds such as S288c. **Table 2.2** shows sporulation efficiencies measured on plates at 30°C and **Table 2.3** shows sporulation efficiencies measured in liquid culture at 30°C. For both plates and liquid cultures, the combination of pre-sporulation and sporulation media that gave the highest sporulation efficiency is identified for each strain (*see columns labeled* "Best" *in* **Tables 2.2 and 2.3**). In addition, the sporulation efficiency is tabulated for each strain in three "consensus" pre-sporulation/sporulation combinations, at least one of which supports good sporulation for each strain tested (*see columns labeled* "Consensus" *in* **Tables 2.2 and 2.3**). Effects of temperature on sporulation efficiencies were also examined on the "Best" medium for each strain (**Table 2.4**). Major conclusions are as follows:

- No single "magic" sporulation protocol supports optimal sporulation for all strains.
- Media supporting high levels of sporulation on plates do not always support high levels in liquid (and vice versa).
- Sporulation efficiency is affected by both pre-sporulation and sporulation conditions.

Table 2.1
Strains compared in this study

Strain name	Strain background	Source of strain or parents
BR2171-3B	Unknown	G.S. Roeder lab
BR2495	Unknown	G.S. Roeder lab
EB1501	S288c	C. Holm lab
NH39	A364a	L. Hartwell lab
NKY278	SK1	N. Kleckner lab
SE1200	AP1	B. Byers lab
SE1202	W303	R. Rothstein lab
YPH274	S288c	P. Hieter lab
Y55	Y55	J.E. Haber lab

Table 2.2
Sporulation efficiencies on plates

	Best[a]		Consensus[b]		
Strain	**Media**	**Sporulation**	**6% YEPD, SPM**	**2% YPA, 1% KAc**	**PSP2, 1%KAc**
BR2171-3B	MSAc, 2% KAc	66.2 ± 0.6	57.2 ± 6.5	37.5 ± 2.6	50.5 ± 9.5
BR2495	MSAc, 1% KAc	71.5 ± 2.2	50.8 ± 5.0	26.8 ± 15.4	68.2 ± 3.4
EB1501	6% YEPD, 1% KAc	29.3 ± 12.1	21.2 ± 12.3	1.8 ± 1.4	3.5 ± 1.7
NH39	2% YPA, 1% KAc	23.4 ± 7.3	8.5 ± 5.6	23.4 ± 7.3	1.4 ± 1.5
NKY278	6% YEPD, 1% KAc	95.8 ± 1.4	83.0 ± 3.8	54.6 ± 2.6	65.8 ± 7.5
SE1200	6% YEPD, SPM	67.6 ± 6.6	67.6 ± 6.6	9.2 ± 4.6	21.0 ± 9.4
SE1202	2% YEPD, 0.3% KAc	58.2 ± 2.2	56.6 ± 3.4	34.2 ± 6.4	23.8 ± 7.3
YPH274	1% YPA, SPM	28.5 ± 1.5	15.5 ± 4.2	24.5 ± 13.1	12.0 ± 1.4
Y55	6% YEPD, SPM	98.3 ± 1.8	98.3 ± 1.8	91.0 ± 5.3	97.2 ± 0.6

[a]"Best" is the combination of pre-sporulation and sporulation media that gave optimal sporulation efficiency (expressed as % of cells forming asci, ± sd).
[b]"Consensus" media are three pre-sporulation/sporulation combinations, at least one of which supports good sporulation for each strain tested. The three columns under this heading show sporulation efficiencies (% of cells, ± sd).

Table 2.3
Sporulation efficiencies in liquid culture

	Best[a]		Consensus[b]		
Strain	**Media**	**Sporulation**	**1% YPA, 1% KAc**	**6%YEPD,CSHSPO**	**PSP2, 1%KAc**
BR2171-3B	1% YPA, CSHSPO	72.5 ± 2.8	55.8 ± 7.4	62.3 ± 9.5	15.0
BR2495	PSP2, 1% KAc	39.0	30.0 ± 5.7	19.5 ± 13.4	39.0
EB1501	1% YPA, 1% KAc	52.0 ± 7.8	52.0 ± 7.8	32.3 ± 8.8	27
NH39	6% YEPD, CSHSPO	28.0 ± 3.5	12.0 ± 2.1	28.0 ± 3.5	0
NKY278	1% YPA, CSHSPO	97.8 ± 1.1	97.5 ± 1.4	92.8 ± 2.5	91
SE1200	1% YPA, 1% KAc	77.3 ± 6.7	77.3 ± 6.7	51.5 ± 4.9	60.5 ± 4.9
SE1202	1% YPA, 1% KAc	51.8 ± 0.4	51.8 ± 0.4	37.3 ± 7.4	39.5
YPH274	PSP2, 1% KAc	44.5	32.8 ± 3.9	23.8 ± 1.8	44.5
Y55	1% YPA, 1% KAc	83.8 ± 0.4	83.8 ± 0.4	82.0 ± 4.2	60.5

[a]"Best" is the combination of pre-sporulation and sporulation media that gave optimal sporulation efficiency (expressed as % of cells forming asci, ± sd).
[b]"Consensus" media are three pre-sporulation/sporulation combinations, at least one of which supports good sporulation for each strain tested. The three columns under this heading show sporulation efficiencies (% of cells, ± sd).

Table 2.4
Effects of temperature on sporulation efficiency

Strain name	Effect of temperature[a]
BR2171-3B	Cold-sensitive
BR2495	Cold- and temperature-sensitive
EB1501	Temperature-sensitive
NH39	Cold- and temperature-sensitive
NKY278	Little or no effect
SE1200	Cold-sensitive
SE1202	Cold- and temperature-sensitive
YPH274	Cold- and temperature-sensitive
Y55	Little or no effect

[a]Cold-sensitive: sporulation was reduced at 15°C and/or 18°C relative to 30°C. Temperature-sensitive: sporulation was reduced at 34.5°C relative to 30°C.

- In general, sporulation media containing dextrose and yeast extract do NOT support high levels of sporulation (defined as >70% of maximal) on plates, although they may in liquid. In general, acetate-based pre-sporulation media support higher levels of sporulation in liquid than do those containing glucose.
- In many strains, the proportion of asci containing three or four spores is higher at 18°C than at 30°C.

2. Materials

Standard equipment for yeast culture (e.g., stationary and shaking incubators, centrifuge with rotors and sterile centrifuge tubes, sterile culture flasks, microscope). The compositions of liquid pre-sporulation and sporulation media tested in this study are given below. For plates, 2% Bacto agar was included.

2.1. Pre-sporulation Media

1. 10% YEPD: 0.8% yeast extract, 0.3% peptone, 10% dextrose.
2. 1% YPA: 1% yeast extract, 2% peptone, 1% potassium acetate.
3. 2% YPA: 1% yeast extract, 2% peptone, 2% potassium acetate.
4. MSAc: 0.67% yeast nitrogen base without amino acids, 2% potassium acetate (*see* **Note 1** below).

5. 2% YEPD: 1% yeast extract, 2% peptone, 2% dextrose.
6. PSP2: 0.1% yeast extract, 0.67% yeast nitrogen base without amino acids, 1% potassium acetate (*see* **Note 1**).
7. SAcCa: 0.67% yeast nitrogen base without amino acids, 2% potassium acetate, 2% casamino acids, 20 μg/mL adenine, 20 μg/mL histidine, 20 μg/mL tryptophan, and 20 μg/mL uracil.
8. 6% YEPD: 1% yeast extract, 2% peptone, 6% dextrose.

2.2. Sporulation Media

1. SPM: 0.3% potassium acetate, 0.02% raffinose, plus amino acid supplements (*see* **Note 2**).
2. 2% KAc: 2% potassium acetate (*see* **Note 2**).
3. 1% KAc: 1% potassium acetate (*see* **Note 2**).
4. CSHSPO: 1% potassium acetate, 0.05% dextrose, 0.1% yeast extract (*see* **Note 2**).
5. SPO: 1.5% potassium acetate, 0.05% dextrose, 0.25% yeast extract (*see* **Note 2**).
6. 0.3% KAc: 0.3% potassium acetate (*see* **Note 2**).
7. SPO7: 1.5% potassium acetate, 0.1% dextrose, 0.25% yeast extract (*see* **Note 2**).
8. SPO8: 0.3% potassium acetate, 0.02% raffinose, 1% yeast extract (no amino acid supplements).

3. Methods

3.1. Standardized Test Protocol on Plates

1. Grow fresh patches on medium of choice.
2. Replica-plate to pre-sporulation medium. Incubate at 30°C for 18–24 h.
3. Replica-plate to sporulation plates. Incubate at 30°C (or lower). Check for spores after 2, 5, or 10 d (*see* **Note 3**).

3.2. Standardized Test Protocol in Liquid

1. Grow fresh overnight in standard rich medium (e.g., YEPD).
2. Inoculate pre-sporulation medium to allow growth in 18–24 h to late log phase (1–2 × 10^7 cells/mL) for acetate media or early stationary phase (1–2 × 10^8 cells/mL) for dextrose-containing media.
3. Harvest cells by centrifugation (3,000*g*, 5 min). Wash cells once or twice with sterile distilled H_2O and resuspend in sporulation medium at 1–2 × 10^7 cells/mL.
4. Incubate in shaking incubator or water bath at 30°C (or lower) for 30–72 h (48 h is usually sufficient).

4. Notes

1. Supplement as needed with 20 μg/mL each of adenine, uracil, histidine, arginine, and methionine; 60 μg/mL each of tyrosine and lysine; and 80 μg/mL each of tryptophan, leucine, and isoleucine.
2. Supplement as needed with 10 μg/mL each of adenine, arginine, histidine, isoleucine, leucine, lysine, methionine, phenylalanine, threonine, tryptophan, uracil, and valine.
3. Strains vary in length of time to complete sporulation as well as optimal temperature. Lower temperatures may require longer sporulation times. Older asci may come apart more easily during dissection, if they have not prematurely dissociated.

Acknowledgments

This work was performed in the Department of Food Science and Genetics Graduate Group, University of California, Davis, CA.

References

1. Primig, M., Williams, R. M., Winzeler, E. A., Tevzadze, G. G., Conway, A. R., Hwang, S. Y., Davis, R. W., and Esposito, R. E. (2000) The core meiotic transcriptome in budding yeasts. *Nat. Genet.* **26,** 415–423.
2. Deutschbauer, A. M. and Davis, R. W. (2005) Quantitative trait loci mapped to single-nucleotide resolution in yeast. *Nat. Genet.* **37,** 1333–1340.
3. Borner, G. V., Kleckner, N., and Hunter, N. (2004) Crossover/noncrossover differentiation, synaptonemal complex formation, and regulatory surveillance at the leptotene/zygotene transition of meiosis. *Cell* **117,** 29–45.
4. Kupiec, M., Byers, B., Esposito, R. E., and Mitchell, A. P. (1997) Meiosis and sporulation in *Saccharomyces cerevisiae*, in *The Molecular and Cellular Biology of the Yeast Saccharomyces: Cell Cycle and Cell Biology,* vol. 3 (Pringle, J. R., Broach, J. R., and Jones, E. W., eds.), Cold Spring Harbor Laboratory Press, Cold Spring Harbor, pp. 889–1036.

Chapter 3

Modulating and Targeting Meiotic Double-Strand Breaks in *Saccharomyces cerevisiae*

Alain Nicolas

Abstract

Meiotic recombination is initiated by DNA double-strand breaks (DSBs) formed by the evolutionary conserved Spo11 protein. Along the *S. cerevisiae* chromosomes, the DSB sites are not evenly distributed and the cleavage frequencies vary 10–100-fold from site to site. Herein are reviewed the methods used in budding yeast to modulate locally and globally the native DSB frequencies, including a powerful method to target Spo11-dependent meiotic DSB in novel chromosomal regions. These methods serve to investigate the control and the mechanism of recombination initiation and modify the natural distribution of meiotic recombination.

Key words: meiosis, recombination, initiation, Spo11.

1. Introduction

In all eukaryotes, meiotic recombination is initiated by DNA double-strand breaks (DSBs) formed by the evolutionarily conserved Spo11 protein and associated factors [reviewed in *(1, 2)*]. The DSB fragments are physically detectable easily in yeasts by gel electrophoresis assays (*see* **Chapters 9, 14, and 15** in this volume) or genome-wide analysis using microarrays (**Chapter 10**) of genomic DNA extracted from well-synchronized cells. In *S. cerevisiae*, the DSB frequencies vary 10–100-fold from site to site and are not evenly distributed, creating large cold and hot regions for meiotic recombination *(3–7)*. This heterogeneity contributes to the large variation of the frequency of recombination per unit of physical distance (cM/Kb ratio). The *cis*- and *trans*-acting factors that determine whether a specific region or site is prone to DSB

Scott Keeney (ed.), *Meiosis, Volume 1, Molecular and Genetic Methods, vol. 557*

DOI 10.1007/978-1-59745-527-5_3 Springerprotocols.com

formation (and hence, recombination) act both locally and globally [reviewed in *(1, 2, 8–10)*]. As reviewed here, perturbation of these mechanistic and regulatory factors has provided ways to modify recombination frequencies. Locally, the deletion and insertions of DNA fragments as well as mutation of specific transcription factors allow reduction or enhancement of DSB frequencies. Globally, some *spo11* mutants allow reduction of DSBs throughout the genome *(11)*, while fusion of Spo11 to a heterologous DNA binding domain allows stimulation of targeted DSBs at a large variety of predetermined sites *(7, 12)*.

2. Local Modification of Spo11-Dependent DSB Frequencies

2.1. Methods to Reduce DSB Formation

Several approaches have been established for reducing DSB formation in particular regions of the genome.

1. *Deleting the cis-acting element controlling DSB formation.* Deletions of the *cis*-acting region controlling DSB formation at the *ARG4, HIS4,* and *HIS2* hot spots were found to reduce DSB frequencies up to 10-fold *(13–15)*. It is important to note, however that a deletion may sometimes create a fortuitous compensatory DSB site *(13, 16)*.
2. *Inducing transcription across the DSB region.* Near the intergenic *ARG4* DSB region, deletion of the transcription terminator of the upstream gene (*DED81/YHR019C*) leading to read-through transcription of the intergenic region was found to completely abolish DSB formation *(17)*.
3. *Modifying chromatin-associated factors.* In budding yeast, most if not all natural hot-spots of DSB formation are located in intergenic-promoter containing regions that correspond to nuclease-hypersensitive regions of chromatin [reviewed in *(1, 8, 10, 18)*] and that are bound by specific transcription factors. At some hot-spot loci, such as *HIS4*, the binding of some specific transcription factors (Bas1, Bas2 or Rap1) is required for hot-spot activity. By deleting the transcription factor binding sites or the gene that encodes these transcription factors, DSB formation and recombination can be reduced *(19)*.

2.2. Methods to Stimulate DSB Formation

One method to stimulate Spo11-dependent DSB formation is the insertion/substitution of ectopic or foreign DNA fragments. Namely, the substitution of a 167-base pair region by a polylinker sequence at *ARG4 (16)*, the insertion of the yeast *LEU2* gene and a fortuitous 77-base pair bacterial DNA in the *HIS4* region *(20)* or the insertion of an array of $(CCGNN)_{12}$ tandem repeats at

HIS4 (21) were found to strongly enhance DSB frequencies above the wild-type level. The stimulatory effect of these sequences likely results, at least in part, from the local increase of chromatin accessibility. Another method is the insertion of a reporter cassette containing a DSB site cleaved by Spo11 or Gal4BD-Spo11 (*see* **Section 3.3**). Cleavage occurs at the expected site but often elsewhere in the inserted fragment or at the junction of the insert and the resident chromosomal region *(7, 22, 23)*. The overall efficiency of DSB formation in the insert will largely depend on the region of insertion: it remains under the influence of the regional control *(7, 23)* and is modulated by a competitor effect between closely spaced DSB sites *(20, 22–24)*.

3. Methods to Globally Modify Spo11-Dependent DSB Frequencies

Several strategies allow one to reduce DSB frequencies in a fairly uniform manner across the genome and to various extents. Most drastically, DSB formation can be abolished by inactivation of the proteins required for DSB formation (for example upon deletion of the *SPO11* gene) but no viable meiotic products are obtained. However, spore viability can be rescued by inactivation of the *SPO13* gene *(25)*. Homozygous *spo11Δ spo13Δ* diploid cells will yield viable diploid ascopores. Less drastically, DSB frequencies can be gradually reduced by using partially defective *spo11* mutants *(11)*. In a homozygous *spo11-HA3His6* diploid, DSB forms at approximately 80% of the wild-type level. DSBs are further reduced to 30% in a strain heterozygous for *spo11-HA3His6* and a tagged allele carrying the *spo11Y135F-HA3His6* mutation, and further reduced to 20% in a strain homozygous for the *spo11-D290A-HA3His6* mutant *(11)*. In these strains, the reduction of meiotic recombination leads to a correlated loss of spore viability but does not reduce the genetic distance in parallel due to the compensatory phenomena of crossover homeostasis, in which crossover numbers are maintained at the expense of non-crossover recombinants *(11)*.

4. Site-Specific Targeting of Spo11 DSBs

How Spo11 selects its natural cleavage sites is unknown but, to target meiotic DSB formation at specific sites, Spo11 can be fused to a sequence-specific DNA binding domain that can be sufficient to induce novel cleavage sites. Fusion of Spo11 with the Gal4 DNA binding domain (Gal4BD) *(7, 12)* or the QQR artificial zinc finger array (N. Uematsu, personal communication) are efficient. In such a method, the fusion construct was stably integrated in the yeast genome at the *TRP1* locus and expressed in meiotic cells using the constitutive promoter (*pADH1*) and the *ADH1* transcription terminator (*tADH1*). In the absence of the wild-type *SPO11* gene, the resulting fusions allow the formation of DSB at the native Spo11 sites plus novel sites, which can be characterized by the physical methods for mapping meiotic DSBs described in **Chapters 9 and 14** in this volume. Stimulation of meiotic recombination frequencies (non-crossovers and crossovers) near the targeted sites can be observed [*(7, 12)*, N. Uematsu, personal communication] using genetical or physical markers as described in **Chapters 4 and 14**.

4.1. DSB Targeting by Gal4BD-Spo11

The genome-wide map of the DSB sites stimulated by Gal4BD-Spo11-myc13 has been established *(7)*. Approximately 50% of the targeted sites contain the consensus Gal4 recognition sequence $CGGN_{11}CCG$, including the well-known Gal4 binding sites such as the *GAL1-10*, *GAL2*, *GAL7* and *GAL80* genes. The strongest stimulation (over 100-fold) was observed in the promoter of the *GAL2* gene that contains five Gal4 recognition sequences (**Fig. 3.1**). Genetic markers placed near the targeted DSBs at the

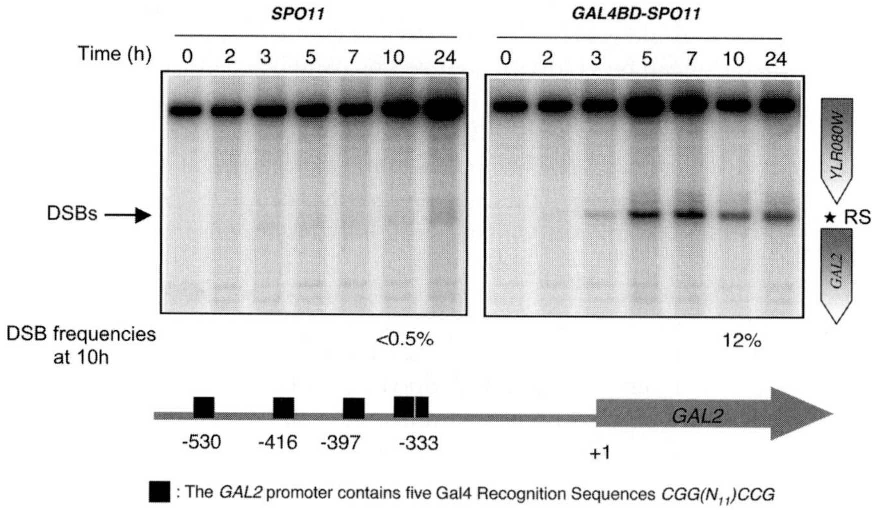

Fig. 3.1. Gal4BD-Spo11 stimulates DSBs in the *GAL2* promoter region near the $Gal4_{UAS}$ recognition sequences. Genomic DNA was prepared from *SPO11* (ORD1181) and *GAL4BD-SPO11* (ORD5807) diploids taken at the indicated time after transfer to sporulation medium. DSB fragments were detected by Southern analysis as described in **Chapter 9** of this volume. The DNA was digested with *Xba*I and *Nco*I, separated on 1% agarose gel and probed with an internal fragment of the *GAL2* gene. At the right of the gels, a map of the chromosome XII region shows ORFs (*open arrows* indicate transcriptional sense), (*) position of the $Gal4_{UAS}$ ($CGGN_{11}CCG$) recognition sequences (coordinates from the +1 *GAL2* translation initiation nucleotide), DSB sites (*arrow*). Quantification of the prominent DSB fragments in the *SPO11* and *GAL4BD-SPO11* cells measured at t = 10 h corresponds to the percentage of the DSB fragment over the sum of all the bands in the lane. [Data from reference *(12)*].

GAL2 locus demonstrated that both non-crossovers *(12)* and crossing-over (Murakami & Nicolas, unpublished results) products are strongly stimulated, creating a novel hot spot. High-resolution mapping indicates that the targeted DSBs occur within 20–30 nucleotides from the Gal4 recognition sequence (Murakami & Nicolas, unpublished results). Targeted DSB stimulation by Gal4BD-Spo11 has also been observed in chromosomal regions that are bound by Gal4 but do not contain the expected Gal4 recognition sequence *(7)*. The reason for Gal4BD-Spo11 targeting to regions lacking canonical Gal4 recognition sequences is not known. The Spo11-Myc13 and Gal4BD-Spo11-Myc13 microarray data are available at Gene Expression Omnibus (http://www.ncbi.nlm.nih.gov/geo) under the accession number GSE5884. Interestingly, Gal4BD-Spo11 binding to meiotic chromatin is not sufficient for Spo11 cleavage (due to chromosomal context) and the occurrence of a novel strong DSB site induces a long-distance effect which reduces DSB frequencies of the native adjacent sites, over 80 kb *(7)*.

4.2. Targeting Spo11 with the Artificial QQR Zinc Finger Array

In principle, synthetic zinc-finger (ZF) DNA binding domains are capable of directing fusion proteins to arbitrarily chosen genomic sequences. One ZF-Spo11 fusion (QQR-Spo11) has recently been shown to stimulate meiotic DSB formation (N. Uematsu, personal communication). QQR is a three-finger ZF which has the recognition sequence (RS), 5′-GGGGAAGAA-3′ *(26)*. As for a Gal4BD-Spo11 fusion, classical gel electrophoresis assays and ChIP-chip analysis allowed us to map and quantify DSB formation. This analysis showed that QQR-Spo11 cleaves the natural Spo11 sites and targets cleavage and crossover frequencies in novel chromosomal regions which contain the expected recognition sequence, for example within the *YML127W-YML128C*, intergenic region (**Fig. 3.2**). However, unexpectedly, stimulation of DSBs in QQR-RS-lacking regions is also very frequent (estimated at 89% of the cases) suggesting that QQR also directs chromatin binding by protein-protein interactions, rather than DNA sequence preferences. The QQR-Spo11-myc13 microarray data are available at Gene Expression Omnibus (http://www.ncbi.nlm.nih.gov/geo) under the accession number GSE8200.

5. Yeast Strains

The laboratories working on *S. cerevisiae* meiosis use different strain backgrounds and therefore appropriate strains should be obtained by request to the authors cited here in reference. A guide to variation in recombination and sporulation among

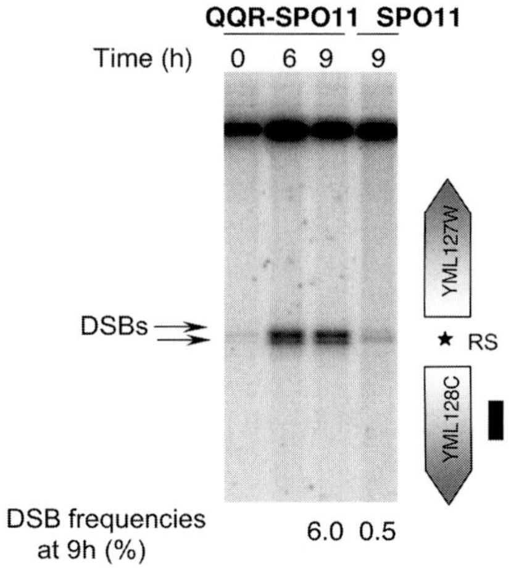

Fig. 3.2. QQR-Spo11 stimulates DSBs in the *YML127W-YML128C* intergenic region near the QQR recognition sequence. Genomic DNA was prepared from *SPO11* (ORD7239) and *QQR-SPO11* (ORD8146) diploids. The DNA was digested with *AseI*, separated on 1% agarose gel and probed with an internal fragment of the *YML128C* gene. At the right of the gel, a map of the chromosome XIII region shows the ORFs, (*) position of the QQR Recognition Sequence (GGGGAAGAA). Other figure components as in **Fig. 3.1**. (Data from N. Uematsu, personal communication).

S. cerevisiae strains is provided in **Chapters 1 and 2** of this volume. The strains of the SK1 background are more widely used for meiotic studies because sporulation is very efficient (>80%) and well-synchronized meiotic samples can be obtained to perform time-course physical analysis of recombination intermediates, including DSB formation, crossover and non-crossover products.

References

1. Keeney, S. (2001) Mechanism and control of meiotic recombination initiation. *Curr. Top. Dev. Biol.* **52**, 1–53.
2. Hunter, N. (2007) *in* "Homologous recombination" (Aguilera, A., Rothstein R., Eds.), Springer-Verlag, Heidelberg.
3. Zenvirth, D., Arbel, T., Sherman, A., Goldway, M., Klein, S., and Simchen, G. (1992) Multiple sites for double-strand breaks in whole meiotic chromosomes of *Saccharomyces cerevisiae. EMBO J.* **11,** 3441–7.
4. Baudat, F., and Nicolas, A. (1997) Clustering of meiotic double-strand breaks on yeast chromosome III. *Proc. Natl. Acad. Sci. USA* **94,** 5213–8.
5. Gerton, J. L., DeRisi, J., Shroff, R., Lichten, M., Brown, P. O., and Petes, T. D. (2000) Inaugural article: global mapping of meiotic recombination hotspots and coldspots in the yeast *Saccharomyces cerevisiae. Proc. Natl. Acad. Sci. USA* **97,** 11383–90.
6. Borde, V., Lin, W., Novikov, E., Petrini, J. H., Lichten, M., and Nicolas, A. (2004) Association of Mre11p with double-strand break sites during yeast meiosis. *Mol. Cell* **13,** 389–401.
7. Robine, N., Uematsu, N., Amiot, F., Gidrol, X., Barillot, E., Nicolas, A., and Borde, V. (2007) Genome-wide redistribution of meiotic double-strand breaks in *Saccharomyces cerevisiae. Mol. Cell. Biol.* **27,** 1868–80.
8. Lichten, M., and Goldman, A. S. (1995) Meiotic recombination hotspots. *Annu. Rev. Genet.* **29,** 423–44.
9. Smith, K. N., and Nicolas, A. (1998) Recombination at work for meiosis. *Curr. Opin. Genet. Dev.* **8,** 200–11.

10. Petes, T. D. (2001) Meiotic recombination hot spots and cold spots. *Nat. Rev. Genet.* **2,** 360–9.
11. Martini, E., Diaz, R. L., Hunter, N., and Keeney, S. (2006) Crossover homeostasis in yeast meiosis. *Cell* **126,** 285–95.
12. Peciña, A., Smith, K. N., Mézard, C., Murakami, H., Ohta, K., and Nicolas, A. (2002) Targeted stimulation of meiotic recombination. *Cell* **111,** 173–84.
13. Nicolas, A., Treco, D., Schultes, N. P., and Szostak, J. W. (1989) An initiation site for meiotic gene conversion in the yeast *Saccharomyces cerevisiae*. *Nature* **338,** 35–9.
14. Fan, Q., Xu, F., and Petes, T. D. (1995) Meiosis-specific double-strand DNA breaks at the *HIS4* recombination hot spot in the yeast *Saccharomyces cerevisiae*: control in *cis* and *trans*. *Mol. Cell. Biol.* **15,** 1679–88.
15. Haring, S. J., Halley, G. R., Jones, A. J., and Malone, R. E. (2003) Properties of natural double-strand-break sites at a recombination hotspot in *Saccharomyces cerevisiae*. *Genetics* **165,** 101–14.
16. de Massy, B., and Nicolas, A. (1993) The control in *cis* of the position and the amount of the *ARG4* meiotic double-strand break of *Saccharomyces cerevisiae*. *EMBO J.* **12,** 1459–66.
17. Rocco, V., de Massy, B., and Nicolas, A. (1992) The *Saccharomyces cerevisiae ARG4* initiator of meiotic gene conversion and its associated double-strand DNA breaks can be inhibited by transcriptional interference. *Proc. Natl. Acad. Sci. USA* **89,** 12068–72.
18. Nicolas, A. (1998) Relationship between transcription and initiation of meiotic recombination: toward chromatin accessibility. *Proc. Natl. Acad. Sci. USA* **95,** 87–9.
19. White, M. A., Dominska, M., and Petes, T. D. (1993) Transcription factors are required for the meiotic recombination hotspot at the *HIS4* locus in *Saccharomyces cerevisiae*. *Proc. Natl. Acad. Sci. USA* **90,** 6621–5.
20. Xu, F., and Petes, T. D. (1996) Fine-structure mapping of meiosis-specific double-strand DNA breaks at a recombination hotspot associated with an insertion of telomeric sequences upstream of the *HIS4* locus in yeast. *Genetics* **143,** 1115–25.
21. Kirkpatrick, D. T., Wang, Y. H., Dominska, M., Griffith, J. D., and Petes, T. D. (1999) Control of meiotic recombination and gene expression in yeast by a simple repetitive DNA sequence that excludes nucleosomes. *Mol. Cell. Biol.* **19,** 7661–71.
22. Wu, T. C., and Lichten, M. (1995) Factors that affect the location and frequency of meiosis-induced double-strand breaks in *Saccharomyces cerevisiae*. *Genetics* **140,** 55–66.
23. Borde, V., Wu, T. C., and Lichten, M. (1999) Use of a recombination reporter insert to define meiotic recombination domains on chromosome III of *Saccharomyces cerevisiae*. *Mol. Cell. Biol.* **19,** 4832–42.
24. Fan, Q. Q., Xu, F., White, M. A., and Petes, T. D. (1997) Competition between adjacent meiotic recombination hotspots in the yeast *Saccharomyces cerevisiae*. *Genetics* **145,** 661–70.
25. Malone, R. E., and Esposito, R. E. (1981) Recombinationless meiosis in *Saccharomyces cerevisiae*. *Mol. Cell. Biol.* **1,** 891–901.
26. Smith, J., Bibikova, M., Whitby, F. G., Reddy, A. R., Chandrasegaran, S., and Carroll, D. (2000) Requirements for double-strand cleavage by chimeric restriction enzymes with zinc finger DNA-recognition domains. *Nucleic Acids Res.* **28,** 3361–9.

Chapter 4

Methods for Analysis of Crossover Interference in *Saccharomyces cerevisiae*

Franklin W. Stahl and Elizabeth A. Housworth

Abstract

Interest in crossover interference in yeast has been spurred by the discovery and characterization of mutants that alter it as well as by the development and testing of models to explain it. This chapter describes methods for detecting and for measuring interference, with emphasis on those that exploit the ability to examine all four products of individual acts of meiosis.

Key words: NPD ratio, coefficient of coincidence, counting model, Erlang distribution, Gamma distribution, Chi-square distribution.

1. Introduction

On most chromosomes there are many sites at which exchange can occur, but, in a typical act of meiosis, only a few exchanges do occur. If these exchanges were to occur independently of each other there would be three identifiable consequences. (i) For a given chromosome, crossovers will be Poisson-distributed among acts of meiosis. (ii) When distances are measured in genetic map units (Morgans or centiMorgans), the frequency distribution of inter-exchange distances would be exponential, but censored in accordance with the linkage map length of the chromosome. (iii) There would be no correlation of exchanges in any given interval with those in any other interval.

"Interference" is any distribution of exchanges ("crossovers") that deviates from independence. "Positive interference," the deviation characteristic of most organisms, refers to the apparent "ability"

Scott Keeney (ed.), *Meiosis, Volume 1, Molecular and Genetic Methods, vol. 557*

DOI 10.1007/978-1-59745-527-5_4 Springerprotocols.com

of an exchange to "discourage" nearby exchanges. Various indicators of interference detect deviations from one or another of the above three manifestations of independence.

Several frequently used indicators of interference (e.g., coefficient of coincidence, NPD ratio) are functions not only of the strength of interference but also of the map length(s) of the interval(s) used in the measurement. We shall deal with these indicators and also with an *index* of interference, an indicator that assesses interference, per se, freed from the map lengths used in the measurement.

Because meiotic crossing over occurs in the four-chromatid stage, i.e., after the premeiotic DNA replication, crossover interference can be of two types: (i) nonindependence in the choice of partners (*chromatid interference*) and (ii) nonindependence of crossover location along the bivalent (*chromosome interference*). (Since exchanges between sister chromatids, should they occur, do not result in crossing over of markers, their only consequence to our analysis is the possibility that they interfere with exchanges between homologs. Absent of evidence that they do so, we will have no more to say about them.) In some of what follows we shall proceed as if the relatively rare phenomenon of chromatid interference did not exist. However, we shall point out some occasions when this assumption might lead us astray and/or where we have the opportunity to verify it.

Studies on interference have made use of "cytological" and "genetical" methods. The former visualizes crossovers either as chiasmata, as "late nodules," or as foci of crossover-specific proteins. If microscopic resolution is adequate, deviations from a Poisson distribution of exchange events can provide evidence of chromosome interference. When genetic distance is (at least roughly) proportional to cytological distance, departure of the interfoci distance-distribution from exponential can provide further evidence *(1)*. Although meiotic yeast chromosomes are inconveniently small, studies of protein foci in yeast have contributed to the interference conversation *(2)*. Cytological studies give no information on chromatid interference.

Our primary focus will be on genetical methods of analyzing interference. The discovery *(3, 4)* of interference was by genetical methods, which have high resolution and, unlike cytological methods, are intrinsically independent of variation in crossover density along the chromosome. The interference of greatest interest is the almost ubiquitous interference at the chromosome level (i.e., chromosome interference, often called crossover interference and sometimes called chiasma interference, although genetic analyses do not visualize chiasmata). The detection of interference as deviation from the Poisson distribution in numbers of exchanges requires that a chromosome be well marked so that few exchanges are missed. Chromosome interference is generally positive, as manifested by a distribution of exchanges among

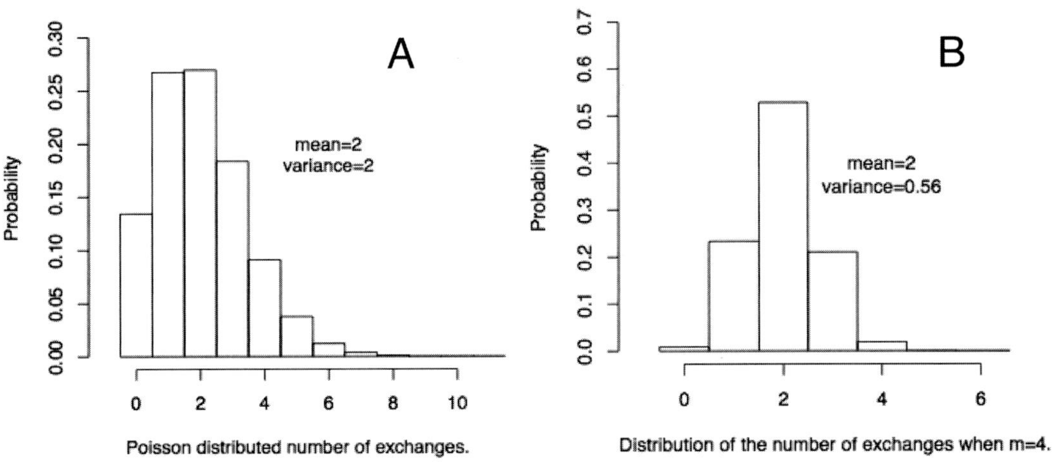

Fig. 4.1. The distribution of exchanges among acts of meiosis. If the abscissa is in units of exchanges per bivalent, this chromosome has a map length of 100 centiMorgans. If the abscissa is in units of crossovers per chromosome, this chromosome has a map length of 200 centiMorgans. (**A**) In the absence of interference, exchanges on a given bivalent (or chromosome) are distributed according to the Poisson distribution, for which the variance is equal to the mean. (**B**) In the presence of positive chromosome interference, the distribution of exchanges will have a variance that is less than the mean. The distribution shown is calculated from a model (Counting Model with $m = 4$) that provides an index of interference (m), which is independent of the length(s) of the intervals involved. The logic underlying the Counting Model is described in **Section 3.4**.

meioses in which the variance is less than the mean (**Fig. 4.1**). Because this interference diminishes with distance, inter-exchange distances, instead of being exponentially distributed, tend to be concentrated near the mean (**Fig. 4.2**). When chromosomes are

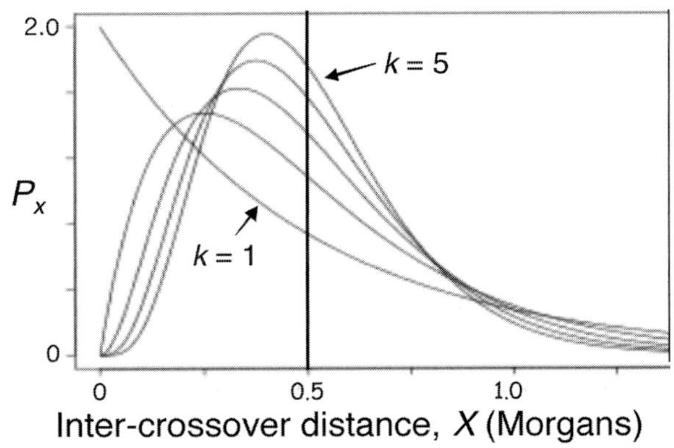

Fig. 4.2. The curves are members of the family of Erlang distributions, which have provided useful characterizations of chromosome interference. The distributions have a single adjustable parameter (k). When $k = 1$, the curve is the exponential expectation for no interference. The curve for $k = 5$ provides a good description of data from the X chromosome of *Drosophila* (*7, 12*). (k of the Erlang distribution corresponds to $m + 1$ of the Counting Model.) Because length is measured in Morgans, the rate parameter in these Erlang distributions is 2k, and the distributions all have the same mean value, 0.5 Morgans, which is the average linkage map distance between crossovers. The probability density function is then $P_X = 2 \cdot k \frac{(2kX)^{k-1}}{(k-1)!} e^{-2kX}$ The logic underlying the Erlang distribution is described in **Section 3.4**.

well enough marked to establish these deviations from independence, powerful methods of analyzing interference are available (**Section 3.4**).

But first, we shall describe the more routinely useful methods for analyzing interference in two-factor data (**Section 3.1**), three-factor data (**Section 3.2**) and four-factor data (**Section 3.3**). Finally, we shall deal with assessing the effects of meiotic mutants on interference (**Section 3.5**).

2. Materials

2.1. Collecting Data

1. Standard microbial genetics laboratory equipment.
2. Dissection microscope and micromanipulator for dissecting meiotic tetrads. Singer Instruments makes computer-assisted equipment that eases the task of dissection.

2.2. Recording, Organizing, and Evaluating Data

1. Computer with Excel spread sheet program.
2. For sorting data, the MacTetrad program, available by Gopher from merlot.wekj.jhu.edu. A more versatile program is available on request from Eric Alani at eea3@cornell.edu.
3. Internet access.

3. Methods

3.1. Two-Factor Data (*see* Note 1)

When all four spores are viable and both marked loci segregate 2:2 (no conversion), the diploid *AB/ab* produces three types of meiotic tetrads. These tetrads are classifiable as parental ditypes (PDs), tetratypes (TTs), or nonparental ditypes (NPDs), respectively.

Tetrad types	Genotypes of the four haploid spores			
PD	*AB*	*AB*	*ab*	*ab*
TT	*AB*	*Ab*	*aB*	*ab*
NPD	*Ab*	*Ab*	*aB*	*aB*

We shall refer to the observed frequencies of those three types of tetrads as fP_{obs}, fT_{obs} and fN_{obs}, respectively.

3.1.1. NPD Ratio

Papazian *(5)* provided an equation for using fT_{obs} (called F by Papazian) to calculate the NPD frequency expected in the absence of interference (fN_{exp}).

$$fN_{exp} = \frac{1}{2}\left[1 - fT_{obs} - 1 - \frac{3fT_{obs}^{\frac{2}{3}}}{2}\right] \qquad [1]$$

valid for $0 \leq fT_{obs} \leq 2/3$

The "NPD ratio" is fN_{obs}/fN_{exp}. Since its introduction the NPD ratio has been widely used by yeast geneticists as an indicator of interference (*see* **Note 2**). Values less then unity suggest interference within the interval examined, while values near unity suggest weak or no interference. As an indicator of interference, the NPD ratio has the following shortcomings:

Imprecision: Because the NPD ratio involves two values that are both characteristically imprecise, it is especially subject to statistical imprecision. (*i*) In many data sets NPDs will be rare and, as a result, the numerator of the ratio is often subject to large statistical uncertainty. (*ii*) The denominator of the ratio, fN_{exp}, is also subject to imprecision – it is sensitive to changes in fT_{obs} when fT_{obs} is close to zero and is even more sensitive to fT_{obs} fluctuations at values close to 2/3 (*5*).

NPD ratios with Standard Errors can be calculated by entering raw tetrad data at "Map Distance, Interference, and Statistical Significance Based on Tetrad Data" at Stahl Lab Online Tools http://molbio.uoregon.edu/~fstahl/.

Limited range: Papazian's NPD ratio cannot be calculated when fT_{obs} exceeds 2/3 because, under the null hypothesis of no interference, fT_{obs} cannot exceed 2/3 (*see* **Note 3**).

Inefficient use of data: Papazian's calculation for the expected frequency of NPDs (fN_{exp}) uses the available data inefficiently. It determines expectations of the no-interference model solely on the basis of the observed TT frequency, fT_{obs}, rather than on the basis of the entire data set.

A questionable use of the Papazian equation for fN_{exp}: Some authors have calculated p values based, somehow, on a Chi-Square comparison of N_{exp} and N_{obs} but have offered no details. Because such calculations make inefficient use of the tetrad data, they can lead to unjustified decisions regarding significance.

Dependent on length of interval: While the NPD ratio is an indicator of interference, it is not a measure of interference per se because it is dependent both on the strength of interference and the linkage map length of the interval employed.

3.1.2. A Better Two-Factor Test

The first four shortcomings (above) are eliminated by a two-factor test for interference that is both more efficient than the NPD ratio test and valid for any fT_{obs} and fN_{obs} *(27)*. The calculation is carried out in six steps, all of which are exact.

1. From a set of tetrad data, calculate the recombinant frequency (R) from fN_{obs} and fT_{obs}:

$$R = fN_{obs} + fT_{obs}/2 \qquad [2]$$

2. Use the value of R from Eq. [*2*] to calculate X, the map length in Morgans, under the assumption of no interference *(6)*:

$$X = \frac{1}{2}\ln(1 - 2R) \qquad [3]$$

3. Use X from Eq. [3] to calculate fT_{exp} using Papazian's equation [*2*] (rewritten):

$$fT_{exp} = \frac{2}{3}\left(1 - e^{-3X}\right) \qquad [4]$$

4. Use the fT_{exp} from Eq. [4] to calculate fN_{exp} using Papazian's equation [*3*] (rewritten):

$$fN_{exp} = \frac{1}{2}\left[\left(1 - e^{-2X}\right) - fT_{exp}\right] \qquad [5]$$

5. Calculate fP_{exp} as $1 - fT_{exp}$ - fN_{exp}
6. Multiply fP_{exp}, fT_{exp}, and fN_{exp} by total tetrads to obtain the expected numbers of the three tetrad classes, P_{exp}, T_{exp}, and N_{exp}.
7. Calculate Chi-Square to obtain a statistic with one degree of freedom:

$$\chi^2 = (P_{obs} - P_{exp})^2/P_{exp} + (T_{obs} - T_{exp})^2/T_{exp} + (N_{obs} - N_{exp})^2/N_{exp}$$

where P_{obs}, T_{obs}, and N_{obs} are the observed numbers of the three tetrad types.

If the probability of such a large deviation from the null hypothesis of no interference (p value) of this statistic is < 0.05 (d.f. $= 1$) then, by convention, the no-interference model fails to fit the data. If the test gives a p value indicative of interference, positive and negative interference can be distinguished by noting whether $N_{obs} < N_{exp}$ (positive interference) or $N_{obs} > N_{exp}$ (negative chromosome interference, positive chromatid interference, or a mitotic clone). This method uses all the tetrad data and does so without approximations.

Example We consider two sets of data. This test is superior to the NPD ratio test for any fT_{obs}. We have chosen $fT_{obs} > 2/3$ to illustrate, as well, the extended range of the test. One set is of 1,000 tetrads, and the other is 200 tetrads. $fT_{obs} = 0.700$ in both. For illustration, we take $fN_{obs} = 0.100$ and $fP_{obs} = 0.200$. Following steps 1–5 above, $R = 0.45$, $X = 1.1513$, $fT_{exp} = 0.64558$, $fN_{exp} = 0.12721$, and $fP_{exp} = 0.22721$. We calculate the expected numbers of tetrads in each class by multiplying each expected frequency by 1,000 for the larger data set and by 200 for the smaller one.

These calculations are automated at the Web site Stahl Lab Online Tools. As long as the expectation $N_{exp} \geq 5$, Chi-Square values are calculated. A link to Vassarstats *Chi-Square to p calculator* is provided. For the larger data set, Chi-Square = 13.66. With d.f. = 1, $p = 0.0002$. With the smaller data set, Chi-Square = 2.73. With d.f. = 1, $p = 0.0985$ (*see* **Note 4**). Absent negative chromatid interference, the larger data set clearly establishes chromosome interference, which is positive since $fN_{obs} < fN_{exp}$.

3.1.3. Quantifying Interference

When the Chi-Square test (**Section 3.1.2**) indicates significance, as it does for the larger data set, the "strength" of positive interference can be quantified within the framework of one or another model that relates interference to genetic distance. As explained in **Section 3.4**, the Counting Model *(7)* is one of a set of "renewal models" which have proven most useful. The magnitude of the Counting parameter (m) that gives the best fit to the data is an estimate of the degree to which the data deviate from the null hypothesis of no interference. The best m value for your data can be determined graphically from **Fig. 4.3**, by noting which m curve lies closest to the intersection of your fN_{obs} and fT_{obs} values. If your straight edge is straight, your hand steady, and your eyes good, you will see that the best m value for the data in the Example of **Section 3.1.2** is about two.

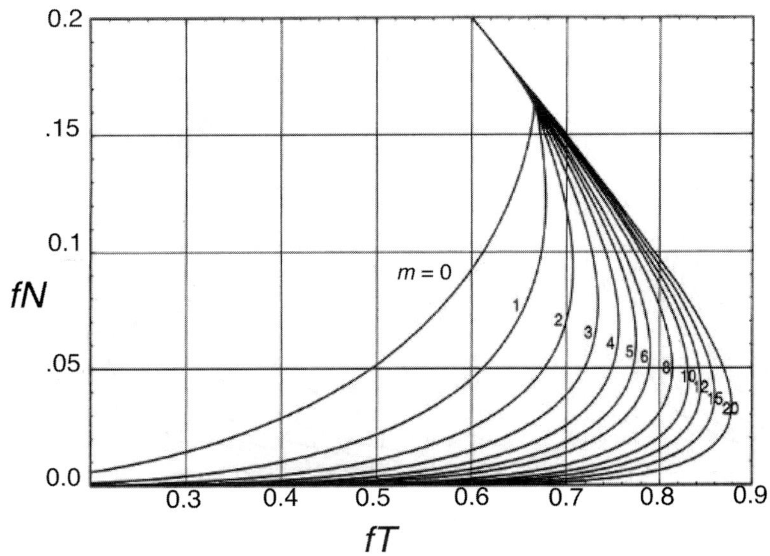

Fig. 4.3. Curves for fN_{exp} plotted vs. fT_{exp} for selected values of m. The value of m that best characterizes an fN_{obs}, fT_{obs} pair is given by the curve that comes closest to the intersection of those observed values. The $m = 0$ curve corresponds to no interference. The other curves correspond to expectations of the Counting Model *(7)*. The figure is from Stahl and Lande *(29)*.

3.1.4. A Faster, Internet-Dependent Way of Detecting and Quantifying Interference

The section of Stahl Lab Online Tools called "Calculation of m Value from Tetrad Data" provides a calculator that identifies values of the parameter m that fall within the 95% or 99% (enter your choice) confidence intervals. The calculator tests the fit of the data to selected values of m from zero (no interference) to 20. For the larger data set above, for which $p = 0.0002$, the "m calculator" gave the results pictured in **Fig. 4.4**. By failing to intersect the $m = 0$ curve, the confidence envelopes indicate significant interference. By intersecting the curves for $m = 1$, 2, and 3 the confidence envelopes indicate that the data are compatible with those values of m. The point, which is at the intersection of fN_{obs} and fT_{obs}, indicates that 2 is the best choice for m.

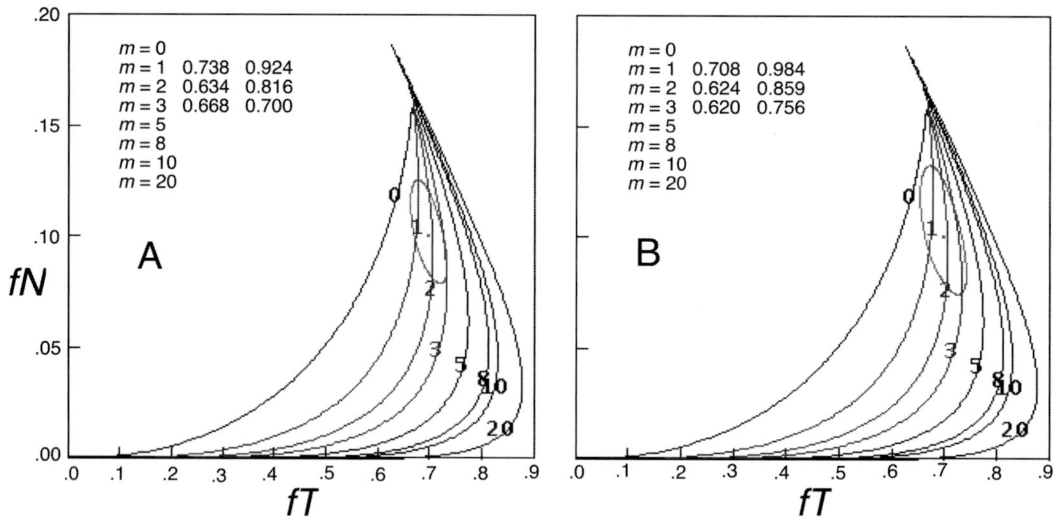

Fig. 4.4. The curves, downloaded from "Calculation of *m* Value from Tetrad Data" at the Stahl Lab Online Tools Web site, are based on the Counting Model *(7, 29)*. **A**. Area for 95% confidence. **B**. Area for 99% confidence. The (fictitious) data entered ($N_{obs} = 100$, $T_{obs} = 700$, $P_{obs} = 200$) manifests significant interference by the failure of the 95% confidence curve to intersect the curve for $m = 0$. Within the framework of the Counting Model (*see* **Section 3.4**), these data indicate $m \sim 2$, but are compatible with $m = 1$ and $m = 3$. Upper and lower limits for map length (X in Morgans) are presented for the three acceptable values of m. The figures were printed from "Calculation of *m* Value from Tetrad Data" at Stahl Lab Online Tools. Examples of real two-factor yeast data tested for interference with the *m* calculator can be found in Stahl and Lande *(29)*.

3.2. Three-Factor Data

Three factor crosses, when the two intervals are short, is a limiting case of the classical four-factor analysis of interference *(4, 7)*, described in the next section. In this section we treat three-factor data without the limitation that both intervals be short.

3.2.1. Random Spore Data

If data on recombinant frequencies are collected by taking a small sample of spores from a large number of acts of meiosis, the classical coefficient of coincidence can be calculated exactly as defined for *Drosophila* data *(3, 4)* (*see* **Note 5**). The coefficient of coincidence is defined for any two intervals as $S =$ (observed double recombinants)/(double recombinants expected on the assumption of independence). This expectation is $(R_1)(R_2)$ where the *R*s refer to the recombinant frequencies for the individual intervals determined without regard to the status of the other interval. The expectation can be tested by Chi-Square. The test does not distinguish chromatid interference from chromosome interference.

Example For a data set of 1,000 random spores with $R_1 = 0.100$ and $R_2 = 0.200$, the expected number of double recombinants is 20.0. If only ten double recombinants were observed, the coefficient of coincidence is 0.5 and one suspects significant positive interference. The user-friendly web site VassarStats ("For a 2 × 2 Table of Cross-Categorized Frequency Data: Version 1," under "Frequency Data") will determine the statistical significance for you with minimal effort on your part if you enter the observed numbers in the 2 × 2 table as suggested by Bailey *(8)* (1961, p. 147). For the cross *ABC* × *abc*:

Observed number of *ABC* + *abc* = 710	Observed number *Abc* + *aBC* = 90
Observed number *ABc* + *abC* = 190	Observed number *AbC* + *aBc* = 10

The Chi-Square value (Yates, corrected for continuity) is 6.27 and $p = 0.012$. If this *p* value isn't good enough for you, you may halve it to get an (approximate) one-tailed value, since you were testing not just for interference, but for positive interference. Note that The Fisher Exact test is simultaneously calculated by VassarStats (for numbers, such as these, that are not too large) and gives a one-tailed *p* value = 0.004.

3.2.2. Testing Three-Factor Tetrad Data for Chromatid Interference

The following test for chromatid- (as opposed to chromosome-) interference is the classical one:

1. Identify tetrads that are TT in both intervals.
2. Test the identified tetrads to see whether the frequencies of two-,three-, and four-strand double crossovers deviate significantly from the expected ratio of 1:2:1.

Example 500 tetrads that were TT in interval 1 were identified. 100 of these were also TT in interval 2. Of those 100, 27 involved the same two chromatids (regressive exchanges), 54 involved one of the same chromatids (progressive exchanges), and 19 involved the two chromatids that were not involved in the interval 1 exchange (digressive exchanges). These observed frequencies can be tested by the Chi-Square "Goodness of Fit" test (under Frequency Data at Vassarstats) against the expectations of the 1:2:1 ratio, which is 25:50:25 for a sample of size 100. These data give a Chi-Square value of 1.92, which, at two degrees of freedom, has a *p* value of 0.38. Thus, these data give no clear evidence of chromatid interference, and, in view of the rarity of such interference, it is justifiable to calculate chromosome interference as if there were no chromatid interference.

The Statistically Savvy may wish to use the more powerful (but less user-friendly) method *(9)* to test for chromatid interference.

3.2.3. A Tetrad-specific "Coefficient of Coincidence" for Chromosome Interference

Absent of chromatid interference, the frequencies of "crossover tetrads" ($fT_{obs}+fN_{obs}$) for the two intervals can be used to calculate a "coefficient of coincidence" for chromosome interference as if they were R values.

Example In 1,000 tetrads, $fT_{obs}+fN_{obs}$ for interval 1 = 0.500 and $fT_{obs}+fN_{obs}$ for interval 2 = 0.400. The frequency of double crossover tetrads for the two intervals is expected, in the absence of interference, to be 0.500 × 0.400 = 0.200. If the observed frequency of double crossover tetrads was 0.100, you could report a "coefficient of coincidence" of 0.100/0.200 = 0.500, indicating positive interference, whose statistical significance could be tested with a 2 × 2 table:

Tetrads that are PD in both intervals = 200	Tetrads that are not PD in interval 1 but are PD in interval 2 = 400
Tetrads that are PD in interval 1 and not PD in interval 2 = 300	Tetrads that are PD in neither interval = 100

According to VassarStats, the Yates Chi-Square statistic for these data is 165, for which the p-value is less than 0.0001

3.2.4. Another Tetrad Test for Chromosome Interference

This test measures the reduction in the Perkins map length *(10)* of one interval as a consequence of exchange(s) in a second interval (the adjacent interval, in a three factor cross).

1. Identify the crossover tetrads (TTs plus NPDs) in one interval.
2. Among those tetrads use Perkins's *(10)* equation, $X = fT_{obs}/2 + 3fN_{obs}$, to calculate the map length of the other interval (*see* **Note 6**).
3. Compare that map length with the Perkins map length of that interval calculated from the noncrossover (PD) tetrads.

Example The *URA3-KAN* interval is adjacent to the *KAN-lys5* interval. The *KAN-lys5* data were divided into (TT+NPD) and PD classes. Each of these two classes was then scored as PD, TT, or NPD in the *URA3-KAN* interval:

	PD	TT	NPD
TT + NPD	521	203	3
PD	598	548	12

Those two sets of tetrad data *(11)* were then entered into the table at "Map Distance, Interference, and Statistical Significance Based on Tetrad Data" on Stahl Lab Online Tools. The Web site calculator reported Perkins map lengths of 15.2 ± 1.1 cM and 26.8 ± 1.1 cM, respectively. They were declared significantly different ($p < 0.05$) on the basis that $2(\mathrm{Var}[X1] + \mathrm{Var}[X2])^{1/2} < |X1 - X2|$. Since the authors were testing for positive interference, they could have used the one-sided test: $1.645\ (\mathrm{Var}[X1] + \mathrm{Var}[X2])^{1/2} < |X1 - X2|$.

Interference is, by definition, reciprocal, so we can expect to find a significant effect of crossovers in *URA-KAN* on the Perkins map length of *KAN-lys*:

	PD	TT	NPD
TT + NPD	560	204	2
PD	598	519	2

These data for the two classes of *URA-KAN* tetrads yield the significantly different Perkins map lengths for *KAN-lys* of 14.1 ± 1 and 23.7 ± 0.8, respectively (*see* **Note 6**).

When fN_{obs} is large, this method may be a more powerful detector of interference than that in **Section 3.2.3.**

3.3. Four-Factor Data

Sturtevant *(3)* and Muller *(4)* characterized interference in *Drosophila* by determining coefficients of coincidence (S_4 in **Fig. 4.5**) for pairs of short intervals whose members were separated from each other by various map distances. S_4 was close to zero when the short intervals were adjacent (i.e., a three-factor cross) and was

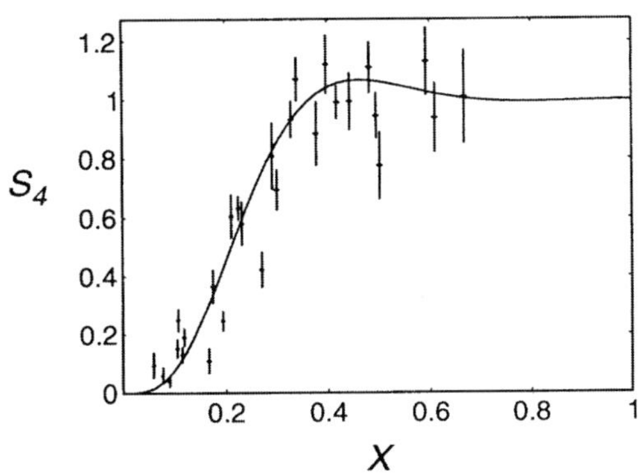

Fig. 4.5. Coefficients of coincidence (S_4) for pairs of short intervals, *AB* and *CD*, at varying linkage map distances (*X* cM) from each other: ***AB------X------CD***. The curve drawn is based on the Counting Model with $m = 4$ (equivalent to $k = 5$ of the Erlang distribution). The fit of the $m = 4$ curve is significantly better than the fits for the $m = 3$ and $m = 5$ curves *(7, 12)*. Data are from crosses involving multiply marked sex chromosomes of *Drosophila*. Figure is from Foss et al. *(7)*.

close to unity when the intervals were about 45 cM apart. The quasi-uniform spacing of crossovers, which characterizes interference, can result in values of S_4 that are greater than 1 when X approaches 0.5, falling below 1 as X increases and stabilizing at 1 for larger map distances. In conducting such tests, exchanges that may occur *between* the two intervals are ignored.

Because the value of S_4, the indicator of interference gained from a four-factor cross, depends on the distance between the two short test intervals, it is, by itself, only an indicator of interference. When S_4 data are graphed over the full range of X, however, they can provide a true *index* of interference (**Fig. 4.5**), which is freed from the particulars of any single cross *(7)*.

3.4. Multifactor Data and Models

A multiply marked chromosome allows for the fitting of statistical models for crossover interference. Of the many such statistical models historically proposed, only the Counting Model, also known as the Chi-Square model, and its close variants seem to fit genetic data *(12, 13, 14)*. A physical model based on stress-induced fractures *(15)*, which does not lend itself to parameter estimation from the kinds of data geneticists typically collect, provides a visually similar distribution for inter-crossover distances. We will concentrate our discussion on statistical models that can be tested with genetic data and for which the estimated parameter(s) provide useful indices of interference.

The Counting Model, as derived from *(7)*, envisions meiotic double-strand breaks as Poisson-distributed attempts with every $(m+1)^{st}$ break becoming a crossover and the intervening m attempts becoming noncrossovers. The noncrossovers are identifiable as gene conversions when they happen to impinge upon a genetic marker. The model was highly successful in its application to *Drosophila* data. Fitting historical genetic datasets from *Drosophila* to the statistical Counting Model gave an estimate of $m = 4$ *(7, 13)*. This corresponds nicely with work showing that, at the *rosy* locus, there are approximately four noncrossovers for every crossover *(16, 17)*.

The Counting Model is also called the Chi-Square model because the theoretical inter-crossover distribution is a scaled Chi-Square distribution. These distributions are also known as Erlang distributions, based on principles explored by the telephone engineer Agner Krarup Erlang (1878–1929). The typical parameterization of an Erlang distribution follows from the assumption that "attempts" are Poisson distributed and that every *k*th attempt in a linear sequence will be a "success." Thus, the parameter *k* in the Erlang distribution corresponds to $m + 1$ in the Counting Model.

The Chi-Square distribution and the Erlang distribution are members of the Gamma distribution family. The density function for a generic Gamma family member involves a rate parameter, λ, and a shape parameter, ν and is

$$f(x|\lambda, \nu) = \lambda^{\nu} x^{\nu-1} e^{-\lambda x}/\Gamma(\nu)$$

The rate parameter, λ, describes the rate of the Poisson events. When distance is measured in terms of the expected number of crossovers along a chromosome, then the rate is twice (for the tetrad) the number of Poisson events needed to ensure a crossover, so $\lambda = 2k = 2(m+1)$. The shape parameter, ν, is the equivalent of the number of events needed to ensure a crossover (k or $m+1$). Statisticians often favor the Gamma distribution when describing the inter-crossover distance distribution because there is no mathematical requirement that ν be an integer.

3.4.1. Data with Relatively Sparse Genetic Markers

Fisher *(18)*, Owen *(19)*, and Payne *(20)*, as described by Bailey *(8)*, laid the groundwork for the statistical analysis of recombination data involving multiple crossovers in a series of intervals along a chromosome, assuming an Erlang distribution. The statistical analysis is anchored in the assumed Poisson distribution of (unobserved) "attempts." For each interval, a matrix is used to keep track of the possibilities for the number of noncrossover "attempts" before the first crossover and after the last crossover in an interval. In a process dubbed the segmental calculus *(8)*, matrices for adjacent intervals are multiplied together, so that the dependencies between intervals are preserved. Cobbs *(21)* and Stam *(22)* corrected Owen's model to involve the crossover process on the four-strand bundle rather than on an individual chromatid, and, to allow $m > 1$, they extended the model for the four-strand bundle along the lines that Payne extended the original model.

Since the observed data are whether an interval is PD, TT, or NPD, rather than the actual number of crossovers that exist in the interval, formulas of Mather *(23)* are used to translate the probability (matrix) of having s crossovers in an interval into the probability (matrix) of observing a PD, TT, or NPD in the interval *(13)*.

Finally, with no information about the starting and ending configuration of crossovers and noncrossovers, one assumes a uniform starting distribution and calculates the probability of the observed data pattern *(13)*.

Such an analysis surpasses the statistics available in most yeast laboratories. However, there is a JAVA program that will analyze raw tetrad data under the Counting Model (and its extension discussed below) at: http://mypage.iu.edu/~ehouswor/InterferenceAnalyzer/

3.4.2. Data with Densely Spaced Markers

The ability to detect sequence polymorphisms that are phenotypically neutral (unlike many other markers) makes possible the collection of data in which viabilities of all types are good and essentially every exchange is detected. If the location of every exchange is known to within a very small margin of error, then the statistical analysis changes in order to make full use of the observed inter-crossover distances. The proper analysis makes use of all tetrads, even those with no exchange or with only one exchange. The inter-crossover distances are modeled as an Erlang or Gamma distribution while the distance from the last crossover to the end of the chromosome is modeled as a censored Erlang or Gamma distribution. The distance from the beginning of the chromosome to the first crossover is modeled to preserve stationarity (so that reversing "first" and "last" or "beginning" and "end" does not change the distribution of crossover positions on the tetrad). The use of such analyses is limited by the current lack of software to implement the procedure, although this general technique has been employed by individuals capable of programming the technique themselves *(14, 24, 25)*.

3.4.3. Complications in Applying the Counting Model to Yeast

There are several complications when fitting the above models to data from yeast. In yeast, as in many other organisms, crossovers appear to occur in two phases with differing interference properties *(26)*. A variant of the Counting Model that allows for an independent mixture of noninterfering ($m = 0$) and interfering ($m > 0$, estimated from the data) crossovers, with noninterfering crossovers making up a proportion p of all crossovers, has improved the fit of the model to tetrad data from *Arabidopsis* and Yeast *(11, 28)*.

Another complication in yeast is the heterogeneity of the crossover process within and among chromosomes in yeast. Highly marked yeast chromosomes provide data that can be broken down into many sets of two-, three-, or four-factor crosses and can be characterized for interference by any or all of the methods described above. The two- and three-factor data in the yeast literature suggests that the strength of interference varies among chromosomes and along chromosomes, defying global characterization by the methods described above *(13, 29)*. The application of the Counting Model and its extension to yeast by *(11)* was limited to one fairly homogeneous chromosome arm.

Crossing over in yeast and other fungi (and possibly other organisms) is initiated by meiosis-specific double-strand breaks, which occur at "hotspots." The assumption of the Counting Model that "attempts" are Poisson-distributed may be compromised by the occurrence of intervals containing only a few such hotspots. This caveat, which is yet to be explored, applies to the no interference model ($m = 0$) as well.

3.5. Interference in Meiotic Mutants (see **Note 7***)*

Meiotic mutants that alter the values of interference indicators are of special interest because they promise gateways to understanding the mechanism(s) of interference. However, most, at least, of such yeast mutants apparently alter interference simply by altering the proportion of crossovers in the two phases described in **Section 3.4**

Deletion of genes in the "*ZMM*" epistasis group *(30)* reduces crossing over and also reduces interference, as manifested by a reduction in the value of all the indicators. The association of these two phenotypes results from a reduction in crossing over specifically in the phase that is subject to interference *(26, 31)*. In some yeast strains, deletion mutants of *NDJ1/TAM1* have higher than normal crossing over and weaker than normal interference *(26)*. In these strains, the association of these two phenotypes is the consequence of a specific increase in crossing over in the noninterference phase *(26)*.

3.5.1. Detection of a Significant Change in Interference

The frequently used indicators of interference are functions of both map length and interference. Consequently, they are not well suited to establishing the biological significance of changes in the value of interference indicators when those changes are accompanied by changes in the rate of crossing over. In the absence of any change in interference, these indicators (NPD ratio, coefficients of coincidence) approach unity as map length increases. Thus, some combinations of map-length/interference phenotypes are less easily interpreted than others. Potentially ambiguous are (*i*) mutants that simultaneously increase map lengths and appear to decrease interference as judged by increase in the NPD ratio or the coefficient of coincidence (*see* **Note 8**) and (*ii*) mutants that decrease map lengths and appear to increase interference by decreasing the NPD ratio or the coefficient of interference. These problems call for the use of *indices* of interference (such as *m* of the Counting Model).

4. Notes

1. In general, chromatid and chromosome interference are indistinguishable in two-factor crosses. Consequently, conclusions regarding chromosome interference are provisional. Papazian *(5)* deals with some two-factor data in which chromatid interference is apparently evidenced by $fN_{obs} > 1/6$, which exceeds the theoretical limit in the absence of chromatid interference. However, as Papazian *(5)* points out, premeiotic crossing over, rather than nonrandom choice of partners at the 4-strand stage, could be responsible for such data. Yeast workers are aware that crossing over occurring during pre-sporulation mitotic growth can contaminate a set of tetrad data with NPDs, leading to false conclusions regarding interference (and map lengths). This seems to be a rare problem with wild-type yeast strains, but might be a considerable problem with some mutants. The problem can sometimes be minimized by "mass matings" of haploid parents followed by sporulation of the resulting diploid population with minimal intervening mitotic growth.
2. Papazian *(5)* often refers to recombinant frequencies (*R*) as "map distances" (e.g., **page 180**). Since the concept of distance in one-dimension implies additivity, equating *R* values, which generally lack additivity, with distances perpetuates a

confusion that seems to originate with Sturtevant *(32)*. Muller *(4)* clarified the distinction between *R* and distance, and Haldane *(6)* clearly identified "map distance" with the mean number of exchanges. Papazian [*(5)*, **page 179**] exacerbates the confusion when he seems to imply that map distances ($m/2$ in his notation) can be only *provisionally* taken to be additive. As a result, the reader may have difficulty following some of the arguments and, consequently, have reservations regarding some of the conclusions. The hazards are compounded by the misprinting of the equation that relates fN_{exp} to fT_{obs} in the absence of interference (**page 182**).

3. Papazian *(5)* offers the following algebraic approximation to his equation for fN_{exp}:

$$fN_{exp} = \frac{fT_{obs}^2}{8}\left[1 + \frac{2}{3}fT_{obs}\right].$$

He states that this approximation "...is valid for long intervals and can be used where other methods of measuring interference cannot." Strickland *(33)* stated that Papazian's approximation was incorrect. The correct expression, he asserted, is

$$fN_{exp} = \frac{fT_{obs}^2}{8}\left[1 + \frac{3}{2}fT_{obs}\right].$$

Strickland stated that this approximation is valid only for small values of fT_{obs}.

We can see who is right, if either. fN_{exp}, calculated exactly from fT_{obs} using Eq. [*1*], can be compared with the values obtained from two approximations. In fact, the Papazian approximation is found to be within 1% of the exact equation at $fT_{obs} = 0.1$, while the Strickland version is off by 7%. At $fT_{obs} = 0.66$, the Papazian approximation is off by 47% and the Strickland approximation, which tracks the exact equation pretty well until $fT_{obs} = 0.57$, is off by 26%. Thus, neither approximation lives up to its billing, and there is really no use for them in this day of electronic calculators. Some authors (who shall remain nameless) have taken Papazian literally and used his approximation when $fT_{obs} > 2/3$. That is a No-No.

When faced with $fT_{obs} > 2/3$, interference can be tested quickly by asking whether fT_{obs} is significantly greater than $2/3$, since such values are impossible in the absence of interference. If fT_{obs} is insignificantly greater than $2/3$, one can ask whether fN_{obs} is significantly less than fP_{obs}, a good indication of interference when $fT_{obs} > 2/3$. However, the best test to use when $fT_{obs} > 2/3$ is that described in **Section 3.1.2**.

4. The difference in the p values in these two examples is (obviously) due to the greater statistical power of the data set with 1,000 tetrads. With this larger data set, the presence of interference is convincingly demonstrated. These calculations were chosen to bring home the importance of not equating a large p value (i.e., $p > 0.05$) with the absence of interference. Such a p value may, instead, indicate only a failure to demonstrate the presence of interference as a result of too few data. By the same token, a small p value gives confidence only regarding the existence of interference, not its magnitude. Should you feel the need, an NPD ratio can be calculated using the fN_{exp} obtained from this calculation. In the example given here, that ratio is $0.100/0.127 = 0.79$, where a ratio of unity would imply no interference. The Papazian approximation (**Note 3**) for these data gives a misleading value of 1.11, which would be indicative of negative interference if taken seriously.
5. Random spore analysis may be an option when tetrad dissection is not. However, since random spores can give no evidence regarding chromatid interference, tetrad analysis is preferable.
6. The Perkins equation *(10)* is valid, in the absence of chromatid interference, for intervals in which the frequency of more than two exchanges is negligible.
7. Shinohara et al. *(34)* described a situation in which overexpression of a meiotic protein in a strain with genetically impaired recombination, due to lack of a different meiotic protein, resulted in restoration of the wild-type recombination rate, but with severely impaired interference. The overexpressed gene was carried on a multicopy plasmid that is notoriously variable in copy number. This concoction is a prime candidate for negative interference (resulting from population heterogeneity) canceling positive interference (intrinsic to the crossover process). The authors may have reached the wrong conclusion as a result of failing to conduct an inter-chromosomal test for negative interference.
8. For deletion of the *NDJ1* gene, the simultaneous increase in map length and decrease in indicators of interference was clearly a result of increase in crossovers identifiable as belonging to the noninterference pathway. Chromosome bisection has the same pairing of phenotypes. The reality of the reduction in interference as a result of bisection was supported by a comparison of interference-indicator values for intervals of about the same (genetic) map length (D. Kaback pers. com.). As with the *ndj1* deletion, the paired phenotypes are probably a consequence of increased crossing over specifically in the noninterference phase *(26)*.

Acknowledgments

F.W.S. thanks John P. Nolan for drawing his attention to the view that the distribution of intercrossover distances in the Counting Model is aptly referred to as an Erlang distribution. Tom Petes notified us that the Web site "Stahl Lab Online Tools" mindlessly calculated fN_{exp} values for $fT_{obs} > 2/3$. It no longer does. Jette Foss provided invaluable editing of the manuscript. The Genetics Community is grateful to Richard Lowry, author of the continually improving, user-friendly web site VassarStats. Dan Graham kindly updated Stahl Lab Online Tools to reflect things we learned during the preparation of this Chapter; in doing so he caught some mistakes in our manuscript. The contribution by E.A.H. was supported by an NSF grant (DMS-0306243) to Indiana University.

References

1. Lhuissier, F. G. P., Offenberg, H. H., Wittich, P. E., Vischer, N. O. E. and Heyting, C. (2007) The mismatch repair protein MLH1 marks a subset of strongly interfering crossovers in tomato. *The Plant Cell* **19**, 862–876.
2. Fung, J. C., Rockmill, B., Odell, M. and Roeder, G. S. (2004) Imposition of crossover interference through the nonrandom distribution of synapsis initiation complexes. *Cell* **116**, 795–802.
3. Sturtevant, A. H. (1915) The behavior of the chromosomes, as studied through linkage. *Zeit. f. ind. Abst. u. Vereb.* **13**, 234–287.
4. Muller, H. J. (1916) The mechanism of crossing over. *Am. Nat.* **50**, 194–221 and *ff.*
5. Papazian, H. P. (1952) The analysis of tetrad data. *Genetics* **37**, 175–188.
6. Haldane, J. B. S. (1919) The combination of linkage values and the calculation of distances between loci of linked factors. *J. Genet.* **8**, 299–309.
7. Foss, E., Lande, R., Stahl, F. W. and Steinberg, C. M. (1993) Chiasma interference as a function of genetic distance. *Genetics* **133**, 681–691. Corigendum: *Genetics* **134**, 997.
8. Bailey, N. T. J. (1961) *Introduction to the Mathematical Theory of Genetic Linkage*, Oxford University Press, London
9. Zhao, H., McPeek, M. S. and Speed, T. P. (1995a) Statistical analysis of chromatid interference. *Genetics* **139**, 1057–1065.
10. Perkins, D. D. (1949) Biochemical mutants in the smut fungus *Ustilago maydis*. *Genetics* **34**, 607–626.
11. Malkova, A., Swanson, J., German, M., McCusker, J. H., Housworth, E. A., Stahl, F. W. and Haber, J. E. (2004) Gene conversion and crossing over along the 405-kb left arm of *Saccharomyces cerevisiae* chromosome VII. *Genetics* **168**, 49–63.
12. McPeek, M. S. and Speed, T. P. (1995) Modeling interference in genetic recombination. *Genetics* **139**, 1031–1044.
13. Zhao, H., Speed, T. P. and McPeek, M. S. (1995b) Statistical analysis of crossover interference using the Chi-Square model. *Genetics* **139**, 1045–1056.
14. Broman, K. W. and Weber, J. L. (2001) Characterization of human crossover interference. *Am. J. Hum. Genet.* **66**, 1911–1926.
15. Kleckner, N., Zickler, D., Jones, G. H., Dekker, J., Padmore, R., Henle, J. and Hutchinson, J. (2004) A mechanical basis for chromosome function. *Proc. Natl. Acad. Sci. USA* **101**, 12592–12597.
16. Hilliker, A. J. and Chovnick, A. (1981) Further observations on intragenic recombination in *Drosophila melanogaster*. *Genet. Res.* **38**, 281–296.
17. Hilliker, A. J., Clark, S. H. and Chovnick, A. (1991) The effect of DNA sequence polymorphisms on intragenic recombination in the *rosy* locus of *Drosophila melanogaster*. *Genetics* **129**, 779–781.
18. Fisher, R. A. (1951) A combinatorial formulation of multiple linkage tests. *Nature.* **167**, 520.
19. Owen, A. R. G. (1949) The theory of genetical recombination. I. Long chromosome arms. *Proc. Roy. Soc. B.* **136**, 67–94.

20. Payne, L. C. (1956) The theoryof genetical recombination: a general formulation for a certain class of intercept length distributions appropriate to the discussion of multiple linkage. *Proc. Roy. Soc. B.* **144**, 528–544.
21. Cobbs, G. (1978) Renewal process approach to the theory of genetic linkage: case of no chromatid interference. *Genetics* **89**, 563–581.
22. Stamm, P. (1979) Interference in genetic crossing over and chromosome mapping. *Genetics* **92**, 573–594.
23. Mather, K. (1935) Reductional and equational separation of the chromosomes in bivalents and multivalents. *J. Genet.* **30**, 53–78.
24. Housworth, E. A. and Stahl, F. W. (2003) Crossover interference in humans. *Am. J. Hum. Genet.* **73**, 188–197.
25. Lam, S. Y., Horn, S. R., Radford, S. J., Housworth, E. A., Stahl, F. W. and Copenhaver, G. P. (2005) Crossover interference on nucleolus organizing region-bearing chromosomes in Arabidopsis. *Genetics* **170**, 807–812.
26. Getz, T. J., Banse, S. A., Young, L. S., Banse, A. V., Swanson, J, Wang, G. M., Browne, B. L., Foss, H. M. and Stahl, F. W. (2007) Differential mismatch repair of heteroduplexes distinguishes interfering from "non"-interfering crossing over in *Saccharomyces cerevisiae*. *Genetics* **178**, 1251–1269.
27. Stahl, F. W. (2008) On the "NPD ratio" as a test for crossover interference. *Genetics* **179**, 701–704.
28. Copenhaver, G. P., Housworth, E. A. and Stahl, F. W. (2002) Crossover interference in Arabidopsis. *Genetics* **160**, 1631–1639.
29. Stahl, F. W. and Lande, R. (1995) Estimating interference and linkage map distance from two-factor tetrad data. *Genetics* **139**, 1449–1454.
30. Börner, G. V., Kleckner, N. and Hunter, N. (2004) Crossover/noncrossover differentiation, synaptonemal complex formation, and regulatory surveillance at the leptotene/zygotene transition of meiosis. *Cell* **117**, 29–45.
31. Zalevsky, J., MacQueen, A. J., Duffy, J. B., Kemphues, K. J. and Villeneuve, A. M. (1999) Crossing over during *Caenorhabditis elegans* meiosis requires a conserved MutS-based pathway that is partially dispensable in budding yeast. *Genetics* **153**, 1271–1283.
32. Sturtevant, A. H. (1913) The linear arrangement of six sex-linked factors in Drosophila, as shown by their mode of association. *J. Exp. Zool.* **14**, 43–59.
33. Strickland, W. N. (1958) An analysis of interference in *Aspergillus nidulans*. Proc. Roy. Soc. Lond. B. **149,** 82–101.
34. Shinohara, M., Sakai, K., Shinohara, A. and Bishop, D. K. (2003) Crossover interference in *Saccharomyces cerevisiae* requires a *TID1/RDH54*- and *DMC1*-dependent pathway. *Genetics* **163**, 1273–1286.

Chapter 5

Measurement of Spatial Proximity and Accessibility of Chromosomal Loci in *Saccharomyces cerevisiae* Using Cre/*loxP* Site-Specific Recombination

Doris Lui and Sean M. Burgess

Abstract

Several methods have been developed to measure interactions between homologous chromosomes during meiosis in budding yeast. These include cytological analysis of fixed, spread nuclei using fluorescence in situ hybridization (FISH) *(1, 2)*, visualization of GFP-labeled chromosomal loci in living cells *(3)*, and Chromosome-Conformation Capture (3C) *(4)*. Here we describe a quantitative genetic assay that uses exogenous site-specific recombination to monitor the level of homolog associations between two defined loci in living cells of budding yeast *(5)*. We have used the Cre/*loxP* assay to genetically dissect nuclear architecture and meiotic homolog pairing in budding yeast. Data obtained from this assay report on the relative spatial proximity or accessibility of two chromosomal loci located within the same strain and can be compared to measurements from different mutated strains.

Key words: budding yeast, meiosis, homolog pairing, chromosome, collision assay, site-specific recombination, Cre/*loxP*.

1. Introduction

The frequency of Cre-mediated recombination events (referred to here as "collisions") between pairs of *loxP* sites engineered into the *S. cerevisiae* genome provides a measure of the relative spatial proximity of those sites to one another in a living cell. A 13-fold level of induction of meiosis specific collision events in wild-type cells indicates a state of close, stable homolog juxtaposition (CSHJ) *(6)*. From mutant analysis, we have shown that CSHJ is

Scott Keeney (ed.), *Meiosis, Volume 1, Molecular and Genetic Methods, vol. 557*

DOI 10.1007/978-1-59745-527-5_5 Springerprotocols.com

achieved primarily through the function of trans-acting factors required for early and intermediate stages of meiotic recombination *(6–8)*.

Cre recombinase is supplied under the control of a galactose-inducible promoter (*pGAL1*). Cre-mediated crossing over between *loxP* sites engineered into specific chromosomal loci gives a genetically detectable product (**Fig. 5.1**) *(6)*. One *loxP* insert bears an upstream constitutive promoter (*pGPD*) and the other *loxP* insert bears a downstream open reading frame from a reporter gene (e.g. *ADE2*, *URA3*, or *GFP*). Collisions between pairs of *loxP* sites can be measured at allelic positions on homologous chromosomes and at ectopic positions on nonhomologous chromosomes in the same strain by using two different reporters. While ectopic collisions are consistently low, allelic collisions increase as cells progress through meiosis *(6)*. Depending on the type of reporter used for the assay, the output is either the frequency of prototroph formation upon return-to-growth (RTG) on media supporting vegetative growth or the frequency of fluorescent spores. For the protocol described here, *pGPD1-loxP* inserted in Chromosome *V* recombines with *loxP:ura3* inserted at an allelic site, or with *loxP:ade2* on Chromosome *VIII* at an ectopic position to generate Ura+ or Ade+ prototrophs, respectively. Strains listed in **Table 5.1** are available upon request.

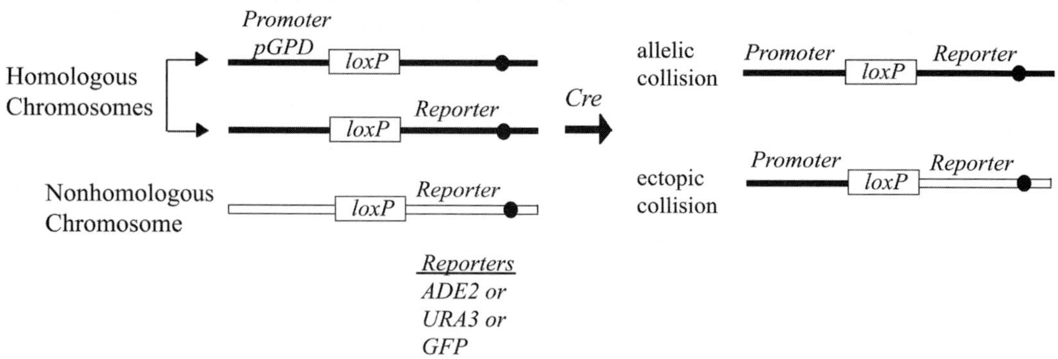

Fig. 5.1. Orientation and location of promoter (*pGPD-loxP*) and reporter constructs (*loxP*-reporter) relative to the centromere of the chromosome prior to Cre-mediated recombination and the resulting allelic and ectopic products following collision events. Reporter genes include *ADE2*, *URA3*, and *GFP*.

2. Materials

2.1. Synchronization of Cells for Entry into Meiosis

1. All media and solutions are made with deionized water.
2. Solid growth media: YPG (1% yeast extract, 2% peptone, 3% glycerol, 0.004% tryptophan, 0.01% adenine sulfate, 2% agar). YPD is the same as YPG except in place of glycerol, dextrose is

Table 5.1
Yeast strains

Name	Genotype[a]
SBY 1438	*MATα lys2::GAL1-Cre-LYS2 flo8::LEU2-pGPD1-loxP-lacZ ndt80::LEU2*
SBY 1160	*MATa lys2 flo8::LEU2-loxP-ura3*
SBY 1338	*MATa lys2 flo8::LEU2-loxP-ura3 ndt80::LEU2-loxP-ade2*
SBY 2877	*MATa lys2 arg4::LEU2-loxP-ade2 ndt80::LEU2*
SBY 2879	*MATα lys2::GAL1-Cre-LYS2 arg4::LEU2-pGPD-loxP ndt80::LEU2*
SBY 2865	*MATa lys2 flo8::LEU2-loxP-ade ndt80::LEU2*
SBY 2867	*MATa lys2 flo8::LEU2-loxP-ade*
SBY 2021	*MATα lys2::GAL1-Cre-LYS2 flo8::LEU2-pGPD1-loxP-lacZ*
SBY 2061	*MATα lys2 arg4::LEU2-loxP-ade2*
SBY 2063	*MATa lys2::GAL1-Cre-LYS2 arg4::LEU2-pGPD-loxP*
SBY 1060	*MATa lys2 arg4::LEU2-loxP-ura3*
SBY 1780	*MATa lys2::GAL1-Cre-LYS2 arg4::LEU2-loxP-ura3 ndt80::LEU-loxP-ade2*
SBY 1755	*MATa lys2 arg4::LEU2-loxP-ura3 ndt80::LEU2-loxP-ade2*
SBY 1192	*MATα lys2 ndt80::LEU2-pGPD1-loxP-lacZ*
SBY 1693	*MATa lys2::GAL1-Cre-LYS2 arg4::LEU2-act1-loxP-GFP ste7-1 cyh2-z MATα lys2 arg4::LEU2-GPD-act1-loxP-GFP ste7-1 CYH2*
SBY 1476	*MATa lys2::GAL1-Cre-LYS2 flo8::LEU2-pGPD1-loxP-lacZ ndt80::LEU2 MATα lys2 flo8::LEU2-loxP-ura3 ndt80::LEU2-loxP-ade2*

[a]All strains are SK1 and contain the following genotype: *ho::hisG ura3Δ::hisG leu2::hisG ade2Δ::hisG trp1::hisG GAL3*. Wild-type SK1 is *gal3 (15)*. *FLO8* is on Chromosome *V* and *NDT80* is on Chromosome *VIII*.

added from a filter-sterilized 40% (w/v) stock to a final concentration of 2% (w/v) after autoclaving. For all media, tryptophan and adenine sulfate are added from filter-sterilized 1% (w/v) stock solutions to autoclaved media cooled to 65°C. Keep 1% tryptophan stock at 4°C and shield from light. Stored 1% adenine sulfate will form crystals that can be melted by heating the solution.

3. Liquid growth media: Liquid YPD is prepared as above except agar is omitted. YPA is 1% yeast extract, 2% peptone, 1% potassium acetate, 0.004% tryptophan, 0.01% adenine sulfate.
4. Synthetic complete (SC) powder: mix together 0.8 g adenine sulfate, 0.8 g arginine, 2 g aspartic acid, 0.8 g histidine, 2.4 g leucine, 1.2 g lysine, 0.8 g methionine, 2 g phenylalanine, 8 g threonine, 0.8 g tryptophan, 1.2 g tyrosine, 0.8 g uracil and grind to a fine powder with a mortar and pestle. SC-URA drop-out powder is made similarly except that uracil is omitted.
5. Liquid sporulation medium (SPM): 1% potassium acetate, 0.02% raffinose, 0.009% SC powder.
6. 18-cm and 25-cm test tubes

2.2. Cre Induction and Return-to-Growth

1. SC-URA medium: 0.67% yeast nitrogen base without amino acids, 0.09 % SC-URA drop-out powder, 2% dextrose, 0.004% tryptophan, 0.01% adenine sulfate, 2% agar. Supplement with dextrose, tryptophan, and adenine sulfate as described above.
2. YPD-ADE solid medium: same as YPD but without supplementation with adenine sulfate.
3. Sugar solutions: 2% (w/v) galactose prepared from a 20% (w/v) filter-sterilized stock solution; 2% (w/v) glucose prepared from a 40% (w/v) filter-sterilized stock solution.
4. 550 Sonic ZD-dismembrator (Fisher Scientific).
5. Glass pipets (1-mL and 0.2-mL).
6. Fixed volume or calibrated liquid dispenser.
7. 16-cm test tubes.

3. Methods

Cells from three independent colonies are cultured and synchronized to enter meiosis upon transfer to sporulation medium. Transfer of cells to SPM marks t = 0 h of the meiotic time course. At t = 1 h (about the time of DNA replication), Cre recombinase expression is induced by the addition of galactose. Sample aliquots are removed from the culture at t = 1 (before induction), 2, 4, 6, 8, and 10 h after transfer to SPM and diluted appropriately for plating on selective and nonselective media for RTG. The *ndt80*Δ mutant arrests in pachytene with full-length synaptonemal complexes and unresolved double-Holliday junction intermediates *(9, 10)*. The use of the *ndt80*Δ mutation allows for recovery of cells in most mutants via RTG that would otherwise be inviable if allowed to complete meiotic divisions and sporulate *(11–13)*. Random spore analysis can be performed with strains that are *NDT80*. Expression of the *GFP* reporter can be visualized using fluorescence microscopy in individual spores contained in an ascus due to the spore-autonomous expression of the *GPD* promoter *(14)*.The number of prototrophs per colony-forming unit (C.F.U.) or the number of fluorescent tetrads per total tetrads for each time point indicates the frequency of collisions. Comparison of C.F.U. at each time point relative to t = 1 h reports on survivability of cells by RTG.

3.1. Synchronization of Cells for Entry into Meiosis

1. From frozen stocks stored in 15% (v/v) glycerol at –80°C, patch cells onto YPG plates. Incubate less than 15 h at 30°C (*see* **Note 1**).
2. From YPG plate, streak for single well-spaced colonies on YPD plates. Incubate 2 d at 30°C.
3. Inoculate 5 mL YPD (18-cm test tube) with a single colony from a YPD plate. In a typical experiment, analysis of each strain is carried out in triplicate and is performed in conjunction with a wild-type control strain, also in triplicate (*see* **Note 2**). Incubate at 30°C on roller drum at 56 rpm for at least 30 h.
4. In a 25-cm test tube, inoculate 10 mL of YPA to a final OD_{600} of 0.23 (*see* **Note 3**). Incubate on roller drum at 30°C at 56 rpm for 14 h.
5. Examine cultures microscopically to determine whether vegetative growth has been arrested at G0/G1. Quantify the fraction of mitotic dividing cells, which are cells with small buds, for one culture of each strain. If more than 5% of cells are small-budded in any culture, consider incubating all of the cultures another 30 min. If more than 10% of cells have small buds, abort the experiment.
6. Pellet cells at 2,400*g* for 3 min.
7. Resuspend cells in 10 mL of sterile deionized water, then pellet at 2,400*g* for 3 min.
8. Resuspend cells in 10 mL of SPM pre-warmed to 30°C and transfer to fresh 25-cm test tubes. Incubate on roller drum at 30°C at 56 rpm. The time at which the SPM cultures are placed back in the roller drum is designated as t = 0 h.

3.2. Cre Induction and Return-to-Growth

1. Prepare one 16-cm test tube that contains 5 mL sterile deionized water and two 16-cm test tubes that contain 4.5 mL for each timepoint for each culture. These tubes will be used to perform the dilution series in **Step** 7 below. They can be prepared in advance using a fixed-volume water dispenser. We typically carry out analysis on four strains in triplicate, so 36 test tubes are required for each time point.

2. At t = 1 h, remove 1 mL of uninduced cells from each culture and transfer to a 1.5 mL microcentrifuge tube.
3. After removing the first aliquot, induce *Cre recombinase* expression by adding 135 μL of 2% galactose into 9 mL of remaining culture (final concentration of 0.03% – *see* **Note 4**).
4. Place cultures back on rollerdrum to incubate at 30°C at 56 rpm. For GFP analysis, *see* **Note 5**.
5. Pellet the 1 mL aliquot of cells for 1 min at 13,000*g* in a microcentrifuge and resuspend pellet in 1 mL of 2% glucose. For random spore analysis, *see* **Note 6**.
6. Briefly vortex resuspended cells.
7. To disperse cells, sonicate for 3 s at setting 1.5 (15% maximum power) using the microtip of a 550 Sonic ZD-dismembrator.
8. Dilution series and plating: Add 25 μL of the sonicated cell suspension to a 16-cm test tube containing 5 mL of sterile deionized water. Vortex briefly. Transfer 0.5 mL using a 1-mL pipet to a 16-cm test tube containing 4.5 mL water. This is the 10^{-3} dilution to be used for counting Ade^+ prototrophs and Ura^+ prototrophs. Transfer 0.5 mL from 10^{-3} dilution to 4.5 mL water. This is the 10^{-4} dilution to be used for determining C.F.U.s.
9. Briefly vortex the test tube before plating. Spread 0.2 mL from the 10^{-3} dilution onto YPD-ADE and SC-URA plates (*see* **Note 7**). From the 10^{-4} dilution, spread 0.2 mL onto YPD-ADE.
10. Incubate plates at 30°C.
11. Repeat **Steps** 5–10 at t = 2, 4, 6, 8, and 10 h.
12. Count the total number of colonies on 10^{-4} dilution plate after 2 d of growth (*see* **Note 8**).
13. Count white colonies (Ade^+ prototrophs) on 10^{-3} dilution plate after 2 d of growth (*see* **Note 9**).
14. Count colonies (Ura^+ prototrophs) on SC-URA plates after 4–5 d of growth.

3.3. Analysis

To calculate the frequency of ectopic collisions for each culture at each time point, divide the number of white colonies (Ade^+ prototrophs) on the 10^{-3} plate by 10 times the total number of colonies on the 10^{-4} plate. To calculate the frequency of allelic collisions for each culture at each time point, divide the number of colonies (Ura^+ prototrophs) on the 10^{-3} plate by 10 times the total number of colonies on the 10^{-4} plate. Viability of a strain is monitored by comparing colony-forming units (C.F.U.s) at each time point to C.F.U.s at the first time point. These calculations give the frequency of events per meiosis. If performing random spore analysis, the frequency is per chromatid.

4. Notes

1. YPG selects against growth of cells that have lost mitochondrial function (i.e., are petite). Cells will sporulate if incubated on YPG longer than 15 h. Typically, cells are incubated on YPG for 14–14.5 h.
2. Variation between cultures from independent colonies from the same strain is relatively low when performed in triplicate. Somewhat more variation is observed when comparing data obtained from experiments performed on different days. Thus, comparison of mutant phenotypes to wild type is more accurate when strains are processed in parallel. Relative differences between wild type and mutant are recapitulated in repeated experiments *(6)*.
3. Strains that have slow vegetative growth can be inoculated here at higher densities to ensure proper density (OD_{600} 2–2.5) prior to transfer to SPM in **Section 3.1**, **Step 8**.
4. A final concentration of 0.03% galactose is sufficient to give high levels of induction of prototroph formation without affecting the kinetics or efficiency of the meiosis I divisions.
5. If performing Cre/*loxP* analysis with the GFP reporter strain, the protocol ends by using fluorescent microscopy to determine the number of tetrads that contain fluorescent spores per total tetrads. Tetrads are analyzed at t = 24 h.
6. If performing random spore analysis with *NDT80* strains, resuspend cells in 1 mL of 20% sorbitol with 0.8 mg zymolyase. Incubate at 37°C for 1 h. Perform four rounds of sonication 15 s at setting 3 (30% maximum power) using the microtip of a 550 Sonic ZD-dismembrator. Rest on ice between rounds. Use a microscope to inspect complete dispersal of spores. If not dispersed, repeat sonication. Continue with **Section 3.2**, **Step 8** but add 6.25 μL of sonicated cell suspension to 5 mL of water instead of 25 μL.
7. When plating, use a 0.2-mL glass pipet to spread the cell suspension evenly across the plate but take care to avoid the edges of the Petri dish. This is achieved by spinning a plate on a plating turntable and slowly releasing the cell suspension as one drags the pipet tip outwards. Leaving at least 0.5 cm of space between the edge of the plated suspension and the Petri dish will make counting colonies easier and more accurate.

8. Take care to distinguish between single round colonies and oddly shaped colonies that may represent two overlapping clones. If colonies are too small to count, slow growing strains can be left to grow for additional days.
9. Ade^+ prototrophs are very distinctly pink or white; however, they may be hard to distinguish from petites or other slow growing colonies. The red color of Ade^- auxotrophs does develop over time.

Acknowledgments

The authors would like to give a special thanks to Tamara Peoples-Holst, Eric Dean, and Joshua Chang Mell for their contribution to the optimization of this protocol over the years. This work has been supported by the National Institutes of Health (NIH) grant NIH R01 GM075119 (S.M.B), the American Cancer Society RSG-01-053-01-CCG (S.M.B), and the NIH-Environmental Health Sciences training grant NIH T 32 ES07059 (D.Y.L).

References

1. Weiner, B. M., and Kleckner, N. (1994) Chromosome pairing via multiple interstitial interactions before and during meiosis in yeast. *Cell* **77**, 977–991.
2. Loidl, J., Scherthan, H., and Kaback, D. B. (1994) Physical association between nonhomologous chromosomes precedes distributive disjunction in yeast. *Proc. Natl. Acad. Sci. USA* **91**, 331–334.
3. Aragon-Alcaide, L., and Strunnikov, A. V. (2000) Functional dissection of *in vivo* interchromosome association in *Saccharomyces cerevisiae*. *Nat. Cell Biol.* **2**, 812–818.
4. Dekker, J., Rippe, K., Dekker, M., and Kleckner, N. (2002) Capturing chromosome conformation. *Science* **295**, 1306–1311.
5. Burgess, S. M., and Kleckner, N. (1999) Collisions between yeast chromosomal loci *in vivo* are governed by three layers of organization. *Genes & Dev.* **13**, 1871–1883.
6. Peoples, T. L., Dean, E., Gonzalez, O., Lambourne, L., and Burgess, S. M. (2002) Close, stable homolog juxtaposition during meiosis in budding yeast is dependent on meiotic recombination, occurs independently of synapsis, and is distinct from DSB-independent pairing contacts. *Genes & Dev.* **16**, 1682–1695.
7. Peoples-Holst, T. L., and Burgess, S. M. (2005) Multiple branches of the meiotic recombination pathway contribute independently to homolog pairing and stable juxtaposition during meiosis in budding yeast. *Genes & Dev.* **19**, 863–874.
8. Lui, D. Y., Peoples-Holst, T. L., Mell, J. C., Wu, H. Y., Dean, E. W., and Burgess, S. M. (2006) Analysis of close stable homolog juxtaposition during meiosis in mutants of *Saccharomyces cerevisiae*. *Genetics* **173**, 1207–1222.
9. Xu, L., Ajimura, M., Padmore, R., Klein, C., and Kleckner, N. (1995) *NDT80*, a meiosis-specific gene required for exit from pachytene in *Saccharomyces cerevisiae*. *Mol. Cell. Biol.* **15**, 6572–6581.
10. Allers, T., and Lichten, M. (2001) Differential timing and control of noncrossover and crossover recombination during meiosis. *Cell* **106**, 47–57.
11. Sherman, F., and Roman, H. (1963) Evidence for two types of allelic recombination in yeast. *Genetics* **48**, 255–261.
12. Esposito, R. E., and Esposito, M. S. (1974) Genetic recombination and commitment to meiosis in Saccharomyces. *Proc. Natl. Acad. Sci. USA* **71**, 3172–3176.

13. Zenvirth, D., Loidl, J., Klein, S., Arbel, A., Shemesh, R., and Simchen, G. (1997) Switching yeast from meiosis to mitosis: double-strand break repair, recombination and synaptonemal complex. *Genes Cells* **2**, 487–498.
14. Mell, J. C., Komachi, K., Hughes, O., and Burgess, S. (2008) Cooperative interactions between pairs of homologous chromatids during meiosis in *Saccharomyces cerevisiae.* Genetics **179,** 1125–1127.
15. Kane, S. M., and Roth, R. (1974) Carbohydrate metabolism during ascospore development in yeast. *J. Bacteriol.* **118**, 8–14.

Chapter 6

Genetic Analysis of Meiotic Recombination in *Schizosaccharomyces pombe*

Gerald R. Smith

Abstract

The fission yeast *Schizosaccharomyces pombe* is well-suited for studying meiotic recombination. Methods are described here for culturing *S. pombe* and for genetic assays of intragenic recombination (gene conversion), intergenic recombination (crossing-over), and spore viability. Both random spore and tetrad analyses are described.

Key words: fission yeast, *Schizosaccharomyces pombe*, meiosis, intragenic recombination (gene conversion), intergenic recombination (crossing-over), spore viability, tetrad analysis.

1. Introduction

Genetic analysis of meiotic recombination is especially facile in the fission yeast *Schizosaccharomyces pombe* for several reasons. Many independent meioses ($>10^8$) can be easily analyzed in one experiment. Having only three chromosomes, *S. pombe* produces many viable meiotic products, called spores, even if the cells are recombination-deficient: strains completely deficient for the early steps of meiotic recombination (DNA double-strand break (DSB) formation) produce about 10–20% as many viable spores as wild type (e.g., *(1)*). This feature has aided the analysis of mutations in more than 50 genes with documented effects on meiotic recombination. These and other studies have led to the formulation of a pathway of meiotic recombination for *S. pombe (2)*. The commonly used strains are isogenic, which facilitates exchange of alleles and comparison of results from different labs. Over 1,600 of the ~5,000

Scott Keeney (ed.), *Meiosis, Volume 1, Molecular and Genetic Methods, vol. 557*

DOI 10.1007/978-1-59745-527-5_6 Springerprotocols.com

genes have been experimentally characterized, which allows crosses between markers covering nearly all of the genome to be analyzed genetically. The genome sequence is essentially complete *(3)* (http://www.sanger.ac.uk/Projects/S_pombe/), which greatly facilitates physical analysis of DNA intermediates as described in the accompanying article (*see* **Chapter 15** in this volume). Cytological analysis of meiotic nuclei, chromosomes, and proteins is described in other articles in this book (**Chapters 22 and 24**). This article describes the genetic analysis of meiosis and recombination. Additional genetic methods for *S. pombe* are described in **Refs.** *(4, 5)* and on the website of Susan Forsburg at the University of Southern California (http://www-rcf.usc.edu/~forsburg/).

1.1. Life Cycle of S. pombe

S. pombe typically grows as haploid cells, which express one of two mating types, called *plus* and *minus*; only cells of opposite mating type can fuse to form diploids and proceed into meiosis. Homothallic (h^{90}) cells switch between *plus* and *minus* approximately once per cell division, by copying information from the unexpressed *mat2* (*plus*) or *mat3* (*minus*) locus into the expressed *mat1* locus *(5)*. Thus, cultures of an h^{90} strain contain cells of both mating types, which can mate and produce spores at high frequency; on appropriate media about 90% of the cells mate and complete meiosis. The *mat* loci are closely linked, and aberrant rearrangements, such as deletions and fusions of *mat2* and *mat3,* can produce heterothallic cells that are stably *plus* (denoted with the genotype h^+) or stably *minus* (h^-). h^+ and h^- strains can mate with each other at high frequency, but strains of the same type do so about 10^{-4} as frequently.

Unlike the budding yeast *Saccharomyces cerevisiae*, *S. pombe* cells mate only when starved of nitrogen. This condition is also needed for entry into meiosis. Thus, when cells of opposite mating type are placed on starvation (sporulation) medium, they mate and undergo meiosis with no intervening mitotic divisions. If mated cells are returned to growth medium before the onset of meiosis, the diploid cells can be propagated indefinitely. Diploids are usually selected by use of closely linked, complementing markers, such as those conferring auxotrophy, and maintained by growth with selection, as diploids give rise to more stable haploids at about 10^{-4} per cell division.

Three kinds of meiosis are recognized. Zygotic meiosis results from cells mating and entering meiosis without intervening mitotic division, and azygotic meiosis results from established diploids entering meiosis. The corresponding asci have different shapes (**Fig. 6.1**), but otherwise meiosis is, to my knowledge, largely the same in the two types (*see* **Note 1**). In addition, the temperature-sensitive mutant *pat1-114* initiates meiosis, even from the haploid state, when the temperature is raised *(6)*. This mutant is

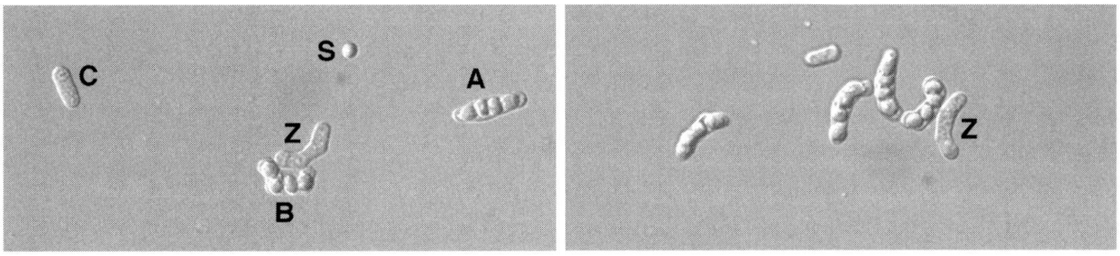

Fig. 6.1. Asci from h^{90} meiosis. Cells of an h^{90} strain were washed and spotted on a supplemented SPA plate, which was incubated for 3 d at 25°. A sample of the cell-ascus mixture was observed under a differential interference contrast microscope and photographed. (*Left*) The linear ascus (A) has the appearance of an azygotic ascus, whereas the curved ascus (B) has that of a zygotic ascus. A zygote (Z), an unmated cell (C), and a spore (S) from spontaneous lysis of an ascus are also visible in this rare field. (*Right*) A more typical field of asci, one of which (Z) is immature.

widely used to induce nearly synchronous meiosis, which is critical for physical analysis of DNA intermediates during meiosis (*see* **Chapter 15** in this volume). Chromosome segregation is slightly deficient during the first meiotic division in *pat1-114* mutants; this deficiency is partially suppressed by ectopic expression of genes from both *mat* loci *(7)*. Otherwise, meiosis seems similar in wild-type and *pat1-114* meiosis.

2. Materials

2.1. Culture Media (*see* Note 2)

1. YEL (yeast extract liquid): 5 g of yeast extract (Difco), 30 g of glucose. Make to 1 L with water and autoclave. The corresponding solid medium containing 2% (w/v) agar is called YEA (yeast extract agar).
2. EMM2 (Edinburgh minimal medium 2): 50 mL of 20 × EMM2 salts, 25 mL of 20% (w/v) NH_4Cl, 25 mL of 0.40 *M* Na_2HPO_4, 25 mL of 40% (w/v) glucose, 1 mL of 1,000 × vitamins, 0.1 mL of 10,000 × trace elements. Make to 1 L with water.
3. 20 × EMM2 salts: 30.6 g of potassium phthalate (monobasic), 10 g of KCl, 5.0 g of $MgCl_2$, 100 mg of Na_2SO_4, 100 mg of $CaCl_2$. Make to 500 mL with water and autoclave.
4. 1,000 × vitamins: 1.0 mg of biotin, 10 mg of calcium pantothenate, 1.0 g of nicotinic acid, 1.0 g of myoinositol. Make to 100 mL with water and autoclave.
5. 10,000 × trace elements: 0.50 g of H_3BO_3, 0.40 g of $MnSO_4$, 0.40 g of $ZnSO_4 \cdot 7H_2O$, 0.20 g of $FeCl_3 \cdot 6H_2O$, 0.15 g of Na_2MoO_4, 0.10 g of KI, 0.040 g of $CuSO_4 \cdot 5H_2O$, 1.0 g of citric acid. Make to 100 mL with water and filter sterilize.
6. NBL (nitrogen base liquid): 50 mL of 20 × NB + $(NH_4)_2SO_4$, 25 mL of 40% (w/v) glucose, water to 1 L. The corresponding solid medium is called NBA (nitrogen base agar).

7. $20\times$ NB + $(NH_4)_2SO_4$: 17 g of yeast nitrogen base without amino acids or ammonium sulfate (Difco), 50 g of $(NH_4)_2SO_4$. Make to 500 mL with water and filter sterilize. Store at 4°C.
8. SPA (sporulation agar): 25 mL of 40% (w/v) glucose, 10 mL of 10% (w/v) KH_2PO_4, 1 mL of $1,000\times$ vitamins. Add to 750 mL of 2.7% (w/v) agar (Difco; autoclaved and tempered). Make to 1 L with water.
9. MEA (malt extract agar): 30 g of malt extract (Difco), 20 g of agar (Difco). Make to 1 L with water and autoclave.
10. YEA + G (*see* **Note 3**): To 1 L of molten YEA, add 4.0 mL of guanine-HCl solution (1.0 g of guanine-HCl plus 17.5 mL of 1.0 *N* NaOH in 50 mL total) and then 1.2 mL of 1.0 *N* HCl. pH should be that of YEA (5.5).

2.2. Spore Preparation and Analysis

1. Sterile distilled water.
2. Sterile flat toothpicks.
3. Glusulase (New England Nuclear).
4. 60% (v/v) ethanol in water.
5. Sterile 1.5 mL microcentrifuge tubes.
6. Sterile velveteen replicating cloths, ~13 cm × 13 cm (**Fig. 6.2**).
7. 70% (v/v) acetone in water.
8. Glass Petri dish cover with inner plastic ring and Whatman 3 MM (or equivalent) filter paper inserts; replicating block (**Fig. 6.2**).
9. Iodine crystals.
10. Microscope with micromanipulator, such as Singer model MSM 300 or Leitz model MTR-27.
11. Microscope and counting chamber (hemacytometer).

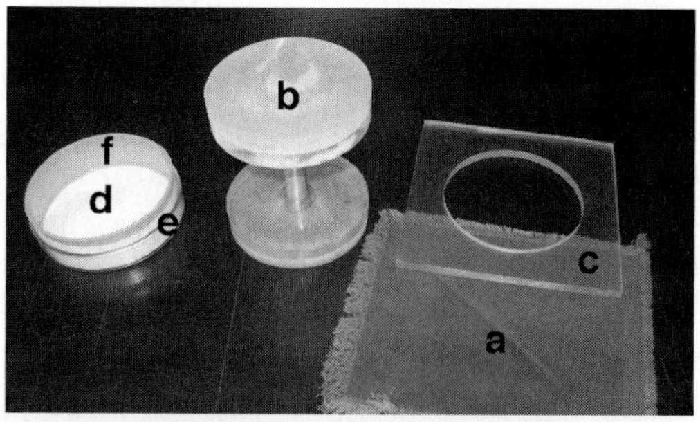

Fig. 6.2. Assembly of a Petri dish with acetone-soaked paper for killing vegetative cells. A replicating velveteen (a) is stretched over the elevated plastic block (b) and held in place with the clear plastic ring (c). A master plate with colonies is replicated onto the velveteen. Acetone is pipetted onto the filter paper (d) held in the glass Petri dish (e) with an internal plastic ring (f). This assembly is then inverted over the velveteen. After exposure to the acetone vapors, the cell-ascus mixtures on the velveteen are replicated onto a fresh YEA plate.

3. Methods

3.1. Analysis of Meiosis and Viable Spore Yield on Plates

This method is useful to identify rare heterozygous diploid spore colonies that could be confused with recombinants (**Section 3.3.4**). It is also useful for screening colonies from heterothallic strains to identify homothallic variants, or vice versa, and for screening homothallic colonies for mutants that produce few viable spores, as expected for mutants that make DSBs but do not repair them ("late" Rec mutants).

1. Use toothpicks to place isolates to be tested on a grid on YEA (**Fig. 6.3**). Incubate 1–2 d at 32°C (*see* **Note 4**). Replicate sequentially to supplemented SPA and to supplemented EMM2. Randomly placed colonies on SPA or EMM2 can also be used.
2. Incubate the SPA plate at 25°C for 2 d (*see* **Note 5**). Pour iodine crystals, about 25 g, into a large glass Petri dish, and invert the SPA plate over the crystals to expose the cell-ascus patches to iodine vapors for about 5–10 min. The ascus-spore mixture stains dark brown to black. Patches of cells that do not mate or that do not enter meiosis stain yellow.

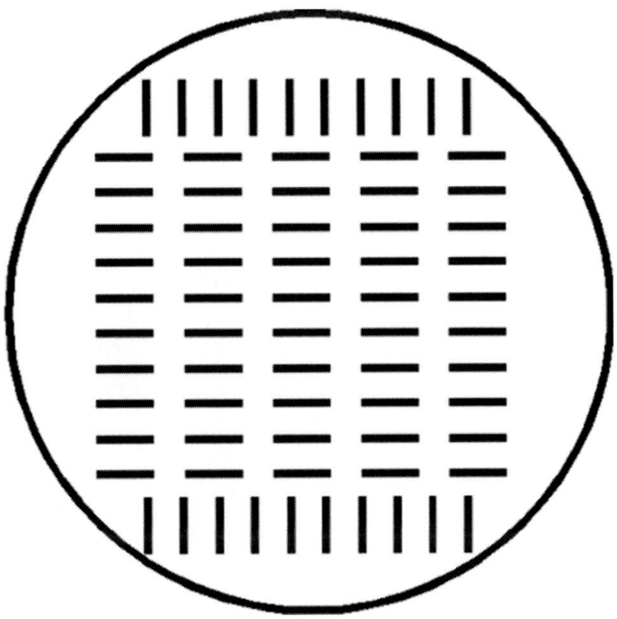

Fig. 6.3. Grid (full size) for picking colonies for replication to test media. A fresh YEA plate is centered over this grid. Cells are transferred from colonies onto the YEA plate, one over each grid mark. After incubation, this plate is replicated to test media.

3. Incubate the EMM2 plate at 25°C for 2 d (*see* **Note 5**). Transfer the cell-ascus mixture from the EMM2 plate onto velveteen (as for ordinary replica plating). Pipette 2.5 mL of 70% acetone onto the filter paper in the glass Petri dish assembly (**Fig. 6.2**), to thoroughly saturate the paper (*see* **Note 6**). Immediately place the glass Petri dish assembly over the velveteen, taking care that the velveteen does not touch the acetone-soaked paper. After 15 min. at room temperature (about 22°C), remove the glass Petri dish assembly. Replicate the cell-ascus mixture onto YEA and incubate at 32°C. Isolates that produce viable spores will yield confluent growth on this YEA plate after 1–2 d. Isolates that do not mate, that do not complete meiosis, or that make inviable spores yield few or no colonies on this YEA plate.

3.2. Growing Cells and Conducting Crosses

1. Streak cells of each parent separately on YEA and incubate at 32°C. After 2–3 d, pick a colony to 5 mL of YEL (*see* **Note 7**) and incubate on a roller drum at 32°C until saturated (overnight for healthy strains). Mix 0.05 mL of each parent in a 1.5 mL microcentrifuge tube. (For h^{90} × heterothallic crosses, use 0.05 mL of the h^{90} parent and 0.5 mL of the heterothallic parent.) Centrifuge 13,000 rpm (16,000*g*) for 10 s. Pour off or aspirate off supernatant. Wash cells twice in 1 mL of water. Suspend cells in the trace of water (about 20 μL) left in the tube. Deposit on a sporulation plate (supplemented SPA, EMM2, or MEA; *see* **Note 8**), making a spot about 5–7 mm in diameter. Incubate at 25°C (*see* **Note 9**). Sporulation is usually complete after 2 d.
2. If quantitation of the viable spore yield is desired, determine the number of viable cells per mL of saturated culture by diluting and plating on YEA (*see* **Note 10**). Incubate at 32°C for 2–4 d. Count the number of colonies and, with the dilution factors, calculate the concentration of viable cells per mL of saturated culture. Use this result in **Section 3.3.2**.
3. With a sterile toothpick remove a small amount of mating mixture from the sporulation plate, and examine it in a water mount under a phase contrast microscope. *See* **Fig. 6.1** for examples of zygotic and azygotic asci and cells. If asci are not abundant, continue to incubate.

3.3. Analysis of Random Spores

With a sterile toothpick remove the entire cell-ascus mixture from each spot on the sporulation plate (**Section 3.2**). Suspend in 0.5 mL of glusulase (1:200 in water) in a 1.5 mL microcentrifuge tube. Vortex gently but thoroughly to make a homogeneous suspension of the cell-ascus mixture. Incubate at 32°C overnight. (*see* **Note 11**) Alternatively, use glusulase at 1:100 for 3–6 h, with occasional mixing. Add 0.5 mL of 60% ethanol, and incubate at room temperature (~22°C) for 10 min. Immediately centrifuge at 13,000 rpm (16,000*g*) for 10 s. Wash twice in 1 mL of water. Suspend in 1 mL of water and store at 4°C. This suspension is stable for many months.

3.3.1. Analysis of Spore Viability

1. Determine the concentration of visible spores in the suspension using a counting chamber (hemacytometer) under a phase contrast microscope.
2. Plate appropriate dilutions of the spore suspension on YEA (*see* **Note 10**) and incubate at 32°C for 3–4 d. Count the number of colonies and, with the dilution factors, calculate the concentration of viable spores per mL of spore suspension. Divide the concentration of viable spores (colony counts) by the concentration of visible spores (microscope counts) to obtain the fraction of spores that are viable.
3. As an alternative to **Steps 1** and **2**, plate ~10^5 spores on a YEA plate and incubate at 32°C overnight. With a dissecting microscope determine, for a sample of the spores, the fraction of spores that germinate and produce a microcolony of at least four cells (*see* **Note 12**). Spore viability can also be determined from tetrad analysis (**Section 3.4**).

3.3.2. Analysis of Viable Spore Yield

Divide the total number of viable spores in the spore suspension (**Section 3.2**) by the total number of cells (not the concentration) of the less-numerous parent put into the mating mixture. This result is the viable spores per viable cell. In theory, this is two for cells that, upon mating and meiosis, produce four viable spores per diploid cell (two per one haploid cell). In practice, the number may be higher, if there is residual mitotic growth on the sporulation plate, or lower, if mating is inefficient. Thus, the number relative to a control, such as wild type, is most meaningful.

3.3.3. Intragenic (Gene Conversion) Recombinant Frequency (see **Note 13***)*

1. Serially dilute spores appropriately in water (*see* **Note 10**) and plate on an appropriate medium for total viable spore determination and on a medium selective for the measured property. For example, plate on YEA for total viable spores and on YEA + G for Ade^+ spores (*see* **Note 3**), or on EMM2 with uracil for total viable spores and on EMM2 without uracil for Ura^+ spores (*see* **Note 14**).
2. Incubate plates at 32°C for 2–4 d, when colonies are usually large enough for all to be readily visible. Count colonies and, with the dilution factors, calculate the concentration of selected types per mL of spore suspension and the concentration of total

viable spores per mL. Divide the concentration of the selected type by the concentration of total viable spores to obtain the recombinant frequency.

3.3.4. Intergenic (Crossover) Recombinant Frequency (see **Note 13***)*

1. Serially dilute the spore suspension and plate on YEA, as in **Section 3.3.3**, **Step 1**. Incubate at a temperature that allows all genotypes to grow. Plate the parents or other appropriate strains as controls.
2. After 2–4 d, pick individual, well-isolated colonies to a gridded pattern (**Fig. 6.3**) on YEA (the "master plate"). Include each parent, or other appropriate strains, as controls. Incubate 1 or 2 d.
3. Using sterile velveteen, replicate from the master plate to plates that will allow a distinction of the two parental types for each marker to be scored. For example, replicate to EMM2 with and without adenine to score Ade^+ and Ade^-, or to YEA at 25°C and 37°C to score temperature-sensitive mutants (*see* **Note 15**). Incubate at the appropriate temperature(s) for 1–2 d. Also, replicate to SPA (*see* **Note 16**), incubate, and score for I_2-reaction (**Section 3.1**) to identify diploids heterozygous for mating type; these diploids should be removed from the analysis if complementing diploids would have the same phenotype as recombinants.
4. Score each spore colony for each phenotype. This is conveniently done with spread sheet software, such as Excel, or by marking a scoring sheet, a piece of paper with many rows (one for each colony) and columns (one for each marker). Count the number of recombinant types and divide by the number of colonies tested. To convert the resultant recombinant fraction into genetic distance, use Haldane's formula *(13)*, $x = -50 \ln(1 - 2R)$, where x is the distance in centimorgans, and R = the recombinant fraction (a value between 0 and 0.5) (*see* **Note 17**).

3.4. Analysis of Meiotic Tetrads

1. Transfer a small amount of cell-ascus mixture (from a 2–3 day-old sporulation plate; **Section 3.2**, **Step 1**) to 1 mL of water. Spread about 20 µL on the edge of a YEA plate (*see* **Note 18**). Under a microscope with a micromanipulator, place an isolated ascus (**Fig. 6.1**) at each of several isolated points on the plate. Mark the plate so it can be repositioned on the microscope stage. Incubate at 37°C for a few hours, or overnight at room temperature, to allow the ascal walls to dissolve (*see* **Note 19**).
2. With the micromanipulator, move each of the four spores from an ascus onto well-defined points separated by about 7 mm on the plate.
3. Incubate the plate for 2–4 d. Score the colony phenotypes by inspection (for example, for the red color of *ade6* or *ade7* mutants) or by transferring the colonies to a grid on

YEA (the master plate) and, after incubation for 1–2 d, replicating to plates appropriate to score each phenotype (as in **Section 3.3.4, Step 3**). Alternatively, replicate the dissection plate without transferring to a master plate (*see* **Note 20**).

4. Score each spore colony in each tetrad for each marker. Tetrads giving 1:3 segregation for a marker had a gene conversion for that marker. Single colonies with mixed phenotype indicate post-meiotic segregation (PMS; *see* **Note 20**). For each pair of linked markers, tetrads with two reciprocal recombinant types and two parental types (a tetratype tetrad) had one crossover (or an odd number of crossovers) between those markers. Tetrads with four reciprocal recombinant types (a nonparental ditype tetrad) had two crossovers (or an even number of crossovers) between the markers. Tetrads with all parental types had no or an even number of crossovers.
5. Calculate the frequency of gene conversion (or PMS) tetrads by dividing the number of tetrads with a conversion (or PMS) event by the total number of four-spore viable tetrads. Calculate the frequency of crossing-over between two linked markers using an appropriate formula that takes account of spores with two recombination events but unseen with the markers used. Perkins's formula *(14)* is often used: $x = 50\,(\mathrm{T} + 6\,\mathrm{N})/\Sigma$, where x is the distance in cM, T is the number of tetratype tetrads, N = the number of nonparental ditype tetrads, and Σ = the total number of four-spore viable tetrads. This formula assumes that only single and double crossovers occur at significant frequency and that each N tetrad, which results from a four-strand (four-chromatid) double crossover event, is accompanied in the population on a statistical basis by two three-strand doubles, which are seen as tetratype tetrads, and one two-strand double, which is nonrecombinant. Alternatively, one can treat the data as random spores and use the formula $\mathrm{R} = (2\mathrm{T} + 4\,\mathrm{N})/\,\Sigma$, where R is the fraction of spores that are recombinant. Then, one can use Haldane's formula *(13)* $x = -50\ \ln(1 - 2\mathrm{R})$, which assumes no crossover interference, as is the case for *S. pombe (15)*.

4. Notes

1. In *tht1* and *tht2* mutants nuclear fusion (karyogamy) is much delayed. Consequently, the mutants have essentially no recombination in zygotic meiosis but normal recombination in azygotic meiosis *(8, 9)*.

2. Sterile distilled water is used throughout. To make solid medium for plates, use 2% agar (Difco). For EMM2 agar and NBA it is convenient to make 2.7% agar in water; after melting, add concentrated components and make to proper volume with water (for example, 300 mL of 2.7% agar is made to 400 mL total). All media, including rich (YEL and YEA) media, are supplemented with all growth requirements (amino acids, purines, and pyrimidines) at 100 μg per mL, except as noted in **Note 3**.
3. Guanine inhibits the uptake of adenine, which is present in YEA at about 5–10 μg/mL. Thus, Ade^- cells cannot grow on YEA + G. On YEA without added adenine, *ade6* and *ade7* mutants make red colonies; the color is enhanced after overnight incubation at 4°C. On YEA with added adenine (>20 μg/mL) Ade^+ and Ade^- cells make white colonies.
4. For convenience, our lab uses 32°C, but 30°C is more commonly used.
5. For reasons unclear, SPA gives a stronger I_2-reaction than EMM2, but EMM2 gives a better discrimination of viable spore yield. For the latter, 34°C is better for some situations, such as those with the temperature-sensitive *rad50S* mutation.
6. If more than one plate is to be treated, replace the filter paper and soak it with 70% acetone, as the acetone preferentially evaporates and killing would be inefficient if the paper were reused. The paper can, however, be dried and reused.
7. YEL liquid cultures can be kept in the refrigerator for about a month. In this case, dilute such a culture 1:50 into 5 mL of YEL and incubate overnight. Freshly saturated cultures mate more efficiently and give higher yields of asci than do old cultures.
8. An alternative method is to transfer with a toothpick a small amount of cells of each parent from a fresh colony or patch on a growth plate to about 20 μL of water on the sporulation plate and mix well. This less-quantitative method is suitable for making diploids or for constructing strains. Different laboratories use one or another sporulation medium (SPA, MEA, or EMM2), without obvious differences in results (but *see* **Note 5**). EMM2 with glutamate replacing NH_4Cl is used by some investigators *(4)*.
9. Meiosis is successful up to ~35°C, but the efficiency is lower. Recombinant frequencies differ less than a factor of ~2 at 35°C or 20°C compared to that at 25°C *(10)*.
10. It is convenient to make 1:100 dilutions by pipetting 10 μL into 1.0 mL of water, or 0.05 mL into 5.0 mL, and to make 1:10 dilutions by pipetting 10 μL into 0.090 mL, or 0.1 mL into 0.90 mL. Plating 0.05, 0.1, or 0.2 mL of an appropriate dilution on each plate makes calculations easy.

11. Glusulase digests the ascal wall, to release spores, and digests vegetative cell walls, to kill most of the vegetative cells. At the indicated dose ethanol kills any remaining cells but leaves spores intact.
12. Some aneuploid spores can germinate and begin mitotic growth but soon cease, although chromosome III disomes make slow-growing colonies *(11)*. A larger number of cell divisions, or longer incubation, may be needed for an appropriate distinction between viable and nonviable spores.
13. Most markers convert at <1% of chromatids; the mean for 31 markers is ~0.07% (P. Munz, pers. comm. in ref. *12*). Thus, if markers are separated by more than a few cM, few of the recombinants arise from conversion; i.e., most of the events are reciprocal (crossovers). Conversely, in the few studies done, intragenic recombination is exclusively the result of conversion.

    Note that selection methods will give genuine recombinant frequencies only if the markers are noncomplementing. While this is generally true for markers in the same gene, intragenic complementation does occur, for example in the *ade6* gene. Diploid spores arise from wild-type meiosis at about 10^{-3}, and more frequently from meiosis of many mutants deficient in recombination or chromosome segregation. Selected colonies can be tested for their I_2-reaction (**Section 3.1**), which will reveal some classes of diploids.
14. For all mitotic growth, EMM2 can be replaced with NBA, which is easier to prepare. Note that mating and meiosis occur frequently on EMM2 but infrequently on NBA and even less frequently on YEA.
15. Phloxin B (magdala red; 20 μg/mL) added to YEA can enhance scoring of temperature-sensitive mutants, since even a few dead cells in a patch of a temperature-sensitive mutant will stain with this dye and give a deeper red color than that for wild-type cells.
16. The SPA plate is not needed if a supplemented EMM2 plate that allows growth of all segregants is included, since it can be used for I_2-staining.
17. Since *S. pombe* does not have crossover interference *(15)*, Haldane's formula is appropriate.
18. For this plate to have uniform thickness, it should be poured on a surface verified to be level.
19. After ascal wall dissolution the plate can be placed at 4°C if the spores cannot be immediately separated on the plate.
20. Sectored colonies, characteristic of post-meiotic segregation, may be more easily seen by direct replication (i.e., without transfer with toothpicks to a master plate. Unlike

S. cerevisiae, however, *S. pombe* does not make half-sectored PMS colonies; rather, the two phenotypes are intermingled in streaks emanating from the center.

Acknowledgments

I am grateful to Sue Amundsen, Gareth Cromie, and Luther Davis for helpful comments on the manuscript. My laboratory is supported by research grants GM031693 and GM032194 from the National Institutes of Health.

References

1. Davis, L., and Smith, G.R. (2003) Nonrandom homolog segregation at meiosis I in *Schizosaccharomyces pombe* mutants lacking recombination. *Genetics* **163**, 857–874.
2. Cromie, G.A., and Smith, G.R. (2008) Meiotic recombination in *Schizosaccharomyces pombe*: A paradigm for genetic and molecular analysis, *in* "Genome Dynamics and Stability" (Egel, R. and Lankenau, D-H., eds.), Springer-Verlag, Berlin, pp. 195–230.
3. Wood, V., Gwilliam, R., Rajandream, M.-A., Lyne, M., Lyne, R., Stewart, A., Sgouros, J., Peat, N., Hayles, J., Baker, S., Basham, D., et al. (2002) The genome sequence of *Schizosaccharomyces pombe*. *Nature* **415**, 871–880.
4. Moreno, S., Klar, A.J.S., and Nurse, P. (1991) Molecular genetic analysis of fission yeast *Schizosaccharomyces pombe*. *Methods Enzymol.* **194**, 795–823.
5. Egel, R. (Ed.) (2004) The Molecular Biology of *Schizosaccharomyces pombe*. Springer-Verlag, Berlin.
6. Iino, Y., and Yamamoto, M. (1985) Mutants of *Schizosaccharomyces pombe* which sporulate in the haploid state. *Mol. Gen. Genet.* **198**, 416–421.
7. Yamamoto, A., and Hiraoka, Y. (2003) Monopolar spindle attachment of sister chromatids is ensured by two distinct mechanisms at the first meiotic division in fission yeast. *EMBO J.* **22**, 2284–2296.
8. Tange, Y., Horio, T., Shimanuki, M., Ding, D.Q., Hiraoka, Y., and Niwa, O. (1998) A novel fission yeast gene, *tht1*$^+$, is required for the fusion of nuclear envelopes during karyogamy. *J. Cell. Biol.* **140**, 247–258.
9. Martín-Castellanos, C., Blanco, M., Rozalén, A.E., Pérez-Hidalgo, L., García, A.I., Conde, F., Mata, J., Ellermeier, C., Davis, L., San-Segundo, P., Smith, G.R., and Moreno, S. (2005) A large-scale screen in *S. pombe* identifies seven novel genes required for critical meiotic events. *Curr. Genet.* **15**, 2056–2062.
10. Li, Y.F., and Smith, G.R. (1997) The *Schizosaccharomyces pombe rec16* gene product regulates multiple meiotic events. *Genetics* **146**, 57–67.
11. Niwa, O., and Yanagida, M. (1985) Triploid meiosis and aneuploidy in *Schizosaccharomyces pombe*: an unstable aneuploid disomic for chromosome III. *Curr. Genet.* **9**, 463–470.
12. Young, J.A., Schreckhise, R.W., Steiner, W.W., and Smith, G.R. (2002) Meiotic recombination remote from prominent DNA break sites in *S. pombe*. *Mol. Cell* **9**, 253–263.
13. Haldane, J.B.S. (1919) The combination of linkage values, and the calculation of distances between the loci of linked factors. *J. Genet.* **8**, 299–309.
14. Perkins, D.D. (1949) Biochemical mutants in the smut fungus *Ustilago maydis*. *Genetics* **34**, 607–626.
15. Munz, P. (1994) An analysis of interference in the fission yeast *Schizosaccharomyces pombe*. *Genetics* **137**, 701–707.

Chapter 7

Analysis of Meiotic Recombination in *Caenorhabditis elegans*

Kenneth J. Hillers and Anne M. Villeneuve

Abstract

Caenorhabditis elegans is an important experimental organism for the study of recombination during meiosis. A variety of techniques have been developed for the measurement of meiotic recombination in *C. elegans*, ranging from traditional genetic measures to direct cytological determination of chiasma frequency. Here, we provide methods for some of the varied approaches used for the study of meiotic recombination in these tiny but powerful worms.

Key words: crossing over, PCR, recombination, chiasmata, snip-SNP, bivalent, univalent.

1. Introduction

Meiotic recombination in *C. elegans*, as in other eukaryotes, follows a characteristic series of steps or stages. Techniques are available for the study of recombination at a number of these steps, and are detailed herein. Meiotic recombination is initiated by meiosis-specific DNA double-strand breaks (DSBs) introduced by the SPO-11 enzyme *(1)*. However, DNA damage induced by ionizing radiation can mimic these spontaneous DSBs *(2, 3)*, providing researchers a means to supplement or replace normal meiotic DSBs (**Section 3.4**). Repair of a subset of meiotic DSBs results in formation of crossovers between homologous chromosomes (*see* **Note 1**). These crossovers, in conjunction with sister chromatid cohesion, form the basis of the chiasmata (singular: chiasma) that hold homologous chromosomes together until Anaphase I of meiosis; these chiasmata play an important role in ensuring proper meiotic chromosome disjunction *(4)*. Chiasma formation can be assayed cytologically, through examination of nuclei in the

Scott Keeney (ed.), *Meiosis, Volume 1, Molecular and Genetic Methods, vol. 557*

DOI 10.1007/978-1-59745-527-5_7 Springerprotocols.com

diakinesis stage of meiosis I prophase; at this stage, homologous chromosomes that have enjoyed a crossover will be held together by the resulting chiasma, and appear as a single entity. Homolog pairs that have failed to successfully form a chiasma will not be connected, and will appear as two separate univalent chromosomes (**Section 3.3**) *(5)*. After completion of meiosis, crossover recombination events can be detected through their effects on the segregation of linked genetic markers using genetic mapping techniques involving either mapped DNA sequence polymorphisms (**Section 3.2**) or morphological markers (**Section 3.1**).

As a result of studies using these approaches, a number of conclusions about meiotic recombination in *C. elegans* can be drawn. Each of the six pairs of chromosomes typically receives only a single crossover per meiosis. This fact, coupled with the observation that E_0 chromosome pairs (lacking a crossover) are very infrequent in wild-type worms, indicates that the process of crossover formation is very tightly controlled in this organism. Another implication of this is that any two genes on the same chromosome are likely to show some degree of linkage.

An additional feature of *C. elegans* biology advantageous for the study of meiosis is the arrangement of nuclei in the germline. Each gonad arm represents a time course of nuclei arranged in a spatiotemporal gradient from premeiotic stages through meiotic prophase and (after fertilization) the meiotic divisions. Thus, within a single fixed animal, nuclei at multiple stages of meiosis and gametogenesis can be examined simultaneously (for more information, *see* **Chapter 31** in this volume).

The following sections describe methods for measuring meiotic crossover formation through the use of traditional genetic markers (**Section 3.1**) and mapped DNA sequence polymorphisms known as snip-SNPs (**Section 3.2**). Methods are also provided for cytological estimation of chiasma frequency (**Section 3.3**), and for exogenous induction of DSBs using ionizing radiation (**Section 3.4**).

Measurement of meiotic crossovers through the use of traditional morphological markers (**Section 3.1**) is relatively straightforward, and allows assaying of large numbers of progeny. Genetic map distances thus measured are therefore relatively accurate. The chief limitation of this approach is that studies are limited in most cases to two markers (due to the relative paucity of morphological phenotypes in *C. elegans*). As a result, each experiment typically measures crossover frequency within a single interval, which prevents detection of chromosomes with multiple crossovers. In addition, some morphological markers can have effects on the viability of homozygotes, complicating determination of crossover frequency. The use of morphological markers to assay crossing over is most frequently applied in situations wherein accurate determination of crossover frequency within a given region is important, or in studies comparing the crossover frequency within a given region in wild-type and mutants.

Measurement of crossing over through the use of DNA sequence polymorphisms (**Section 3.2**), on the other hand, allows the use of multiple markers in the same experiment. This allows measurement of crossing over in numerous intervals along the length of a chromosome (or chromosomes) in a single experiment, and also allows direct detection of chromosomes with multiple crossovers. In addition, the markers are codominant (simplifying analysis) and usually phenotypically neutral. However, the analysis necessary is more difficult and time-consuming, which limits the number of individuals typically assayed. Sequence polymorphism markers are generally used in experiments looking at the distribution of crossovers along the length of a chromosome or chromosomes, and also in studies looking specifically at the incidence of chromosomes with multiple crossovers.

As an alternative to genetic approaches, cytological examination of oocyte nuclei in the diakinesis stage of meiosis (**Section 3.3**) allows estimation of overall levels of crossover formation. In particular, this approach allows estimation of the frequency of E_0 chromosome pairs (those lacking chiasmata) without the confounding effect of progeny inviability due to aneuploid gametes. This approach is most commonly used to study the effects of mutations upon meiosis; it is insensitive to alterations in crossover distribution, but serves as a ready way to detect reductions in overall crossover/chiasma levels.

The methods described above are typically applied to endogenous meiotic DSB repair. However, under some circumstances it is useful to assay the ability of mutant germ cells to repair DSBs provided exogenously. Exposure of worms to ionizing radiation results in DNA damage that mimics DSBs (**Section 3.4**); the fate of these radiation-induced DSBs can be followed both cytologically and genetically, providing information concerning the process of DSB repair in mutant backgrounds.

2. Materials

2.1. Measuring Crossing Over Using Recessive Phenotypic Markers

1. 1 M potassium phosphate buffer pH 6.0: 108.3 g KH_2PO_4, 35.6 g K_2HPO_4, H_2O to 1 L; autoclave.
2. 5 mg/mL cholesterol in 95% ethanol (do not autoclave).
3. NGM plates:
   a. combine and autoclave: 3 g NaCl, 17 g agar, 2.5 g peptone, 975 mL H_2O.

b. cool to 55°C.

c. add and mix well: 1 mL 1 M $CaCl_2$, 1 mL 5 mg/mL cholesterol in 95% ethanol (see above), 1 mL 1 M $MgSO_4$, 25 mL 1 M potassium phosphate buffer (see above).

d. Dispense into 60 mm Petri dishes, using sterile technique.

4. *E. coli* OP-50.
5. *C. elegans* strains homozygous for linked morphological markers. *C. elegans* strains of interest are available for a nominal fee from the Caenorhabditis Genetics Center (CGC); a searchable CGC portal is available at http://wormbase.org/db/gene/strain. For general information concerning care and feeding of *C. elegans*, *see (6)*.

2.2. Measuring Crossing over Using snip-SNP Markers

1. 10 mM Tris-HCl, pH 8.0.
2. 10 mg/mL Proteinase K in H_2O.
3. 2x Single-Worm Lysis buffer: 100 mM KCl, 20 mM Tris-HCl pH 8.3, 5.0 mM $MgCl_2$, 0.9% NP-40, 0.9% Tween-20, 0.02% gelatin. Immediately before use, add Proteinase K to 120 μg/mL (using 10 mg/mL stock).
4. Reagents for polymerase chain reaction: Taq DNA polymerase and PCR buffer (any supplier); dNTPs (any supplier).
5. Primers: a large and growing number of snip-SNPs have been identified, mostly through the efforts of the Genome Sequencing Center at Washington University at Saint Louis and the efforts of Exelixis. Both data sets are available on the web: http://genome.wustl.edu/genome/celegans/celegans_snp.cgi (Washington University) and http://www.exelixis.com/discovery_acad_c_ele.shtml (Exelixis). For further information and suggestions on primer design, *see (7)* and references therein.
6. Restriction digestion master mix: the restriction master mix contains the appropriate restriction enzyme (specific for each snip-SNP marker) and 10x buffer, plus H_2O. To each 15 μL PCR reaction to be digested, add 5 μL of a solution containing 2 μL of the appropriate 10x restriction buffer, 3–5 units of restriction enzyme, and water to make 5 μL.

2.3. Determination of Chiasma Frequency Through Analysis of Bivalents at Diakinesis

1. M9 buffer: 3 g KH_2PO_4, 6 g Na_2HPO_4, 5 g NaCl, 1 mL 1 M $MgSO_4$, H_2O to 1 L; autoclave.
2. 95% ethanol.
3. DAPI (diaminophenolindole) / Vectashield solution: 1:1 mixture of 4 μg/mL DAPI in M9 buffer (diluted from a concentrated DAPI stock of 5 mg/mL in H_2O or dimethylformamide) with Vectashield solution (Vector Laboratories; Burlingame, CA).

2.4. Induction of Meiotic Recombination Through Ionizing Irradiation

1. Source of ionizing radiation (e.g. a ^{137}Cs source).

3. Methods

Sections 3.1 and **3.2** describe methods for measuring crossing over during meiosis in *C. elegans*. Each of these methods follows the same general logic: (1) animals heterozygous for linked markers are generated; (2) crossing over occurring in the germline of these heterozygous animals is assessed by scoring the markers inherited by their progeny. The chief difference between these approaches lies in the nature of the markers used, and thus in the analyses necessary to score the markers. Each approach has advantages and disadvantages. The choice of which approach to use will depend upon the specific circumstances. For more information, see the individual protocols.

Section 3.3 describes a method that allows cytological evaluation of crossing over, by assaying chiasma formation at diakinesis (the last stage of Prophase I). This analysis has the advantage of being independent of reproductive success; thus, crossing over can be studied in mutant strains with compromised reproductive capability (a feature of many mutations affecting meiotic DSB formation or repair). However, this approach is limited in that it can only detect a reduction in the overall number of crossovers; it is blind to changes in crossover distribution.

The methods provided herein are often applied to the study of meiosis in animals carrying mutations affecting meiosis; in any such case, it is important to ensure that the animals in which recombination is measured have a normal diploid chromosome complement. This is necessary because mutations affecting meiosis often result in chromosome missegregation during gametogenesis, with resulting aneuploidy among the progeny of homozygous animals. Thus, particular care must be taken during experimental design to avoid the problems that can result from attempting to analyze recombination in aneuploid animals. For further details, *see* **Note 2**.

3.1. Measuring Crossing Over Using Recessive Phenotypic Markers

Section 3.1.1 describes a protocol for measuring crossing over during both oogenesis and spermatogenesis in hermaphrodites. This approach is usually the simplest way to measure crossover frequencies, allowing analysis of the products of many meioses (and thus relatively accurate determination of crossover frequencies). This is the most common means of measuring crossover frequencies in *C. elegans*.

Under some circumstances, however, it is desirable to assess crossovers formed solely during formation of oocytes, or during spermatogenesis in males. For example, some mutations affect oocyte meiosis and spermatocyte meiosis differently. **Section 3.1.2** describes a protocol for measuring crossing over occurring solely during oogenesis in hermaphrodites. This approach has the advantage of allowing detection of chromosomes that have enjoyed multiple crossovers, something not easily accomplished when assaying self progeny from hermaphrodites (**Section 3.1.1**). However, this approach is labor intensive, as scoring the markers inherited by the progeny of the heterozygous hermaphrodite requires progeny testing (analysis of the phenotypes present in the following generation). **Section 3.1.3** describes a protocol for measuring crossing over during meiosis in males. As males are not self-fertile, this method requires determination of the genotype of the male gamete through analysis of outcross progeny.

Mutations affecting meiosis and gametogenesis typically reduce the number of progeny produced, often drastically. Thus, when measuring crossing over in meiotic mutants, it is often necessary to assay crossover formation in many individuals heterozygous for linked genetic markers. In addition, fewer progeny resulting from these heterozygous parents are typically assayed (often $\leq$200 in total), due to the low number of progeny produced by animals homozygous for meiotic mutations. In such cases, it is important to assay crossover formation in a large genetic interval, to maximize the number of crossovers detected.

3.1.1. Measuring Crossovers During Spermatogenesis and Oogenesis in Hermaphrodites Using Recessive Morphological Markers

When measuring crossing over in self progeny of hermaphrodites, all self progeny from six to eight heterozygous F1 hermaphrodites are typically scored. As each wild-type hermaphrodite normally produces 250–300 self progeny, typical experiments of this sort involve scoring of >1,000 F2 progeny.

1. Generation of heterozygous hermaphrodites: on a small (60 mm) NGM plate seeded with *E. coli*, mate wild-type males with hermaphrodites homozygous for two linked recessive phenotypic markers (*see* **Note 2**).
2. Pick heterozygous (phenotypically wild-type) F1 hermaphrodites (at the L4 stage) individually to small seeded NGM plates. When assaying recombination in animals homozygous for a meiotic mutation, it is often necessary to score all self progeny from many F1 hermaphrodites; in such cases, multiple F1 hermaphrodites can be placed on each plate.
3. Move F1 hermaphrodites to new plates every 24 h until they cease progeny production (*see* **Note 3**).
4. Scoring markers transmitted to self progeny: As F2 progeny reach adulthood, score each for the phenotypic markers used in the cross by determining its phenotype and then removing

it from the plate. Wild-type and double-mutant individuals are parental; individuals expressing either mutant phenotype in isolation are recombinant (*see* **Note 4**).

5. Calculate the recombination frequency (p) using the following equation: $p = 1-(1-2R)^{1/2}$, where R = fraction of progeny with recombinant phenotypes = (M1+M2) / (WT + M1 + M2 + M1M2), where M1 and M2 are the numbers of individuals displaying each of the single mutant phenotypes, M1M2 is the number of individuals with the double mutant phenotype, and WT is the number of individuals with wild-type phenotype *(8)* (*see* **Note 5**).

3.1.2. Measuring Crossovers During Oogenesis in Hermaphrodites Using Recessive Morphological Markers

Use of recessive morphological markers to assay crossover formation during oogenesis is labor-intensive, as *C. elegans* males homozygous for morphological markers usually are unable to mate. Since wild-type males are used, outcross progeny (each the product of one meiosis in the heterozygous hermaphrodite parent) must be picked separately to individual plates for progeny testing. Thus, studies of this type typically only analyze a few hundred progeny, limiting the accuracy of the measured crossover frequencies. As snip-SNP markers are codominant and can thus be directly detected in heterozygotes, use of snip-SNP markers is preferred for measurement of crossover frequencies during oogenesis (*see* **Section 3.2.2**). The exception to this is when crossing over is assessed using *X*-linked morphological markers. *C. elegans* males are *XO*, inheriting their single *X* from the hermaphrodite parent. Thus, by scoring male outcross progeny from a hermaphrodite heterozygous for *X*-linked markers, one can directly assess crossing over that occurred during oogenesis in the hermaphrodite.

1. Generation of heterozygous hermaphrodites: on a small (60 mm) NGM plate seeded with *E. coli*, mate wild-type males with hermaphrodites homozygous for two linked phenotypic markers (*see* **Note 2**).
2. Pick six to eight heterozygous (phenotypically wild-type) hermaphrodite progeny (as late L4) individually to small seeded NGM plates along with five to eight wild-type (N2) males. When assaying recombination during oogenesis in animals homozygous for a meiotic mutation, it is often necessary to score outcross progeny from many heterozygous hermaphrodites; in such cases, multiple hermaphrodites can be placed on each plate.
3. After 24 h, each heterozygous hermaphrodite should have mated with the N2 males present on the plate; thus, progeny produced after 24 h of mating are likely to be outcross progeny (allowing measurement of crossing over that occurred solely during oogenesis). Move heterozygous

hermaphrodites individually to new plates. Each 24 h thereafter for several days (or until they cease producing outcross progeny), move individually to fresh plates (*see* **Note 3**).

4. As the outcross hermaphrodite progeny of the heterozygous hermaphrodites reach adulthood, pick them individually to small NGM plates for progeny testing. When possible, assay all outcross progeny from a given time window or plate (from **Step 3**). This will help avoid introduction of bias due to differential growth rates among progeny genotypes (*see* **Notes 3** and **6**). If scoring crossing over between *X*-linked morphological markers among male progeny, this step and the following step are unnecessary.
5. Score markers transmitted to outcross hermaphrodite progeny through progeny testing. Outcross progeny will fall into four classes: homozygous wild-type (nonrecombinant: + +/+ +), heterozygous for both markers (nonrecombinant: + +/m1 m2), heterozygous for marker 1 (recombinant: + +/m1 +), or heterozygous for marker 2 (recombinant: + +/+ m2). All are phenotypically wild-type, but can be distinguished through the types of progeny they produce. Homozygous wild-type (nonrecombinant) animals will produce only wild-type self progeny. Nonrecombinant animals heterozygous for both markers will produce self progeny with both mutant phenotypes. Recombinant animals heterozygous for marker 1 (+ +/m1 +) will produce self progeny with mutant phenotype 1, but not phenotype 2. Likewise, recombinant animals heterozygous for marker 2 (+ +/+ m2) will produce self progeny with mutant phenotype 2, but not phenotype 1.
6. Scoring *X*-linked markers transmitted to outcross male progeny: as the outcross male progeny of the heterozygous hermaphrodites reach adulthood, score each for the phenotypic markers used in the cross and then remove it from the plate. Wild-type and double-mutant individuals are parental; individuals expressing either mutant phenotype in isolation are recombinant (*see* **Note 4**). When assaying recombination in meiotic mutants, patroclinous *XO* male progeny can be produced at an appreciable frequency. These are outcross male progeny that received an *X* chromosome from the male parent, and no *X* from the hermaphrodite parent (due to meiotic chromosome missegregation). Such patroclinous males will be wild-type for *X*-linked markers, and can thus lead to an underestimation of the recombination frequency. To correct for this, calculate R using the following equation: $R = (M1 + M2)/(2*(M1M2 + M1))$, where M1 and M2 are the

numbers of individuals displaying each of the single mutant phenotypes, and M1M2 is the number of individuals with the double mutant phenotype.

7. The recombination frequency p = R, where R = fraction of outcross progeny carrying recombinant chromosomes.

3.1.3. Measuring Crossing over in Males Using Recessive Morphological Markers

Measurement of crossing over during gametogenesis in males requires mating heterozygous males with hermaphrodites homozygous for the linked morphological markers present in the male parent. In this way, the genotype of the gamete produced by the male can be directly inferred from the phenotype of the outcross progeny. In studies of this sort, several hundred outcross progeny from each of five to seven heterozygous male parents are typically scored.

1. Generation of heterozygous males: on a small (60 mm) NGM plate seeded with *E. coli*, mate wild-type males with hermaphrodites homozygous for two linked phenotypic markers (*see* **Note 2**).
2. Pick four to five heterozygous (phenotypically wild-type) F1 males individually to small seeded NGM plates with several late L4-stage hermaphrodites homozygous for the two phenotypic markers being used. When assaying recombination in animals homozygous for a meiotic mutation, it is often necessary to score outcross progeny from many heterozygous males. In such cases, special care must be taken in identification of outcross progeny (*see* **Note 7**).
3. After 24 h of mating, transfer the mated hermaphrodite partners (but not the heterozygous males) individually to fresh plates. Each of these animals should have mated with the heterozygous males, and will thus produce outcross progeny. Transfer these mated hermaphrodites to fresh plates every 24 h for several days (or until they cease production of outcross progeny) (*see* **Note 3**).
4. Scoring markers transmitted to progeny: as the outcross progeny of the F1 males reach adulthood, score each for the phenotypic markers used in the cross and then remove it from the plate. Wild-type and double-mutant individuals are parental; individuals expressing either mutant phenotype in isolation are recombinant (*see* **Notes 4** and 7).
5. The recombination frequency p = R, where R = fraction of progeny with recombinant phenotypes = (M1+M2)/(WT + M1 + M2 + M1M2), where M1 and M2 are the numbers of individuals displaying each of the single mutant phenotypes, M1M2 is the number of individuals with the double mutant phenotype, and WT is the number of individuals with wild-type phenotype (*see* **Note 8**).

3.2. Measuring Crossing over Using snip-SNP Markers

The wild-type *C. elegans* strain CB4856 (the Hawaiian strain) differs from wild-type N2 Bristol at approximately 0.1% of bases. These differences are broadly dispersed throughout the genome, and provide a dense array of potential genetic markers for use in measurement of recombination. These markers have the advantage of being phenotypically neutral (in general) and codominant, thus avoiding potential complications due to viability and simplifying scoring. In addition, multiple markers can be followed in a single cross (limited only by the number of PCRs one can carry out on the DNA sample obtained). A subset of these polymorphisms alter (create or destroy) cleavage sites for restriction endonucleases. Such polymorphisms, referred to as snip-SNPs, have been exploited for use in a PCR-based approach for mapping genes and measuring meiotic crossing over *(9)*. The basic approaches are similar to those detailed in **Section** 3.1, but analysis of marker segregation involves molecular approaches rather than examination of morphological characters. For more detailed background information and additional technical information, *see (7)* and references therein.

A major advantage of this approach is that multiple intervals can be simultaneously assayed for crossing over, allowing determination of the distribution of crossover events along chromosomes, and also allowing detection of chromosomes that have enjoyed multiple crossovers. However, the nature of the analysis (requiring PCR, restriction enzyme digestion, and gel electrophoresis for each marker scored in each individual) necessarily limits the number of individuals that can be assayed, reducing the accuracy of map distances generated. When analyzing crossing over using snip-SNP markers, 200–300 individuals are typically analyzed (as opposed to the thousands typically analyzed in experiments using morphological markers).

Section 3.2.1 describes a method for measuring crossing over during both oogenesis and spermatogenesis in hermaphrodites using snip-SNP markers. The major advantage of this approach is its simplicity - recombination is assayed by determining the genotype of self-progeny of heterozygous individuals (as in **Section 3.1.1**). The chief disadvantage of this approach, as in **Section 3.1.1**, is that crossing over can occur during both sperm and egg production; thus, only a subset of double-crossover chromosomes can be unambiguously detected *(10)*.

As an alternative, crossing over can be assayed during meiosis in a single germline; in this case, all double crossover chromosomes can be detected. **Section 3.2.2** describes a method for measuring crossing over during oogenesis in hermaphrodites (as in **Section 3.1.2**). This approach has the advantage that each progeny worm assayed represents the product of a single meiosis from the heterozygous hermaphrodite parent; this allows unambiguous detection of all multiply recombinant chromosomes. In addition, the codominant nature of snip-SNP markers means that crossovers can be detected without the additional complication of progeny testing (which is necessary to assay recombination during oogenesis using recessive markers; *see* **Section 3.1.2**). Therefore, use of snip-SNP markers to assay recombination during oogenesis is preferable to use of traditional recessive morphological markers. **Section** 3.2.3 describes a method for measuring crossing over during spermatogenesis in males.

As in **Section 3.1**, when measuring crossing over in meiotic mutants, it is often necessary to assay crossover formation in many individuals heterozygous for linked genetic markers. This is because mutations affecting meiosis and gametogenesis typically reduce the number of progeny produced, often drastically. Thus, when measuring recombination in meiotic mutants, the following protocols should be modified to involve increased numbers of heterozygous parents.

3.2.1. Measuring the Incidence of Crossing over During Both Spermatogenesis and Oogenesis in Hermaphrodites Through the Use of snip-SNP Markers

1. Generation of heterozygous hermaphrodites: on a small (60 mm) NGM plate seeded with *E. coli*, mate Bristol N2-derived hermaphrodites homozygous for a selected morphological marker to homozygous Hawaiian CB4856 males. After 48 h, remove both male and hermaphrodite parents from the plate and allow progeny to develop (*see* **Notes 9** and **10**).
2. Pick heterozygous (phenotypically wild-type) F1 hermaphrodites (as L4 or younger) individually to small seeded NGM plates.
3. Move F1 hermaphrodites to new plates every 12–24 h until they cease producing progeny (*see* **Note 3**).
4. Scoring markers transmitted to self progeny: As F2 progeny reach adulthood, pick individually into 0.2 mL thin-walled tubes containing 10 μL of 10 mM Tris-HCl, pH 8.0 (*see* **Notes 4**, **11**, and **12**).
5. To each tube, add 10 μL of 2x Single-worm Lysis Buffer and mix well.
6. Lyse worms: freeze at –80°C; incubate at 65°C 1 h; 95°C 15 min (*see* **Note 11**).
7. PCR analysis: Each snip-SNP marker is amplified using a specific primer pair. Thus, PCR conditions should be empirically optimized for each marker to be analyzed. However, the following general conditions have worked well in our hands: use 0.5 μL of worm lysate in each 15 μL reaction. PCR cycling: 94° 2 min; 35 cycles of {94° 20 s; 60° 30 s; 72° 40 s}; 72° 10 min (*see* **Note 13**).
8. Restriction Digestion: add an appropriate volume of restriction enzyme master mix to each PCR reaction, and digest 4 h to overnight.

9. Agarose gel analysis: Restriction-enzyme-digested PCR products can be analyzed through agarose gel electrophoresis. Because expected DNA fragments are often small (<300 bp), we use 2.5% agarose gels in 0.5 × TBE.
10. After electrophoresis, score each sample for the presence or absence of the N2- and CB4856-specific band(s). In cases of ambiguity, PCR analysis and restriction enzyme digestion should be repeated.
11. Identifiable recombinant progeny will fall into two types: (a) Recombinants in which crossing over between the assayed markers occurred during production of either sperm or egg but not both. This case results in progeny heterozygous for one marker and homozygous for the other, (e.g. [b1 h2/b1 b2], where b1 and b2 represent N2-derived alleles at loci 1 and 2, and h1 and h2 represent the CB4856 alleles). (b) Recombinants in which crossing over between the assayed markers occurred during production of both sperm and eggs. Detectable recombinants in this case will be homozygous for recombinant chromosomes (e.g. [b1 h2/b1 h2]). Note that an equal number of progeny resulting from this case will be heterozygous for both alleles (e.g. [b1 h2/h1/b2]), and thus indistinguishable from nonrecombinants.
12. The recombination frequency (p) is calculated using the following equation: $p = 1-(1-R)^{1/2}$, where R = ((number of animals heterozygous for one marker and homozygous for the other) + 2*(number of animals homozygous for recombinant chromosomes))/total number of animals scored.

3.2.2. Measuring the Incidence of Crossing over During Oogenesis in Hermaphrodites Through the Use of snip-SNP Markers

1. Generation of heterozygous hermaphrodites: on a small (60 mm) NGM plate seeded with *E. coli*, mate Bristol N2-derived hermaphrodites homozygous for a selected phenotypic marker to homozygous Hawaiian CB4856 males. After 48 h, remove both male and hermaphrodite parents from the plate and allow progeny to develop (*see* **Notes 9** and **10**).
2. Pick heterozygous (phenotypically wild-type) F1 hermaphrodites (as L4) individually to small seeded NGM plates along with five to eight males of N2 background. To aid in identification of outcross progeny, it is often convenient to use GFP-expressing males (*see* **Note 6**).
3. After 24 h, each heterozygous hermaphrodite should have mated with the N2 males present on the plate. Thus, progeny produced after 24 h of mating are likely to be outcross progeny (allowing measurement of crossing over

that occurred solely during oogenesis). Move heterozygous hermaphrodites to new plates. Each 24 h thereafter for several days (or until they cease producing outcross progeny), move individually to fresh plates (*see* **Note 3**).

4. Scoring markers transmitted to progeny: as the outcross progeny of the heterozygous hermaphrodite reach adulthood, pick individually into 0.2 mL thin-walled tubes containing 10 μL of 10 mM Tris-HCl, pH 8.0 (*see* **Notes 4, 6, 11,** and **12**).
5. To each tube, add 10 μL of 2x Single-worm Lysis Buffer and mix well.
6. Carry out worm lysis, PCR, restriction analysis, electrophoresis, and scoring as in **Section** 3.2.1 **Steps 6–10**.
7. For each interval assayed, outcross progeny will fall into four classes: homozygous N2 (nonrecombinant; b1 b2/ b1 b2), heterozygous N2/CB4856 (nonrecombinant; b1 b2/h1 h2), heterozygous for marker 1 (recombinant; b1 b2/h1 b2), heterozygous for marker 2 (recombinant; b1 b2/ b1 h2). (b1 and b2 represent N2-derived alleles, and h1 and h2 represent CB4856-derived alleles).
8. The recombination frequency p = R, where R = fraction of progeny with recombinant genotypes.

3.2.3. Measuring Crossing over in Males Using snip-SNP Markers

1. Generation of heterozygous males: on a small (60 mm) NGM plate seeded with *E. coli*, mate Bristol N2-derived hermaphrodites to homozygous Hawaiian CB4856 males (or vice versa). After 24 h of mating, remove all the male parents from the plate, which will facilitate detection of progeny males in **Step 2** (*see* **Note 10**).
2. Pick heterozygous F1 males individually to small seeded NGM plates with several N2-derived late L4 stage hermaphrodites homozygous for some phenotypic mutation (e.g. *unc-3*).
3. After 24 h of mating, transfer the mated hermaphrodite partners (but not the heterozygous males) individually to fresh plates. Each of these animals should have mated with the heterozygous males, and will thus produce outcross progeny. Transfer these mated hermaphrodites to fresh plates every 24 h for several days (or until they cease production of outcross progeny) (*see* **Note 3**).
4. Scoring markers transmitted to progeny: outcross progeny from mated hermaphrodites will consist of phenotypically wild-type hermaphrodites and males [if the hermaphrodite partners are homozygous for an *X*-linked marker such as *unc-3*, outcross males will be mutant (and thus distinguishable from their phenotypically wild-type fathers)]. As outcross

progeny reach adulthood, pick individually into 0.2 mL thin-walled tubes containing 10 μL of 10 mM Tris-HCl, pH 8.0 (*see* **Notes 4**, **11**, and **12**).

5. To each tube, add 10 μL of 2x Single-worm Lysis Buffer and mix well.
6. Carry out worm lysis, PCR, restriction analysis, electrophoresis, and scoring as in **Section** 3.2.1 **Steps 6–10**.
7. For each interval assayed, outcross progeny will fall into four classes: homozygous N2 (nonrecombinant; b1 b2/b1 b2), heterozygous N2/CB4856 (nonrecombinant; b1 b2/h1 h2), heterozygous for marker 1 (recombinant; b1 b2/h1 b2), heterozygous for marker 2 (recombinant; b1 b2/b1 h2). (b1 and b2 represent N2-derived alleles, and h1 and h2 represent CB4856-derived alleles)
8. The recombination frequency p = R, where R = fraction of progeny with recombinant genotypes.

3.3. Determination of Chiasma Frequency Through Analysis of Bivalents at Diakinesis

As a complement to the above-described genetic methods for assaying meiotic crossover incidence in *C. elegans*, the following protocol allows for estimation of the frequency of chiasma formation through cytological analysis of oocytes at diakinesis (the last stage of meiosis I prophase). Most chromosome pairs in *C. elegans* receive only one crossover per meiosis under normal circumstances. Mutations that reduce the overall levels of crossing over will thus result in an increased incidence of E_0 chromosome pairs (pairs that have failed to cross over). The resulting univalent chromosomes can be detected cytologically by determining the number of chromosomal bodies present in diakinesis nuclei. In nuclei in which each pair of *C. elegans* chromosomes is connected by a chiasma, six bivalents (representing the six pairs of chromosomes) will be present at diakinesis. In nuclei wherein one pair of chromosomes has failed to form a chiasma, five bivalents and two univalents will be seen at diakinesis (for a total of seven chromosomal bodies). If two pairs fail to form chiasmata, then eight chromosomal bodies (four bivalents and four univalents) will be seen, and so on. Thus, the number of chromosomal bodies present at diakinesis provides an indirect readout for the number of chiasma formed during oocyte meiosis. Note that this approach is chiefly useful in situations where the incidence of crossing over is reduced (for example, by mutation); this approach is largely insensitive to the occurrence of multiple chiasmata between a homolog pair [but *see (11)*]. An additional meiotic recombination phenotype can also be detected through analysis of diakinesis chromosome morphology. The diakinesis chromosomes of some meiotic mutants (e.g. *rad-51, brc-2*) are aggregated and poorly condensed; this phenotype is associated with mutants in which meiotic DSBs are formed but not repaired *(12–14)*.

Because of the transparent nature of *C. elegans*, the number of chromosomal bodies present in diakinesis nuclei can be determined simply by fixing whole hermaphrodites and staining with DAPI. In a few minutes, 25–35 individuals can be moved onto a microscope slide and fixed. In the majority of animals, at least one diakinesis nucleus (and commonly as many as four or more) can be visualized. In this way, it is a relatively simple matter to determine the number of chromosomal bodies present in hundreds of diakinesis nuclei, thus providing an estimate of the incidence of achiasmate (E_0) chromosome pairs. Note that this approach provides a minimum estimate of the incidence of E_0 chromosomes. In some nuclei, individual univalents or bivalents may lie too close to each other to be resolved unambiguously. In such nuclei, the number of chromosomal bodies will be underestimated.

1. Pick 50–100 hermaphrodites at the late L4 stage to a small NGM plate seeded with *E. coli*. Grow for an additional 48 h; adults of this age accumulate diakinesis nuclei, simplifying analysis (*see* **Note 14**).
2. Pick 25–35 worms into a drop of M9 buffer on a glass microscope slide (*see* **Note 15**).
3. Wick away as much of the buffer as possible, using a Kim-Wipe or Whatman paper (*see* **Note 16**).
4. Fix worms by adding 15 μL of 95% ethanol. Allow to air dry.
5. Repeat **Step 4** twice more. At this point, the sample has been fixed and can be stored at 4°C until ready for analysis.
6. Stain worms by covering with 7–10 μL of DAPI/Vectashield solution (1:1). Cover sample with a coverslip, and seal with nail polish (if desired).
7. Visualize worms using standard fluorescence microscopy. Focusing up and down through all focal planes, determine the number of DAPI-stained bodies in the nuclei of diakinesis-stage oocytes. For consistency of quantitation, restrict analysis to only the most mature oocytes in a given gonad arm, as chromosomes in these nuclei tend to be well dispersed. Some investigators restrict scoring to oocytes in positions –1 and –2 (numbers refer to position relative to the spermatheca, with –1 being the oocyte immediately adjacent to the spermatheca), while others score oocytes in positions –1, –2, and –3. In some animals, both anterior and posterior gonad arms may be readily visible and scorable; in other cases, only one may be oriented properly for visualization.

3.4. Induction of Meiotic Recombination Through Ionizing Irradiation

DNA breaks induced by ionizing radiation have been shown to mimic the endogenous DNA double-strand breaks (introduced by the meiosis-specific SPO-11 enzyme) that are the initiating lesions of meiotic recombination *(2, 3, 15, 16)*. Thus, ionizing radiation can be used as a supplement to, or replacement for, normal meiotic DSBs. This allows, for example, determination of the DSB repair ability of mutants that do not form spontaneous DSBs [e.g. *(17)*].

Here, we provide a protocol for the most common application, which assesses the effects of irradiation (delivered to nuclei in mid-pachytene) on the cytological appearance of chromosomes at the diakinesis stage of meiotic prophase.

1. Pick hermaphrodites at the late L4 stage to a small NGM plate seeded with *E. coli*. Grow for an additional 20 h; this will result in animals in early adulthood (*see* **Note 14**).
2. Irradiate adult animals with 1,000–5,000 rad of gamma-rays from a ^{137}Cs source. Ionizing radiation from other sources (e.g. X-rays) should have similar effects (*see* **Note 17**).
3. For cytological determination of the ability of irradiation to induce chiasmata, fix animals 18 h after irradiation. This time point has been empirically determined as a time by which a maximal response to irradiation can be detected in the DSB-defective *spo-11* mutant. For quantitation of the effects of irradiation on chiasma formation, fix and stain whole animals as in **Section 3.3**. For visualization of the response to irradiation in more detail, use the dissection, fixation, and immunofluorescence (if desired) protocols detailed in **Chapter 31** of this volume.
4. In some cases it may be desirable to determine the ability of irradiation to induce crossover formation genetically. To do so, irradiate hermaphrodites heterozygous for linked genetic markers and assay crossover formation as in **Sections 3.1** and **3.2**.

4. Notes

1. Repair of DSBs occurs through a process that involves the recombinase RAD-51. RAD-51 protein/DNA complexes that form during this repair process can be visualized through indirect immunofluorescence, providing researchers with an indirect readout of ongoing recombination/repair events *(12, 18)*. For more information concerning use of immunofluorescence in the study of meiosis in *C. elegans*, *see* **Chapter 31** of this volume. For a recent application of RAD-51 immunofluorescence in the study of meiosis, *see (19)*.
2. Measurement of recombination in animals homozygous for mutations affecting meiosis requires construction of worms homozygous for the meiotic mutation under study and heterozygous for linked genetic markers. However, many

meiotic mutants become aneuploid after only a few generations (due to the chromosome missegregation induced by many mutations affecting meiosis). This can greatly complicate both genetic and physical measures of recombination. Thus, it is vitally important to assay recombination in the germlines of euploid mutant animals derived from parents that were heterozygous for the meiotic mutation in question. The simplest approach for doing so involves use of balancer chromosomes marked with a GFP insertion. One way to do so is shown in **Fig. 7.1**. Note that the scheme shown in **Fig. 7.1** is specifically designed to generate males and hermaphrodites homozygous for a meiotic mutation and heterozygous for two phenotypic markers. This scheme can be easily modified to generate animals heterozygous for snip-SNP markers. Note that animals heterozygous for balancer chromosomes should not be used as "wild-type" controls for experiments measuring crossing over in meiotic mutant backgrounds. In balancer chromosome heterozygotes, nonhomologous chromosome synapsis occurs, with subsequent effects on meiotic recombination [e.g. *(20)*]. For more information about balancer chromosomes in *C. elegans*, *see (21)*. In cases where a suitable balancer chromosome is not available, worms of the appropriate genotype should be derived as in *(22)*.

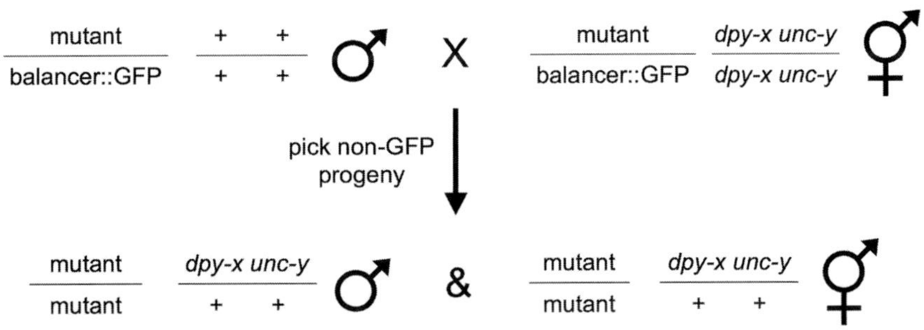

Fig. 7.1. Scheme for production of animals that are both (1) homozygous for a meiotic mutation of interest and (2) heterozygous for linked genetic markers. Males heterozygous for the mutation of interest ("mutant") and a balancer chromosome marked by a gene insertion which leads to GFP expression ("balancer::GFP") are mated to hermaphrodite partners heterozygous for the mutation of interest (balanced by the GFP-marked balancer chromosome) and homozygous for two recessive phenotypic markers (*dpy-x* and *unc-y*) that are linked to each other but unlinked to the meiotic mutation of interest. Male and hermaphrodite progeny from this cross that do not express GFP will be homozygous for the meiotic mutation of interest, and heterozygous for the linked phenotypic markers.

3. A single hermaphrodite produces 250–300 progeny over a 3–4 day period. For measurement of recombination frequencies, it is important to assay all progeny produced by the animal under study during a given time period. By moving hermaphrodites every 24 h, "broods" of roughly

100 progeny are collected. As all of these animals hatched from eggs produced during a single 24-h period, they will all reach adulthood within a relatively narrow time window (but *see* **Note 4**); this greatly simplifies subsequent analyses.

4. As different genotypes may have different growth rates, it is important to score all progeny produced during a given time period. Failure to do so may result in undercounting the number of individuals in certain genotypic class(es), and thus reduce the accuracy of the map distance measurement. Thus, each plate of progeny (each "brood"; *see* **Note 3**) should be scored multiple times over a span of several days; this will increase the likelihood that all progeny will be scored.
5. Under some circumstances, individuals homozygous for two recessive phenotypic markers (M1M2) will show reduced viability. In these cases, the fraction of recombinant progeny (R) from a heterozygous hermaphrodite is calculated using the following equation: R = (M1 + M2)/(4/3*(WT + M1)) *(23)*.
6. To measure the frequency of recombination in the oocyte germline, it is important to only score outcross progeny from the heterozygous hermaphrodite. In crosses of this sort, outcross progeny can be identified in a number of ways.
   - Only score hermaphrodite progeny picked from plates with roughly equal numbers of males and hermaphrodites; these should represent outcross offspring. However, if the animals being assayed are mutant for meiotic function, then self progeny may also have a high proportion of male offspring (the Him phenotype). In such a case, use the following approach.
   - Generate outcross progeny using males homozygous for a third, dominant, marker. One example that has been successfully used is the transgene insertion *ccIs4251*, which expresses GFP under control of the *myo-3* promoter *(24)*. In this case, outcross progeny can be distinguished due to GFP expression.
7. In crosses of this sort, some of the outcross progeny will be homozygous for both markers, and thus indistinguishable from self progeny. Miscounting of self progeny as nonrecombinants will lead to erroneous underestimation of the recombination frequency. To avoid this, three different approaches can be taken.
   - Only score male progeny because male offspring from this cross will arise only as outcross progeny.
   - Calculate R without using the number of doubly-mutant individuals. In this case, R = (M1 + M2)/(2*(WT + M1)).

- Use males homozygous for a third, dominant, marker. One example that has been successfully used is the transgene insertion *ccIs4251*, which expresses GFP under control of the *myo-3* promoter *(24)*. In this case, outcross progeny can be distinguished due to GFP expression.

8. Under some circumstances, individuals homozygous for two recessive phenotypic markers will show reduced viability. In these cases, the fraction of recombinant progeny (R) from a heterozygous male is calculated using the following equation: R = (M1 + M2)/(2*(WT + M1)) *(23)*.
9. The N2-derived parent in this cross is homozygous for a recessive morphological marker to facilitate identification of outcross progeny, which will be wild-type. Self-progeny will be homozygous mutant and thus morphologically distinguishable. This is not necessary, but simplifies identification of outcross progeny. Alternative approaches for identification of outcross progeny are detailed in **Note 6**.
10. Measuring recombination using SNP markers in mutant backgrounds (e.g. meiotic mutants) requires introgression of whole chromosomes from the Hawaiian strain into a strain carrying the mutation under study. This can be done in a similar fashion to the approaches described in **Note 2**, but with the important difference that the SNP markers are codominant, and so can be detected in heterozygous individuals. Strain construction details are left as an exercise for the reader. For an example, *see (11)*.
11. At this point, samples can be stored at –80°C until ready for further analysis.
12. Analysis can also be carried out in 96-well plates.
13. Always amplify N2 and CB4856 controls for amplification and digestion.
14. Age-matching of animals to be analyzed is important, as the proportions of cells in various meiotic stages change as animals age. To synchronize animals, pick late L4 hermaphrodites to a fresh plate and then age to desired degree of maturity. For analysis of diakinesis, animals should be analyzed 48 days post-L4 (older animals accumulate diakinesis nuclei, simplifying the analysis). In a given analysis, always use animals of the same age.
15. *E. coli* also contains DNA, and thus stains with DAPI. The presence of *E. coli* on the slide will thus increase background. To minimize the amount of *E. coli* present on the slide, pick worms to be analyzed to an unseeded NGM plate, and then pick to the slide. In addition, some bacterial contamination can be removed during the wicking process (**Step 3**).

16. If the worms dry out completely before fixation, gonad morphology can be damaged. Thus, it is important to add the ethanol rapidly after removal of the M9 buffer by wicking.

17. In our experience, a dose of either 1,000 or 5,000 rad of ionizing radiation is sufficient to efficiently restore chiasma formation to animals lacking SPO-11 (and thus incapable of forming chiasmata under normal circumstances). Irradiation with 5,000 rad results in cytologically visible chromosome fragmentation and aggregation of chromosomes in animals lacking RAD-50; irradiation with 1,000 rad does not elicit this response *(19)*.

References

1. Keeney, S., Giroux, C. N., and Kleckner, N. (1997) Meiosis-specific DNA double-strand breaks are catalyzed by Spo11, a member of a widely conserved protein family. *Cell* **88**, 375–84.
2. Dernburg, A. F., McDonald, K., Moulder, G., Barstead, R., Dresser, M., and Villeneuve, A. M. (1998) Meiotic recombination in *C. elegans* initiates by a conserved mechanism and is dispensable for homologous chromosome synapsis. *Cell* **94**, 387–98.
3. Kim, J. S., and Rose, A. M. (1987) The effect of gamma radiation on recombination frequency in *Caenorhabditis elegans. Genome* **29**, 457–62.
4. Petronczki, M., Siomos, M. F., and Nasmyth, K. (2003) Un menage a quatre: the molecular biology of chromosome segregation in meiosis. *Cell* **112**, 423–40.
5. Villeneuve, A. M. (1994) A cis-acting locus that promotes crossing over between *X* chromosomes in *Caenorhabditis elegans. Genetics* **136**, 887–902.
6. Stiernagle, T. (2006) *WormBook*, ed. the *C. elegans* Research Community, http://www.wormbook.org.
7. Davis, M. W., and Hammarlund, M. (2006) Single-nucleotide polymorphism mapping. *Methods Mol. Biol.* **351**, 75–92.
8. Brenner, S. (1974) The genetics of *Caenorhabditis elegans. Genetics* **77**, 71–94.
9. Wicks, S. R., Yeh, R. T., Gish, W. R., Waterston, R. H., and Plasterk, R. H. (2001) Rapid gene mapping in Caenorhabditis elegans using a high density polymorphism map. *Nat. Genet.* **28**, 160–4.
10. Hammarlund, M., Davis, M. W., Nguyen, H., Dayton, D., and Jorgensen, E. M. (2005) Heterozygous insertions alter crossover distribution but allow crossover interference in *Caenorhabditis elegans. Genetics* **171**, 1047–56.
11. Nabeshima, K., Villeneuve, A. M., and Hillers, K. J. (2004) Chromosome-wide regulation of meiotic crossover formation in *Caenorhabditis elegans* requires properly assembled chromosome axes. *Genetics* **168**, 1275–92.
12. Alpi, A., Pasierbek, P., Gartner, A., and Loidl, J. (2003) Genetic and cytological characterization of the recombination protein RAD-51 in *Caenorhabditis elegans. Chromosoma* **112**, 6–16.
13. Martin, J. S., Winkelmann, N., Petalcorin, M. I., McIlwraith, M. J., and Boulton, S. J. (2005) RAD-51-dependent and -independent roles of a *Caenorhabditis elegans* BRCA2-related protein during DNA double-strand break repair. *Mol. Cell. Biol.* **25**, 3127–39.
14. Rinaldo, C., Bazzicalupo, P., Ederle, S., Hilliard, M., and La Volpe, A. (2002) Roles for *Caenorhabditis elegans rad-51* in meiosis and in resistance to ionizing radiation during development. *Genetics* **160**, 471–9.
15. Liu, H., Jang, J. K., Kato, N., and McKim, K. S. (2002) *mei-P22* encodes a chromosome-associated protein required for the initiation of meiotic recombination in *Drosophila melanogaster. Genetics* **162**, 245–58.
16. Thorne, L. W., and Byers, B. (1993) Stage-specific effects of X-irradiation on yeast meiosis. *Genetics* **134**, 29–42.
17. Chin, G. M., and Villeneuve, A. M. (2001) *C. elegans mre-11* is required for meiotic recombination and DNA repair but is dispensable for the meiotic G(2) DNA damage checkpoint. *Genes Dev.* **15**, 522–34.
18. Colaiacovo, M. P., MacQueen, A. J., Martinez-Perez, E., McDonald, K., Adamo,

A., La Volpe, A., and Villeneuve, A. M. (2003) Synaptonemal complex assembly in *C. elegans* is dispensable for loading strand-exchange proteins but critical for proper completion of recombination. *Dev. Cell* **5**, 463–74.

19. Hayashi, M., Chin, G. M., and Villeneuve, A. M. (2007) *C. elegans* germ cells switch between distinct modes of double-strand break repair during meiotic prophase progression. *PLoS Genet.* **3**, 2068–84.
20. McKim, K. S., Howell, A. M., and Rose, A. M. (1988) The effects of translocations on recombination frequency in *Caenorhabditis elegans. Genetics* **120**, 987–1001.
21. Edgley, M. L., Baillie, D. L., Riddle, D. L., and Rose, A. M. (2006) Genetic balancers. *WormBook*, ed. the *C. elegans* Research Community, http://www.wormbook.org.
22. Kelly, K. O., Dernburg, A. F., Stanfield, G. M., and Villeneuve, A. M. (2000) *Caenorhabditis elegans msh-5* is required for both normal and radiation- induced meiotic crossing over but not for completion of meiosis. *Genetics* **156**, 617–30.
23. Zetka, M. C., and Rose, A. M. (1990) Sex-related differences in crossing over in *Caenorhabditis elegans. Genetics* **126**, 355–63.
24. Fire, A., Xu, S., Montgomery, M. K., Kostas, S. A., Driver, S. E., and Mello, C. C. (1998) Potent and specific genetic interference by double-stranded RNA in *Caenorhabditis elegans. Nature* **391**, 806–11.

Chapter 8

Visual Markers for Detecting Gene Conversion Directly in the Gametes of *Arabidopsis thaliana*

Luke E. Berchowitz and Gregory P. Copenhaver

Abstract

Measuring meiotic gene conversion is important both because of its role in the fundamental mechanisms of meiotic recombination and because of its influence on linkage relationships and allelic diversity in the genome. Historically, gene conversion has been most thoroughly examined in fungal organisms through the use of tetrad analysis. Here we describe a method for using tetrad analysis in the model plant *Arabidopsis thaliana* to detect and quantify gene conversion events – a resource unavailable in most other higher eukaryotic model systems.

Key words: meiosis, recombination, crossing-over, heteroduplex DNA, mismatch repair.

1. Introduction

Meiotic gene conversion (GC) results from a nonreciprocal exchange of genetic information at polymorphic loci and is classically defined by non-Mendelian segregation ratios in the products of meiosis *(1)*. Geneticists have had a consistent interest in GC because competing mechanistic models of meiotic recombination often make specific predictions involving GC. The ability to detect GC events provides a basis for testing these models. GC also is thought to play an important role in structuring genomes by mediating the spread of variants in repetitive DNA arrays, and their subsequent homogenization including repeats with critical functions like ribosomal RNA genes and centromeric satellite DNA *(2, 3)*. Several studies have proposed a role for GC in directing the evolution of multi-gene families including the

Scott Keeney (ed.), *Meiosis, Volume 1, Molecular and Genetic Methods, vol. 557*

DOI 10.1007/978-1-59745-527-5_8 Springerprotocols.com

aldehyde dehydrogenase, phosphoribosylanthranilate isomerase, ribulose bisphosphate carboxylase, and r-gene families *(4–7)*. More recently, interest in GC has experienced a broader renaissance due to rising popularity of linkage disequilibrium (LD) mapping *(8, 9)*. LD mapping uses the frequency of historic crossover events detected by genotyping haplotypes to establish linkage relationships. GC however, can disrupt linkage and influence the results of LD studies. Therefore, there has been significant interest in measuring and accounting for GC in order to optimize LD mapping algorithms *(10)*. Despite its importance, GC has been historically difficult to measure in complex eukaryotes primarily due to the inability to simultaneously examine all of the products of a meiosis – a process known as tetrad analysis *(11, 12)*. Here we will describe a method for quickly, accurately, and inexpensively measuring GC in the multi-cellular model organism *Arabidopsis thaliana*.

The German botanist Hans Winkler first published the phrase "gene conversion" in 1930, though in his book he makes reference to a lecture on the topic at the 1927 International Genetics Conference in Berlin. Winkler, incidentally, also coined the word "genome" in 1920 *(13, 14)*. The modern view of gene conversion has changed since Winkler, who forcefully defended the idea that it was the basis for all recombination.

Today, most GC is thought to arise through the repair of meiotic DNA double-strand breaks that are catalyzed by an evolutionarily conserved topoisomerase homologue called Spo11 (**Fig. 8.1**). The 5′ ends of the DNA double-strand break are resected to leave 3′ single-stranded tails. One of these ends invades a homologous chromatid and primes new DNA synthesis. The other end then captures the D-loop and uses it as a template for new synthesis. Ligation of the free ends generates a double-Holliday junction intermediate *(15)*. This intermediate can be resolved to produce crossovers or noncrossovers. Some researchers now believe that noncrossovers come from an alternative pathway, called synthesis-dependent strand annealing *(16)*. In synthesis-dependant strand annealing, the invading strand dissociates from the template after some synthesis. The newly synthesized region can then anneal to the other resected end of the break. Subsequent DNA synthesis and ligation results in a noncrossover chromatid. It is important to note that resolution of double Holliday junctions to generate crossovers or noncrossovers and synthesis-dependant strand annealing can all result in regions of heteroduplex DNA that can subsequently give rise to gene conversion. In addition, the DNA synthesis that occurs during double-strand break repair can also give rise to GC (**Fig. 8.1**).

In heteroduplex DNA, the individual strands are derived from different parental DNA molecules. DNA polymorphisms between the parental molecules result in base/base mismatch and

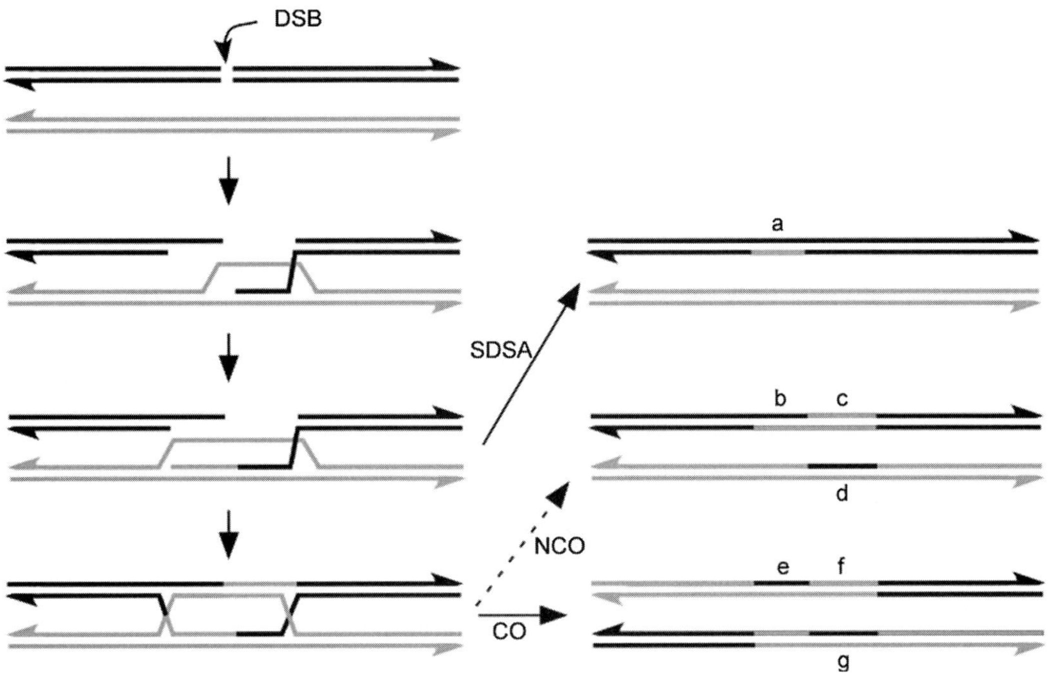

Fig. 8.1. Repair of double-strand breaks (DSBs) is accomplished during meiosis by recombination mechanisms that act on two homologous chromatids (*black and gray bars*). Synthesis-dependant strand annealing (SDSA) results in noncrossovers (NCOs). The resolution of double Holliday junctions forms either NCOs or crossovers (COs) depending on how both Holliday junctions are processed relative to one another (only two of the four possible resolution products are shown for simplicity). During CO or NCO formation DNA synthesis can result in regions of GC (c and f, note that for GC to occur here the heteroduplex DNA at d and g must be repaired in favor of the "gray" information). Formation of heteroduplex DNA (a, b, d, e, and g) can also give rise to GC if the DNA mismatch repair system corrects the mismatch so that the information from the homologous chromatid is retained.

insertion/deletion heterologies in the heteroduplex DNA. The DNA mismatch repair system can recognize and repair these heterologies *(17)*. If the repair system restores the original parental sequence then no GC event occurs. If, however, the repair system corrects the mismatch so that the DNA sequence from the homologous chromatid is used then a GC event is observed. GC, due to repair of heteroduplex DNA or DNA synthesis (**Fig. 8.1**), results in a 3:1 segregation of alleles at the locus instead of the expected Mendelian 2:2 segregation within the four products of meiosis (**Fig. 8.2**).

A meiotic gene conversion assay will optimally be useful directly in the gametes, have an inexpensive and rapid output such as visual scoring, and be applicable at multiple loci. Here we describe an assay for detecting GC events in *Arabidopsis* that meets these criteria. The assay combines a collection of transgenic lines (**Fig. 8.3**), each with a trans-gene construct that encodes a fluorescent protein, and a mutant background (*qrt1/qrt1*) that enables

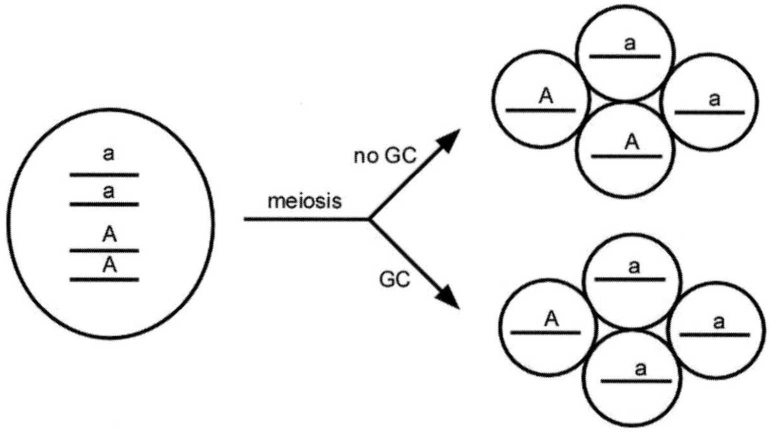

Fig. 8.2. The segregation of heterozygous alleles during meiosis typically results in tetrads with 2:2 "Mendelian" pattern (no GC). If GC occurs during meiosis then a "non-Mendelian" segregation pattern can result in 3:1 (or 1:3) tetrads.

tetrad analysis in *Arabidopsis*. Lesions in *QRT1* prevent the normal separation of developing pollen grains from one another *(18)*. This results in fusion of the pollen exine walls, but leaves *qrt* pollen viable and fertile in every other respect *(19)*. Because the fusion of *qrt* pollen grains does not involve the inner, intine wall, there is no mixing of cytoplasmic or nuclear components between the meiotic products. Pollen from *qrt* plants which are segregating markers can be used for tetrad analysis *(20)*. Using *qrt/qrt* plants we created a library of transgenic lines that express yellow, red, or cyan fluorescent protein (**Fig. 8.3**) that are called fluorescent-tagged lines (FTLs) *(21)*. The marker proteins are expressed using the *LAT52* post-meiotic pollen-specific promoter so that visualizing their expression makes it possible to track the segregation of the genes encoding them using fluorescence microscopy *(22)*. The trans-genes have been precisely mapped by sequencing the DNA flanking their insertion sites. It is possible to create mutant nonfluorescent alleles of these trans-genes through EMS mutagenesis. A plant that is heterozygous for a fluorescent allele and a nonfluorescent allele will typically yield pollen tetrads that have two fluorescing grains and two nonfluorescing grains (i.e., expected 2:2 Mendelian segregation). If GC occurs at the trans-gene locus then a pollen tetrad with 3:1 or 1:3 segregation is observed (**Fig. 8.2**).

The following method describes: (i) mutagenesis of existing fluorescent lines, (ii) screening for nonfluorescent alleles, (iii) genetically verifying nonfluorescent mutant lines, (iv) sequencing the nonfluorescent allele, (v) backcrossing, (vi) scoring GC events in the pollen tetrads produced by the heterozygote, and (vii) using GC test loci in mutant backgrounds.

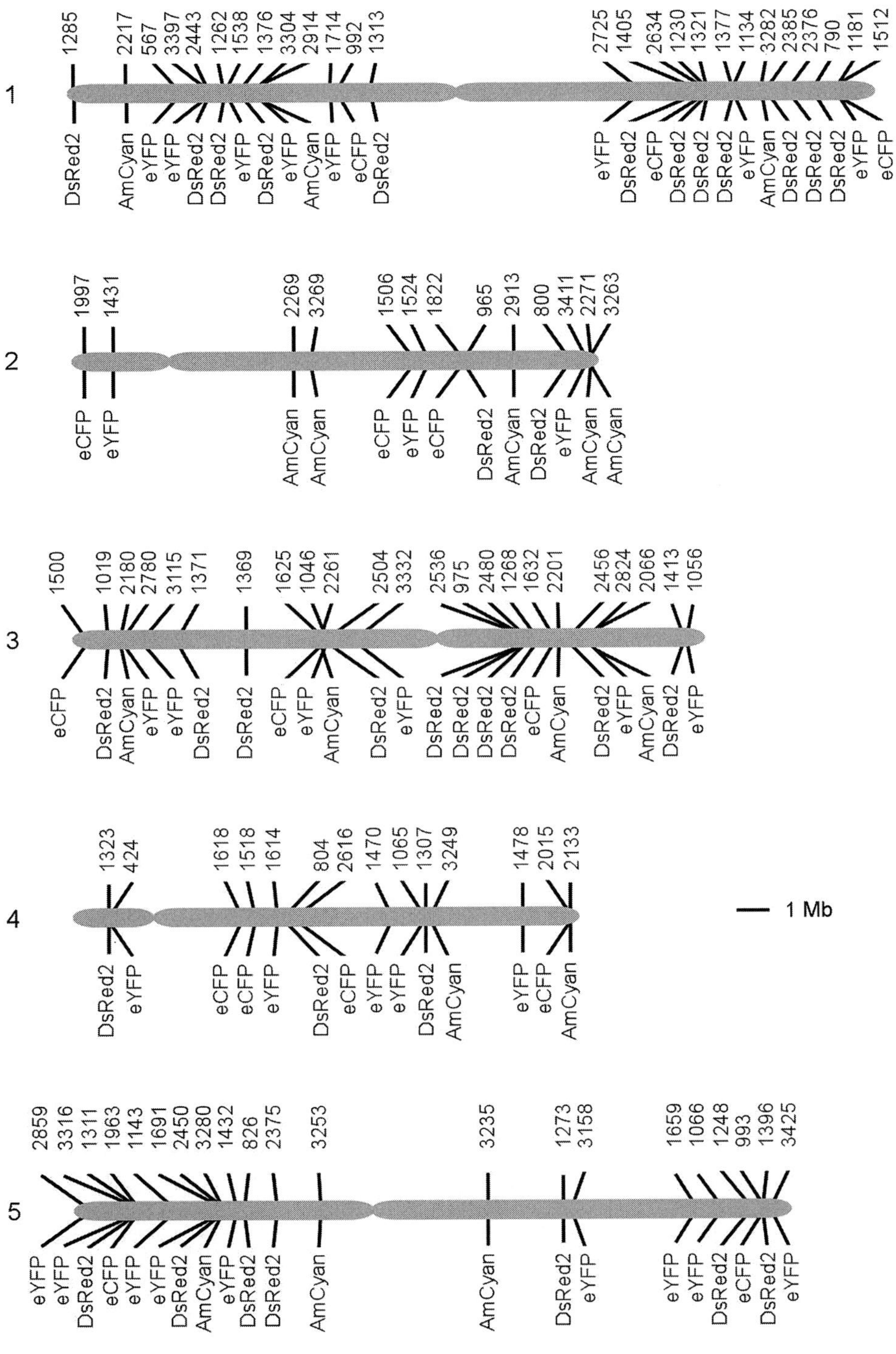

Fig. 8.3. Map of FTL lines in *Arabidopsis*. Each of the five *Arabidopsis* chromosomes is shown as a gray bar (1–5). The position of each transgenic line is indicated by a *black bar*. Above each chromosome is the identification number of the FTL line. Below each chromosome is the fluorescent protein encoded by each trans-gene construct (DsRed2, eYFP, eCFP, or AmCyan).

2. Materials

2.1. Plant Material and Plant Growth

1. *Arabidopsis thaliana* FTL seed (**Fig. 8.3**), available on request from corresponding author.
2. Metro-Mix 400 soil (Sun Gro Horticulture Inc., Bellevue, WA) (*see* **Note 1**).
3. Plastic planting flats with bottom holes, 21 × 11 inches (Wyatt-Quarles, Garner, NC).
4. Plastic planting flats without holes, 21 × 11 inches (Wyatt-Quarles, Garner, NC).
5. Plastic planting inserts, 24-cell (Wyatt-Quarles, Garner, NC).
6. Clear plastic humidome (Wyatt-Quarles, Garner, NC).
7. Plastic plant marker stakes (Wyatt-Quarles, Garner, NC).

2.2. Mutagenesis

1. Ethylmethane sulfonate (EMS) (*see* **Note 2**).
2. 0.1% agarose solution, after autoclaving can be stored at room temperature.

2.3. 2.3.Microscopy

1. Glass microscope slides, 75 × 25 × 1 mm.
2. Soft cotton cloth, like a T-shirt.
3. Pollen growth medium (PGM) solution: 17% (w/v) sucrose, 2 mM $CaCl_2$, 1.625 mM boric acid, 0.1% Triton-X100.
4. Epi-fluorescence microscope (e.g., Nikon Eclipse E1000).
5. CFP, YFP DsRed filters.

2.4. DNA Preparation for Sequencing

1. Extraction buffer: 50 mM Tris-HCl (pH 8.0), 200 mM NaCl, 0.2 mM EDTA, 0.5% SDS, 100 μg/mL proteinase K (add proteinase K just before use to a working aliquot).
2. Plastic grinding pestles.
3. Buffer saturated phenol, pH 6.6.
4. Chloroform/isoamyl alcohol 24:1.
5. 3 M sodium acetate solution.
6. 100% ethanol.
7. 70% ethanol.
8. TE: 10 mM Tris-HCl (pH 8.0), 1 mM EDTA.
9. PCR primers to amplify *LAT52* promoter and fluorescent gene (*see* **Note 3**).

3. Methods

3.1. EMS Mutagenesis of Fluorescent Transgenic Lines

1. Refer to **Fig. 8.3** and select an FTL line (*see* **Note 4**). Lines can be obtained by contacting the corresponding author.

2. Plant 12 seeds for the FTL line (*see* **Note 5**). Sow seeds on autoclaved Metro-mix 400 soil. Cover the flats with a humidome.
3. Stratify the seeds in a 4°C cold chamber for 4 days.
4. Transfer the seeds to an environmentally controlled growth chamber with long-photoperiod (16 h light/8 h dark) at 20°C. Leave the humidome on.
5. After the plants germinate and the first four true leaves emerge, slide the humidome back to create a gap along one edge of the flat but do not remove it entirely.
6. After 1 day with a gapped humidome remove the humidome entirely and let the plants grow until they flower. If the FTL stock was shipped to you as homozygous seed let the plant reach senescence and proceed directly to **Step 15** below. If the FTL stock was shipped to you as a segregating population proceed to the next step.
7. Clean a microscope slide with a soft cotton cloth. This helps the medium that the pollen will be suspended in stay in a neat droplet.
8. Put six evenly spaced 10 μL spots of PGM onto the slide.
9. Place a numbered plastic plant marker next to a FTL plant.
10. With fine-tipped forceps remove an open flower from the FTL plant and place it face down in the first drop of PGM. Let it soak for 20 s.
11. Grasp the base of the flower with the forceps and use a gentle up and down motion to release the pollen into the drop of PGM (*see* **Note 6**).
12. Repeat **Steps 9** through **11** with a different FTL plant until the slide is full.
14. Visually screen the pollen using a fluorescence microscope (*see* **Note 4**). Plants that are homozygous for the trans-gene construct will produce pollen tetrads that have four fluorescing grains (4:0). Hemizygous plants will produce tetrads with two fluorescing grains and two nonfluorescing grains (2:2). Discard hemizygous plants and let the homozygous plants grow until senescence.
15. Collect seeds from plants with dried siliques. Using a small pair of scissors cut the plant away from its roots. Hold the plant over a clean sheet of paper and gently rub it between your hands to release the seeds.

16. Separate the seeds from the chaff and place the seeds in a labeled microfuge tube with a small hole in the lid or in a small coin envelope. Store the seed in a cool, dry place.
17. Place approximately 40 mg of seeds (roughly 800 seeds) in a 50 mL conical tube and add 40 mL of sterile water. Let the seeds imbibe for 1 h.
18. Add 125 μL of freshly opened EMS. Seal the tube and mix immediately. Secure the seal on the tube with parafilm (*see* **Note 2**).
19. Incubate seeds with agitation at room temperature for 16 h (overnight).
20. Place the tube containing the seeds upright in a tube rack, allow the seeds to settle to the bottom for 5 min and pour the EMS solution into an appropriate waste container.
21. Add 40 mL of sterile water to the tube and rinse the seeds by sealing the tube and inverting it 5 times. Place the tube in the tube rack and allow the seeds to settle to the bottom for 5 min and pour the EMS solution into an appropriate waste container.
22. Repeat **Step 21** six more times.
23. Add 40 mL of sterile water and let the seeds soak for 1 h before pouring the rinse water off.
24. Resuspend the rinsed seeds in 40 mL of 0.1% agarose solution.
25. Prepare eight flats of soil (*see* **Note 5**).
26. Use a 5 mL pipette to evenly disperse 5 mL of the seeds suspended in the 0.1% agarose solution across a flat of soil. Repeat for the remaining flats of soil. Cover the flats with humidomes.
27. Grow the plants as described in **Steps 3–6** until they flower.

3.2. Screening M1 FTLs for Nonfluorescent Alleles

1. Clean a microscope slide with a soft cotton cloth.
2. Put six evenly spaced 10 μL spots of PGM onto the slide.
3. Place a numbered plastic plant marker next to an M1 plant.
4. With fine-tipped forceps remove an open flower from the first M1 plant and place it face down in the first drop of PGM. Let it soak for 20 s.
5. Grasp the base of the flower with the forceps and use a gentle up and down motion to release the pollen into the drop of PGM (*see* **Note 6**).
6. Repeat **Steps 4** and **5** with another flower from the same plant (*see* **Note 7**).
7. Repeat **Steps 3** through **6** for two more plants.

8. Visually screen the pollen using a fluorescence microscope (*see* **Note 4**). If neither allele of the fluorescent trans-gene experienced a mutation that disrupts function then the pollen tetrads will have four fluorescing grains (4:0). Conversely, if one of the alleles suffered a mutation that disrupts function then the pollen tetrads will have two fluorescing grains and two nonfluorescing grains. A true candidate line will produce only 2:2 tetrads. Discard plants yielding 4:0 tetrads. Save plants that consistently yield 2:2 tetrads and allow them to set self-fertilized seeds.
9. Collect individual siliques from the mature plants as in **Section 3.1**, **Steps 15–16** (*see* **Note 8**).

3.3. Confirming Putative Nonfluorescent Mutants

1. Select three siliques from **Section 3.2**, **Step 9** and plant eight seeds from each on Metro-mix 400, keeping track of which seeds come from which siliques (*see* **Note 5**). Grow according to the technique described in **Section 3.1**, **Steps 3–6**. Let the plants grow until their first few flowers begin to open.
2. Examine the pollen from each plant as described in **Section 3.2**, **Steps 1–8** with the exception that it is only necessary to examine one flower from each plant. Siliques from **Section 3.2**, **Step 9** that have seeds that are segregating a bona fide nonfluorescent allele should produce three kinds of progeny: those that make tetrads with all four grains fluorescing (4:0), those that make tetrads with only two grains fluorescing (2:2) and those that make tetrads with no fluorescent grains (0:4). These progeny should be produced in a 1:2:1 ratio. Lines that meet this expectation should be saved by collecting self-fertilized seed from the plants that make 2:2 tetrads (*FTL/ftl*) and from plants that make 0:4 tetrads (*ftl/ftl*). Siliques from **Section 3.2**, **Step 9** that have seeds that do not contain the mutation due to mosaicism (*see* **Notes** 7 and **8**) will produce only one kind of progeny, i.e., those that make tetrads with no fluorescent grains. These should be discarded.
3. Harvest four small cauline leafs from the plants that make 0:4 tetrads (*ftl/ftl*) for DNA extraction. These can be stored at –80°C if necessary.

3.4. Preparing DNA from the Nonfluorescent Mutant

This plant DNA mini-prep is adapted from McKinney et al. *(23)*. It produces DNA that is stable for long periods at 4°C and can be used for most down-stream molecular applications including PCR.

1. Place one cauline leaf from **Section 3.3**, **Step 3** in a microfuge tube and add 100 μL of extraction buffer.
2. Grind the leaf using a plastic pestle until it is a homogeneous slurry.
3. Add an additional 100 μL of extraction buffer, vortex briefly and incubate at 37°C for 30 min.

4. Add 200 μL of buffer saturated phenol to each sample, vortex for 15 s and spin in the microfuge at maximum speed for 2 min.
5. Transfer the top aqueous phase to a new microfuge tube.
6. Add 200 μL of chloroform/isoamyl alcohol (24:1), vortex and spin in the microfuge at maximum speed for 2 min.
7. Transfer the top (aqueous) phase to a new microfuge tube.
8. Add 18 μL of 3 M sodium acetate and precipitate the nucleic acids by adding 400 μL (two volumes) of 100% ethanol, mix by inverting and incubate for 10 min at 4°C.
9. Pellet the precipitated DNA by spinning in a microfuge for 10 min at maximum speed.
10. Pour off the supernatant being careful to retain the nucleic acid pellet.
11. Add 500 μL of 70% ethanol, vortex vigorously for 15 s and spin in the microfuge for 2 min at maximum speed.
12. Pour off the supernatant being careful to retain the nucleic acid pellet.
13. Repeat **Steps 11** and **12**.
14. Air dry the nucleic acid pellet.
15. Re-suspend the pellet in 100 μL of T.E.

3.5. Sequencing the Nonfluorescent Allele

1. Design PCR primers to amplify the fluorescent marker gene (*see* **Note 3**).
2. PCR amplify the nonfluorescent allele using template DNA from **Section 3.4, Step 15**.
3. Sequence the PCR amplified DNA. A useful allele will have some DNA sequence change in the promoter or coding region of the fluorescent marker gene (*see* **Note 9**).
4. Sequence information of the mutant allele can be used to genotype plants that descend from this line (*see* **Note 10**).

3.6. Backcrossing Nonfluorescent Alleles

Undesirable background mutations can be eliminated from the nonfluorescent mutant line by back crossing. The nonfluorescent mutant plant from **Section 3.3, Step 2** that produces 0:4 tetrads will be back crossed to the original parental line from **Section 3.1, Step 16**.

1. Plant four seeds of the original parental line from **Section 3.1, Step 16** and four seeds of the nonfluorescent mutant from **Section 3.3, Step 2** and grow them on Metro-mix 400 until they are flowering using the procedure described in **Section 3.1, Steps 3–6**.

2. Use clean fine-tipped forceps to emasculate a flower bud just prior to opening (stage 12) by removing the anthers *(24)*. Remove a few flower buds above and below the emasculated bud.
3. Use fine-tipped forceps to harvest mature anthers that are shedding pollen from the original parental line.
4. Rub the anthers from **Step 2** onto the stigma of the emasculated flower from **Step 1**.
5. Tag the cross with a piece of time tape just below the emasculated flower.
6. Harvest the seed when the silique matures. All the progeny should be heterozygous fluorescent:nonfluorescent. Save these as BC1 (backcross generation 1). Confirm the genotypes by using sequence information from **Section 3.5, Step 4**.
7. Plant four BC1 seeds and four seeds of the original parental line from **Section 3.1, Step 16** and grow them on Metro-mix 400 until they are flowering using the procedure described in **Section 3.1, Steps 3–6**.
8. Use the procedure described in **Steps 2–5** to cross the BC1 plants to the original parental line. 50% of the seeds from this cross will be heterozygous and 50% will be homozygous for the fluorescent allele.
9. Plant 12 seeds from **Step 8** and four seeds of the original parental line from **Section 3.1, Step 16**.
10. Identify the heterozygous individuals from **Step 8** by examining the pollen they produce under the fluorescent microscope using the technique described in **Section 3.2, Steps 1–6**. Heterozygous individuals will produce pollen tetrads that have two fluorescing grains and two nonfluorescing grains (2:2). Save these as BC2. Confirm the genotypes by using sequence information from **Section 3.5, Step 4**.
11. Cross the BC2 individuals to the original parental line from **Step 9** using the technique described in **Steps 2–5**. 50% of the seeds from this cross will be heterozygous and 50% will be homozygous for the fluorescent allele.
12. Plant 12 seeds from **Step 11** and identify the heterozygous individuals using the same technique as **Step 10**. Save these as BC3. Confirm the genotype by using information from **Section 3.6, Step 4**.
13. Allow the BC3 individuals to set self-fertilized seed. These seed will segregate the fluorescent and nonfluorescent tetrad phenotypes in 1:2:1 ratio as described in **Section 3.3, Step 2**. Save seed from individuals with 2:2 and 0:4 fluorescent:nonfluorescent tetrads. Confirm the genotype by using sequence information from **Section 3.5, Step 4**.

3.7. Scoring GC Events in the Pollen Tetrads

1. Plant 12 seed from individuals that produced 2:2 tetrads from **Section 3.6, Step 12** (these will be a segregating population since the parent was heterozygous) and identify heterozygous individuals as described in **Section 3.6, Step 10** as well as individuals that are homozygous for the fluorescent allele (for a control).
2. Clean a microscope slide with a soft cotton cloth.
3. Put two 10 µL spots of PGM onto the slide.
4. With fine-tipped forceps remove an open flower from the heterozygous plant identified in **Step 1** and place it face down in the first drop of PGM. Let it soak for 20 s.
5. Grasp the base of the flower with the forceps and use a gentle up and down motion to release the pollen into the drop of PGM (*see* **Notes 6** and **11**).
6. Repeat **Steps 4–5** for another flower.
7. Gently place a cover slip on top of each drop of PGM.
8. Visually screen the pollen with a fluorescence microscope using the 10X objective (*see* **Note 4**). Meioses that have not experienced a gene conversion event at the fluorescent marker locus will produce pollen tetrads that segregate 2:2. Meioses that have experienced a gene conversion event at the fluorescent marker locus will produce pollen tetrads that segregate either 3:1 or 1:3 (fluorescent:nonfluorescent). Because nonfluorescent pollen can result from nonviable pollen 1:3 tetrads should be ignored. The frequency of GC tetrads can be calculated using the formula:

$$2(\text{number of } 3:1 \text{ tetrads})/(\text{number of } 2:2 \text{ tetrads} + 2(\text{number of } 3:1 \text{ tetrads}))$$

3.8. Using a GC Test Locus for Mutant Analysis

1. Plant four seeds that are homozygous (*see* **Note 12**) for a mutation in your favorite gene (*yfg/yfg*) and four seeds of the original parental homozygous FTL line from **Section 3.1, Step 16** on Metro-mix 400 and grow them until they flower using the procedure described in **Section 3.1, Steps 3–6**.
2. Cross the mutant to the FTL line using the procedure described in **Section 3.6, Steps 2–5** and harvest seeds from the resulting siliques when they reach senescence. These seed will be heterozygous for *quartet* and the mutant, and hemizygous for the FTL insert (*qrt/QRT; FTL/+; yfg/YFG*).
3. Plant four seeds from **Step 2** (*qrt/QRT; FTL/+; yfg/YFG*) and four seeds from **Section 3.6, Step 13** that are homozygous for the nonfluorescent allele (*ftl/ftl*) on Metro-mix 400 and grow them until they flower using the procedure described in **Section 3.1, Steps 3–6**.

4. Cross the two lines from **Step 3** using the procedure described in **Section 3.6**, **Steps 2–5** and harvest seeds from the resulting siliques when they reach senescence.
5. Plant 24 seed from **Step 4** on Metro-mix 400 and grow them until they flower using the procedure described in **Section 3.1**, **Steps 3–6**.
6. Examine the pollen under the microscope using the procedure described in **Section 3.7**, **Steps 2–7**. 50% of the plants will be heterozygous for quartet (*qrt/QRT*) and will have monad pollen. Discard these plants. 50% of the remaining plants will be hemizygous for the nonfluorescent FTL allele (*ftl/+*) and will have tetrads with no fluorescent grains. Discard these as well.
7. Genotype (*see* **Note 10**) the remaining plants for the *YFG* locus. 50% will be *qrt/qrt*; *ftl/FTL*; *YFG/YFG*. Discard these. 50% will be *qrt/qrt*; *ftl/FTL*; *yfg/YFG*. Save these and allow them to set self-fertilized seed.
8. Plant 48 seeds from **Step 7** on Metro-mix 400 and grow them until they flower using the procedure described in **Section 3.1**, **Steps 3–6**.
9. Examine the pollen under the microscope using the procedure described in **Section 3.7**, **Steps 2–7**. 50% of the plants will be homozygous for either the fluorescent or nonfluorescent FTL allele (*FTL/FTL* or *ftl/ftl*) and will produce pollen tetrads that have either all fluorescing grains or no fluorescing grains (4:0 or 0:4). Discard these.
10. Visually screen the pollen tetrads produced by the remaining plants from **Step 9** with a fluorescence microscope using the 10X objective (*see* **Note 4**). Keep separate data records for each plant. Meioses that have not experienced a gene conversion event at the fluorescent marker locus will produce pollen tetrads that segregate 2:2. Meioses that have experienced a gene conversion event at the fluorescent marker locus will produce pollen tetrads that segregate either 3:1 or 1:3 (fluorescent:nonfluorescent). Because nonfluorescent pollen can result from nonviable pollen 1:3 tetrads should be ignored. The frequency of GC tetrads can be calculated using the formula:

$$2(\text{number of } 3:1 \text{ tetrads})/(\text{number of } 2:2 \text{ tetrads} + 2(\text{number of } 3:1 \text{ tetrads}))$$

11. Genotype (*see* **Note 10**) each plant scored in **Step 10** for the *YFG* locus. 25% will be homozygous for the wild-type allele (*YFG/YFG*) and can be used as a control population. 50% will be heterozygous (*yfg/YFG*) and can also be used as a control population. The remaining 25% will be homozygous for

the mutant allele (*yfg/yfg*). Genotyping after data collection (i.e., blind scoring) is good laboratory practice and can reduce unintentional experimenter bias.

4. Notes

1. Commercial soil often comes contaminated with pathogens and pests that can infest growth chambers and alter experimental data. All soil used for planting should be wetted with de-ionized water, covered with foil, autoclaved and cooled to room temperature before use.
2. EMS is mutagenic and should be handled using appropriate lab safety procedures including wearing a lab coat, safety glasses, double gloves and working in a chemical fume hood. EMS can be inactivated by mixing with sodium hydroxide to a final concentration of 1 N.
3. Primers useful for amplifying the genes encoding the fluorescent proteins and the promoter. The Y/CFP primer pair can be used to amplify either the CFP or the YFP gene.

   LAT52 – F:
   5′TCCTATGAGCTCGCTCTAGAACTAGTGGATCCC3′

   LAT52 – R:
   5′AAGCATGCTAGCCGAGGTCGACGGTATCGATAAG3′

   DSRED2 – F: 5′CACCATGGCCTCCTCCGAGAACGT3′

   DSRED2 – R: 5′CTACAGGAACAGGTGGTGGCGGCC3′

   Y/CFP – F: 5′CACCATGGTGAGCAAGGGCGAGGA3′

   Y/CFP – R: 5′TTACTTGTACAGCTCGTCCATGCC3′

   amCYAN – F: 5′ATGGCTCTTTCAAACAAGTTTATCGG′3

   amCYAN – R: 5′AAAGGGACAACAGAGGTTATATGTGC′3
4. The choice of FTL line will determine what optical filters will be used with the fluorescence microscopy. The available FTL lines have trans-gene inserts with either DsRed, eYFP, eCFP, or AmCyan (Clontech, Mountain View, CA).
5. A variety of pots or flats can be used for planting. We recommend 21 × 11 inch plastic flats. For sowing bulk seed a flat with holes is inserted into a flat without holes and filled with soil. This enables bottom watering. For planting individual seeds we recommend using 24-cell inserts in a flat without holes.
6. When loading pollen onto microscope slides it helps to have a black background (such as a composite lab bench) underneath the slide so that it is easier to see when the pollen has been efficiently released into the medium.

7. EMS mutagenesis of *Arabidopsis* seeds can yield mosaic M1 plants. Scoring two or more flowers from the same plant can reduce the chance that the desired mutant will be missed due to sectoring.
8. Because M1 plants may be mosaic (*see* **Note 7**) some of the siliques they produce may contain seed that are segregating the desired mutation and other siliques may contain only seeds that do not have the mutation.
9. The sequence for the *LAT52* promoter can be found in Twell et al. *(25)*. The sequence for DsRed2, eYFP, eCFP, and AmCyan can be found in Clontech Laboratories protocols PT3601-5 (version PR651771), PT3175-5 (version PR08369), PT3258-5 (version PR08391), and PT3476-5 (version PR651824), respectively, which can be found on the Clontech website (http://www.clontech.com/).
10. Genotyping can be achieved by sequencing or by developing an allele-specific PCR-based marker *(26)*.
11. We recommend scoring only tetrads from the 5th to the 30th flower produced by the plant. Flowers outside of this range can have lower pollen yields and higher pollen lethality. We also recommend selecting flowers from only one shoot (preferably the primary shoot).
12. If the mutant phenotype of yfg involves fertility defects or genome stability problems it is advisable to make this initial cross with a heterozygote. Crossing with a heterozygote will generate progeny that are either *YFG/yfg* or *YFG/YFG* (presumably both with wild-type phenotype) therefore some sort of genotyping (*see* **Note 10**) will be required at this stage to select the heterozygotes.

Acknowledgments

We would like to thank Jeff Sekelsky (UNC at Chapel Hill) for critical reading of the manuscript. We would also like to thank NSF (MCB-0618691) and DOE (DE-FGO2-05ER15651) for financial support.

References

1. Roman, H. (1985) Gene conversion and crossing-over. *Environ. Mutagen.* 7, 923–32.
2. Copenhaver, G. P., Doelling, J. D., Gens, J. S., and Pikaard, C. S. (1995) Use of RFLPs larger than 100 kbp to map the position and internal organization of the nucleolus organizer region on chromosome 2 in *Arabidopsis thaliana*. *Plant J.* 7, 273–86.
3. Hall, S. E., Kettler, G., and Preuss, D. (2003) Centromere satellites from Arabidopsis populations: maintenance of conserved and variable domains. *Genome Res.* **13**, 195–205.

4. Cooley, M. B., Pathirana, S., Wu, H. J., Kachroo, P., and Klessig, D. F. (2000) Members of the Arabidopsis HRT/RPP8 family of resistance genes confer resistance to both viral and oomycete pathogens. *Plant Cell* **12**, 663–76.
5. Skibbe, D. S., Liu, F., Wen, T. J., Yandeau, M. D., Cui, X., Cao, J., Simmons, C. R., and Schnable, P. S. (2002) Characterization of the aldehyde dehydrogenase gene families of Zea mays and Arabidopsis. *Plant Mol. Biol.* **48**, 751–64.
6. Meagher, R. B., Berry-Lowe, S., and Rice, K. (1989) Molecular evolution of the small subunit of ribulose bisphosphate carboxylase: nucleotide substitution and gene conversion. *Genetics* **123**, 845–63.
7. Melquist, S., Luff, B., and Bender, J. (1999) Arabidopsis PAI gene arrangements, cytosine methylation and expression. *Genetics* **153**, 401–13.
8. Haubold, B., Kroymann, J., Ratzka, A., Mitchell-Olds, T., and Wiehe, T. (2002) Recombination and gene conversion in a 170-kb genomic region of *Arabidopsis thaliana*. *Genetics* **161**, 1269–78.
9. Nordborg, M., Hu, T. T., Ishino, Y., Jhaveri, J., Toomajian, C., Zheng, H., Bakker, E., Calabrese, P., Gladstone, J., Goyal, R., Jakobsson, M., Kim, S., Morozov, Y., Padhukasahasram, B., Plagnol, V., Rosenberg, N. A., Shah, C., Wall, J. D., Wang, J., Zhao, K., Kalbfleisch, T., Schulz, V., Kreitman, M., and Bergelson, J. (2005) The Pattern of Polymorphism in *Arabidopsis thaliana*. *PLoS Biol.* **3**, e196.
10. Frisse, L., Hudson, R. R., Bartoszewicz, A., Wall, J. D., Donfack, J., and Di Rienzo, A. (2001) Gene conversion and different population histories may explain the contrast between polymorphism and linkage disequilibrium levels. *Am. J. Hum. Genet.* **69**, 831–43.
11. Copenhaver, G. P., Keith, K. C., and Preuss, D. (2000) Tetrad analysis in higher plants. A budding technology. *Plant Physiol.* **124**, 7–16.
12. Mather, K., and Beale, G. (1942) The calculation and precision of linkage values from tetrad analysis. *J. Genet.* **43**, 1–30.
13. Winkler, H. (1920) Verbreitung und Ursache der Parthenogenesis im Pflanzen- und Tierreiche, Verlag von Gustav Fischer, Jena.
14. Winkler, H. (1930) Die Konversion der Gene, Verlag von Gustav Fischer, Jena.
15. Szostak, J. W., Orr-Weaver, T. L., Rothstein, R. J., and Stahl, F. W. (1983) The double-strand-break repair model for recombination. *Cell* **33**, 25–35.
16. Allers, T., and Lichten, M. (2001) Differential timing and control of noncrossover and crossover recombination during meiosis. *Cell* **106**, 47–57.
17. Kunkel, T. A., and Erie, D. A. (2005) DNA mismatch repair. *Annu. Rev. Biochem.* **74**, 681–710.
18. Preuss, D., Rhee, S. Y., and Davis, R. W. (1994) Tetrad analysis possible in Arabidopsis with mutation of the QUARTET (QRT) genes. *Science* **264**, 1458–60.
19. Francis, K. E., Lam, S. Y., and Copenhaver, G. P. (2006) Separation of Arabidopsis pollen tetrads is regulated by QUARTET1, a pectin methylesterase gene. *Plant Physiol.* **142**, 1004–13.
20. Copenhaver, G. P., Browne, W. E., and Preuss, D. (1998) Assaying genome-wide recombination and centromere functions with Arabidopsis tetrads. *Proc. Natl. Acad. Sci. U S A* **95**, 247–52.
21. Francis, K. E., Lam, S. Y., Harrison, B. D., Bey, A. L., Berchowitz, L. E., and Copenhaver, G. P. (2007) Pollen tetrad-based visual assay for meiotic recombination in Arabidopsis. *Proc. Natl. Acad. Sci. U S A* **104**, 3913–8.
22. Twell, D., Yamaguchi, J., and McCormick, S. (1990) *Development* **109**, 705–13.
23. McKinney, E. C., Ali, N., Traut, A., Feldmann, K. A., Belostotsky, D. A., McDowell, J. M., and Meagher, R. B. (1995) *Plant J.* **8**, 613–22.
24. Smyth, D. R., Bowman, J. L., and Meyerowitz, E. M. (1990) *Plant Cell* **2**, 755–67.
25. Twell, D., Yamaguchi, J., Wing, R. A., Ushiba, J., and McCormick, S. (1991) Promoter analysis of genes that are coordinately expressed during pollen development reveals pollen-specific enhancer sequences and shared regulatory elements. *Genes Dev.* **5**, 496–507.
26. Neff, M. M., Neff, J. D., Chory, J., and Pepper, A. E. (1998) dCAPS, a simple technique for the genetic analysis of single nucleotide polymorphisms: experimental applications in *Arabidopsis thaliana* genetics. *Plant J.* **14**, 387–92.

Section II

Molecular Analysis of Recombination and Protein-DNA Interactions During Meiosis

Chapter 9

Gel Electrophoresis Assays for Analyzing DNA Double-Strand Breaks in *Saccharomyces cerevisiae* at Various Spatial Resolutions

Hajime Murakami, Valérie Borde, Alain Nicolas, and Scott Keeney

Abstract

Meiotic recombination is triggered by programmed DNA double-strand breaks (DSBs), which are catalyzed by Spo11 protein in a type II topoisomerase-like manner. Meiotic DSBs can be detected directly using physical assays (gel electrophoresis, Southern blotting, and indirect end-labeling) applied to samples of genomic DNA from sporulating cultures of budding and fission yeast. Such assays are extremely useful for quantifying and characterizing many aspects of the initiation of meiotic recombination, including the timing of DSB formation relative to other events, the distribution of DSBs across the genome, and the influence on DSB formation of mutations in recombination factors and other gene products. By varying the type of gel electrophoresis and other parameters, the spatial resolution of DSB analysis can range from single nucleotides up to whole yeast chromosomes.

Key words: yeast, meiotic recombination, Southern blotting, pulsed-field gel electrophoresis.

1. Introduction

To detect meiotic double-strand breaks (DSBs) in *S. cerevisiae*, genomic DNA is extracted from synchronously sporulating cultures, then digested with restriction enzyme(s) as necessary to yield an appropriately sized fragment. Subsequently, these DNA fragments are separated by gel electrophoresis, and both the parental length (unbroken) DNA and the DSB fragments are detected by Southern blotting and indirect end-labeling by hybridization to an appropriate probe (**Fig. 9.1**).

Scott Keeney (ed.), *Meiosis, Volume 1, Molecular and Genetic Methods, vol. 557*

DOI 10.1007/978-1-59745-527-5_9 Springerprotocols.com

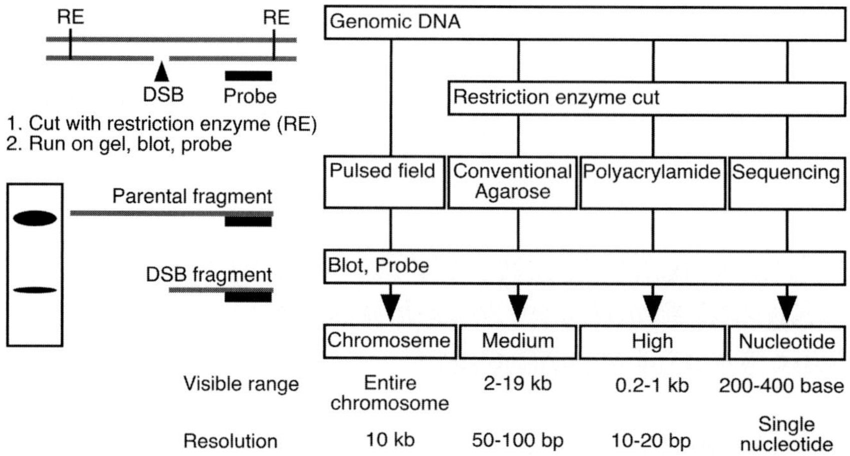

Fig. 9.1. The procedure of DSB mapping at four different resolutions. *Left*: General procedure of DSB mapping. Genomic DNA is digested with restriction enzyme(s) to yield appropriately sized DNA fragments (for chromosome-level mapping, this step is skipped). The DNA fragments are separated by electrophoresis on an appropriate gel according to the size of the fragment, then detected by Southern blot hybridization. *Right*: Comparison of methods and resolution of chromosome, medium, high, and nucleotide-level resolution mapping.

Meiotic DSBs do not occur randomly throughout the genome. Instead, DSBs (and the resulting recombination products) form preferentially within small (≤2 kbp) regions called hotspots *(1)*. DSBs show specific features of distribution according to the resolution of the methods used to detect them. When observed at the resolution of a whole chromosome (100s to 1,000s of kbp in *S. cerevisiae*), alternating hot and cold domains are observed that are typically on the order of 50–100 kbp wide (*see* **Fig. 9.2** for an example). When examined at the resolution of individual genes (i.e., examining ~5–20 kbp at a time), individual hotspots can be observed that are separated by several kbp of DNA in which few, if any, DSBs are formed (**Fig. 9.3A**). Most hotspots in *S. cerevisiae* are located within transcriptional promoter regions *(2)*. Mapping at yet higher resolution reveals that each hotspot consists of multiple DSB sites clustered within regions of ~75–250 bp (**Fig. 9.3B**). Finally, mapping at the resolution of individual nucleotides reveals that Spo11 cleaves the DNA to yield a two-nucleotide 5′ overhang and that some positions within hotspots are cut more frequently than others, although no obvious DNA sequence preference has yet emerged *(3–6)*.

Because there is no single method for separating and detecting DNA fragments across all of the size ranges outlined above, it is necessary to choose DSB mapping strategies appropriate to the purpose of the experimental study. Here we describe procedures for DSB analysis at four levels of spatial resolution, which we define as chromosome level, medium resolution, high resolution, and nucleotide-level mapping. These differ from one another primarily with respect to DNA sample preparation and the method of gel

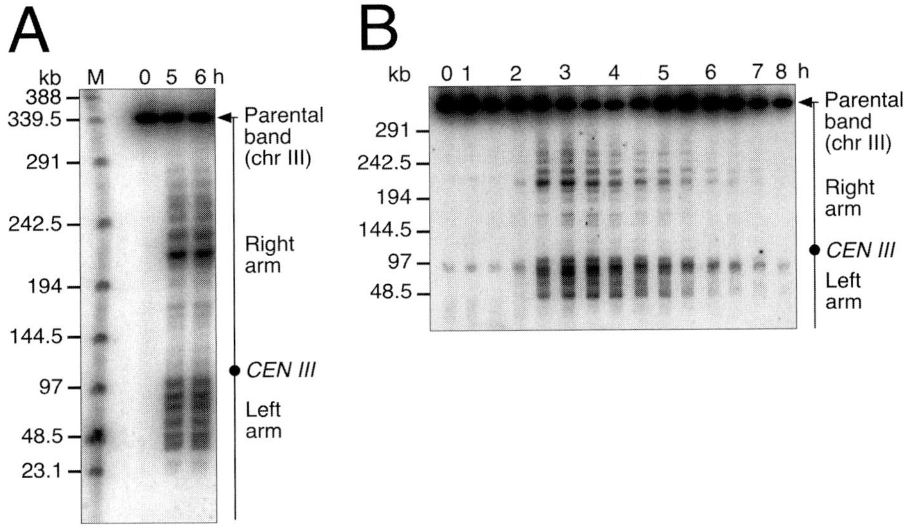

Fig. 9.2. Chromosome-level mapping. Cells from a *sae2Δ* strain (MJL 2305) **(A)** or a wild-type strain (MJL 1071) **(B)** were harvested at the indicated times in meiosis and processed for genomic DNA plug preparation and pulsed-field gel electrophoresis as described in the text. After Southern blotting, the membrane was probed with a chromosome III left end fragment (*CHA1*), revealing, in addition to the unbroken chromosome III (340 kb), the DSB fragments of chromosome III produced during meiosis. In **(A)**, a 14 cm wide by 21 cm long gel was used, and electrophoresis time was for 46 h. In **(B)**, a 21 cm wide by 14 cm long gel was used and electrophoresis time was 30.5 h. The size standard (M) is a *Hind* III digest of bacteriophage λ DNA and λ DNA concatamers. **(B)** is adapted with permission from *(8)*.

electrophoresis (pulsed-field, conventional agarose, polyacrylamide, or sequencing gel, respectively). These protocols are modified from established methods *(5, 7, 8)*.

In many studies, the position and the frequency of meiotic DSBs are measured using *rad50S* or *sae2Δ* mutants, which are deficient in a process of endonucleolytic release of covalently attached Spo11 from DSB ends *(5, 6, 9, 10)*. These mutants accumulate unrepaired DSBs, making observation of the normally transient meiotic DSB much easier. The frequency of DSBs varies from one hotspot to another and, in general, DSB frequencies measured in *rad50S* or *sae2Δ* strains correlate well with overall recombination frequencies in corresponding *RAD50+* or *SAE2+*. However, it is important to note that DSB frequencies are substantially reduced in *rad50S* or *sae2* mutants in regions where replication is experimentally delayed *(8)*, and recent genome-wide studies demonstrate that DSBs in certain regions are specifically underreported in *rad50S*-like mutants *(11, 12)*.

To obtain well-synchronized meiotic samples, cells are precultured in presporulation medium (SPS), and then transferred to sporulation medium (SPM). In the SK1 strain background, the DSB frequency reaches maximum levels about 5–6 h after transfer to SPM. Premeiotic samples are harvested at 0 h (i.e., immediately upon transfer to SPM), and meiotic samples are harvested at appropriate time points thereafter (*see* **Note 1**).

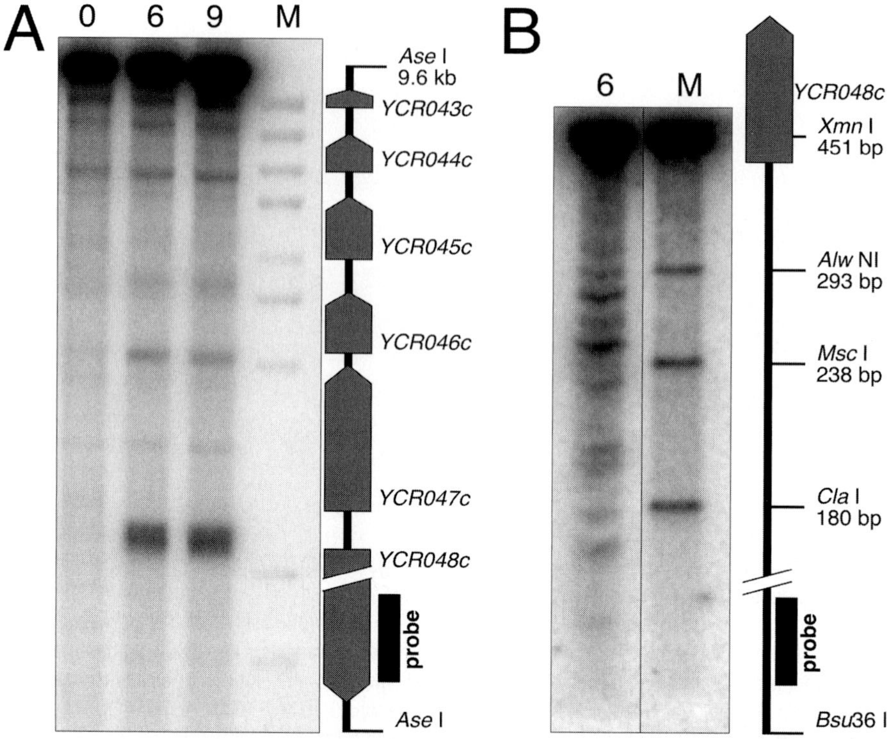

Fig. 9.3. Medium- and high-resolution mapping of DSB in a *rad50S* strain. (**A**) Medium-resolution mapping of DSBs around *YCR048w*. Numbers above the panel indicate the time (in h) after transfer to SPM. The lane marked M includes 200 ng of λ *Bst*E II digest. Genomic DNA was digested with *Ase* I. The parental fragment is 9.6 kb. Both the genomic DNA and the marker were separated on a 0.8% agarose gel and detected using a radiolabeled probe containing 100 ng of the probe DNA fragment and 0.1 ng of λ DNA. (**B**) High-resolution mapping of DSBs in the *YCR048w* promoter region. DSBs, which appeared as a single band in normal resolution mapping, are observed as multiple break sites when analyzed at this resolution. Genomic DNA extracted from meiotic cells (6 h) was digested with *Xmn* I and *Bsu*36 I. For the molecular weight marker, genomic DNA purified from 0 h cells was digested with *Xmn* I and *Bsu*36 I. Aliquots (1% of the digested DNA each) were further digested with either *Alw*N I, *Msc* I, or *Cla* I. After heat inactivation of the restriction enzymes, the differently digested DNA fragments were mixed together and loaded in the lane marked M. These DNA fragments were separated on a 6% polyacrylamide gel containing 8 M urea and detected by Southern blotting and indirect end labeling.

The frequencies of meiotic DSBs are low, so it is important that genomic DNA is extracted gently to avoid mechanical shearing. Excessive shearing generates a high background of hybridizing signal in lanes from the gel electrophoresis, and can obscure weaker DSB signals and/or make DSBs difficult to quantify. DNA samples in solution should be mixed gently, and vigorous pipetting and vortexing should be avoided.

To observe DSBs at different levels of spatial resolution, the lengths of the parental and DSB fragments need to be considered, and an appropriate probe needs to be chosen for indirect end-labeling of Southern blots. Parental and DSB fragments should be well separated by gel electrophoresis in order to accurately estimate DSB frequencies. It is recommended that probes are designed within open reading frames so that the region to which the probe hybridizes is unlikely itself to be a site of DSB formation. Additional specifications are discussed in the individual methods sections below.

2. Materials

2.1. Media for Yeast Culture

1. YPD: 1% Bacto yeast extract, 2% Bacto peptone, 2% dextrose, 2% Bacto agar. Autoclave.
2. YPG: 1% Bacto yeast extract, 2% Bacto peptone, 0.1% dextrose, 3% glycerol, 2% Bacto agar. Autoclave.
3. 0.1 M potassium biphthalate stock solution. Adjust to pH 5.5 with KOH.
4. Presporulation medium (SPS): 0.5% Bacto yeast extract, 1% Bacto peptone, 0.17% Bacto yeast nitrogen base (without amino acids and without ammonium sulfate), 0.5% ammonium sulfate, 1% potassium acetate, 0.05 M potassium biphthalate. Autoclave.
5. Sporulation medium (SPM): 1% potassium acetate. Autoclave.

2.2. Chromosome-Level DSB Mapping

2.2.1. Plug DNA Preparation

1. 0.5 M EDTA pH 7.5 (adjust pH with NaOH). Autoclave. Store at room temperature.
2. SCE solution: 1 M sorbitol, 0.1 M sodium citrate, 0.06 M EDTA, pH 7.0. Filter-sterilize.
3. 1% LMP agarose mix: 1% low melting point agarose (Seaplaque GTG, Lonza), 0.125 M EDTA, pH 7.5. Microwave and equilibrate to 40°C. This solution should be made fresh the day of the experiment to prevent agarose hydrolysis.
4. Solution 1: SCE plus 5% β-mercaptoethanol plus 1 mg/mL zymolyase 100T. This solution must be made fresh and kept on ice until use.
5. Solution 2: 0.45 M EDTA pH 7.5, 0.01 M Tris-HCl pH 7.5, 7.5% β-mercaptoethanol, 10 μg/mL RNase A. This solution should be made fresh the day of the experiment from stock solutions (for preparation of 10 mg/mL RNase A, *see* **Section 2.3** below).
6. Solution 3: 0.25 M EDTA, pH 7.5, 0.01 M Tris-HCl pH 7.5, 1% sarkosyl (from a 10% sarkosyl solution, filter-sterilized), 1 mg/mL proteinase K (directly added as a powder). This solution may be prepared in advance without the proteinase K, filter-sterilized and kept at room temperature. Add Proteinase K just before use.

7. Plug storage solution: 0.05 M EDTA, pH 7.5, 50% (w/v) glycerol. Filter-sterilize and keep at room temperature.
8. Agarose plug molds (e.g., Biorad, catalog no. 170-3622).

2.2.2. Pulsed-Field Gel Electrophoresis (PFGE)

1. TBE electrophoresis buffer, 10 × stock solution: 890 mM Tris base, 890 mM boric acid, 20 mM EDTA pH 8.0. Store at room temperature.
2. Gel casting stand 14 cm by 21 cm (e.g., BioRad, catalog no.170-3704)
3. 14 cm wide, 15-well, 1.5 mm thick comb
4. PFGE system (BioRad CHEF-DRII or CHEF Mapper XA Chiller System, or equivalent).
5. Circulating variable-speed pump (BioRad catalog no. 170-3644 or equivalent)
6. Cooling module (BioRad catalog no. 170-3654 or equivalent)
7. 1.3% agarose gel: Weigh 1.95 g SeaKem agarose in 150 mL of 0.5 × TBE. Microwave. Equilibrate at 55°C (*see* **Note 2**).
8. *S. cerevisiae* pulsed-field size marker (Mid-range I or II PFGE size markers from New England Biolabs)
9. 10 mg/mL ethidium bromide solution

2.2.3. Southern Blotting

1. UV crosslinker (Stratagene Stratalinker or equivalent); vacuum blotter; nylon membrane (positively charged, Hybond N+, GE Healthcare, or equivalent)
2. PFGE transfer solution: 0.6 M NaCl, 0.4 N NaOH. Store at room temperature.
3. 20 × SSC: 3 M NaCl, 0.3 M citrate. Adjust to pH 7.0 with 1 N HCl. Store at room temperature.

2.2.4. Southern Blot Hybridization and Washing

1. ^{32}P-labeled double-stranded DNA probe, prepared by random priming using an appropriate commercial kit according to manufacturer's instructions (e.g., Amersham Rediprime^{TM} II DNA Labeling System, GE Healthcare). The DNA template for the labeling reaction should be generated by PCR from genomic DNA and should be gel-purified by agarose gel electrophoresis and extracted using an appropriate commercial gel extraction kit. It is sometimes convenient to use the gel-purified material from this initial PCR to seed a secondary PCR with the same primers as a way of generating larger quantities of highly purified template. The amplified template DNA can be stored indefinitely at –20°C.
2. 1 M phosphate buffer pH 7.2: 38.6 g NaH_2PO_4•H_2O, 193 g Na_2HPO_4•$7H_2O$, distilled water to 1 L. Autoclave. Store at room temperature.

3. Hybridization solution: 1% bovine serum albumin (BSA), 0.5 M phosphate buffer pH 7.2, 7% SDS, 1 mM EDTA pH 8.0. Dissolve BSA in the phosphate buffer, then add SDS (high SDS concentration prevents BSA from dissolving). Filter sterilize. Store at −20°C.
4. Wash solution: 40 mM phosphate buffer pH 7.2, 1% SDS, 1 mM EDTA pH 8.0.
5. Hybridization oven and bottles (35 × 300 mm).

2.3. Medium-Resolution Mapping

2.3.1. Genomic DNA Extraction

1. 1.2 M sorbitol. Autoclave. Store at room temperature.
2. Zymolyase solution: Dissolve 100 mg of zymolyase 20T in 1 mL of 1 M sorbitol. Store at 4°C up to 2–3 months.
3. β-mercaptoethanol, isopropanol, 100% ethanol, all stored at room temperature. 70% ethanol, stored at −20°C.
4. Extraction solution: 50 mM EDTA pH 8.0, 50 mM Tris-HCl pH 8.0, 0.5% SDS. Autoclave. Store at room temperature.
5. Proteinase K: Prepare 10 mg/mL solution in water. Store at −20°C.
6. 5 M potassium acetate. Autoclave. Store at room temperature.
7. TE buffer: 10 mM Tris-HCl pH 7.5, 1 mM EDTA pH 8.0. Autoclave. Store at room temperature.
8. RNase A: Prepare 10 mg/mL solution in 10 mM Tris-HCl pH 7.5, 15 mM NaCl, 1 mM EDTA pH 8.0. Incubate in boiling water bath for 15 min, then at room temperature for 1 h. Store at −20°C.
9. 3 M sodium acetate pH 5.2: adjust pH with acetic acid. Autoclave. Store at room temperature.

2.3.2. Agarose Gel Electrophoresis

Restriction enzymes; electrophoresis power supply; agarose gel electrophoresis apparatus suitable for running 30 cm-long gels; agarose; TBE buffer (*see* **Section 2.2.2**), 6 × loading buffer (0.25% bromophenol blue, 0.25% xylene cyanol FF, 15% Ficoll PM400); 10 mg/mL ethidium bromide solution.

2.3.3. Southern Blotting

1. Vacuum blotter; nylon membrane (positively charged); 20x SSC (*see* **Section 2.2.3**)
2. Depurination solution: 0.25 N HCl. Store at room temperature.
3. Denaturing solution: 1.5 M NaCl, 0.5 N NaOH. Store at room temperature.
4. Transfer solution: 1.5 M NaCl, 0.25 N NaOH. Store at room temperature.
5. ^{32}P-labeled double-stranded DNA probe (*see* **Section 2.2.4**).

2.4. High-Resolution Mapping

2.4.1. Polyacrylamide Gel Electrophoresis

1. Polyacrylamide gel electrophoresis apparatus and power supply; gel casting plates and combs (8 × 8 cm, 1 mm thick); 40% acrylamide solution (19:1 acrylamide:bis-acrylamide); urea; TEMED; 10 × TBE (*see* **Section 2.2.2**).
2. 3 × loading buffer: 95% formamide, 10 mM EDTA pH 8.0, 0.1% bromophenol blue, 0.1% xylene cyanol FF. Store at −20°C.
3. APS: 10% ammonium persulfate. Store at −20°C.

2.4.2. Southern Blotting

Whatman 3 MM filter paper; transfer apparatus (BioRad Trans-Blot® SD Semi-Dry Electrophoretic Transfer Cell, or equivalent); Nylon membrane (uncharged, GeneScreen™ Hybridization Transfer Membrane (PerkinElmer) or equivalent); UV crosslinker; 20 × SSC (*see* **Section 2.2.3**).

2.5. Nucleotide-Resolution Mapping

2.5.1. Sample Preparation

1. Phusion™ High-Fidelity DNA Polymerase (Finnzymes).
2. SequiTherm EXEL™ II DNA Sequencing Kit (Epicentre).
3. NucleoSpin Extract II (Macherey-Nagel).
4. Primers a, b, c, d in **Fig. 9.4**.

2.5.2. Electrophoresis

High-voltage electrophoresis power supply; sequencing gel electrophoresis apparatus; sequencing gel plates (30 × 40 cm); plate coating reagent (Sigmacote); shark tooth combs and spacers (0.4 mm thick). *See* **Section 2.4.1** for acrylamide gel solutions and loading buffer.

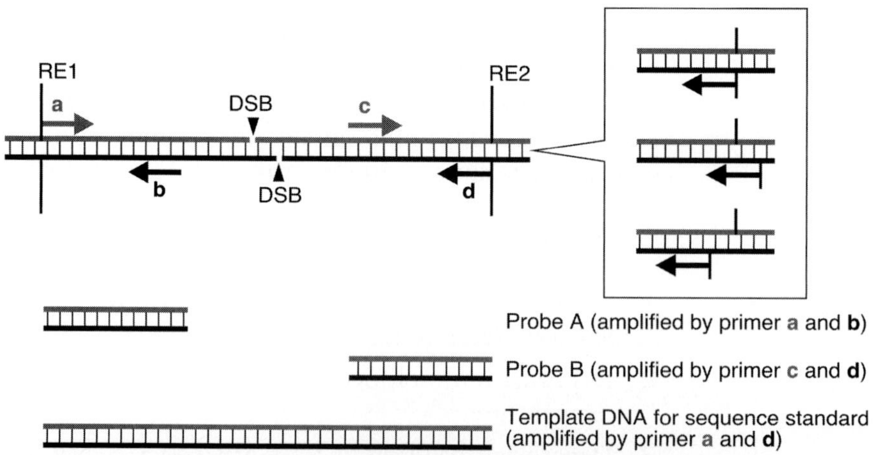

Fig. 9.4. Primers required for nucleotide-level mapping. For nucleotide-resolution mapping, genomic DNA is digested with an appropriate restriction enzyme (RE1 or RE2), which cleaves 150–200 bp from DSB sites. Four primers (a, b, c, and d) are used for the following procedures: to amplify probes A and B and template DNA for sequence standards; to prepare a strand-specific probe (primers b and c); to prepare sequence standards (primers a and d). Primers a and d are designed to be suitable to the shape of the restriction enzyme cut as illustrated in the box on the right.

2.5.3. Southern Blotting

UV crosslinker; 20 × SSC (*see* **Section 2.2.3**); 35 × 45 cm blotting paper (BioRad); Whatman 3 MM filter paper; electro transfer apparatus [TE 90 - GeneSweep™ Sequencing Gel Transfer Unit (Hoefer Scientific Instruments)].

2.5.4. Probe Preparation and Hybridization

Taq polymerase, 333 μM dNTP mix without dCTP, [α-^{32}P] dCTP (6,000 Ci/mmol), Taq amplification buffer, 50 mM $MgCl_2$, probe template DNA (*see* **Section 2.2.4**), 20 μM PCR primer b (**Fig. 9.4**).

3. Methods

3.1. Synchronous Meiotic Cultures

This section is based on *(13)* with minor modifications.

1. Day −4: Patch *rad50S* (or *sae2Δ*) diploid on YPG plate and incubate at 30°C for at least 6 h for the selection of functional mitochondria. Streak diploid on YPD plate to generate single colonies and incubate at 30°C for 2 days.
2. Day −2: Pick a single colony and inoculate in 1 mL of SPS. Shake overnight at 250 rpm, 30°C.
3. Day −1: Calculate the cell density either by measuring OD_{600} or directly counting cells in a hemacytometer. Inoculate cells in 5 mL of SPS at cell density of 0.5 × 10^7 cells/mL and shake at 250 rpm, 30°C for 6 h. Inoculate cells in SPS at cell density of 2–4 × 10^5 cells/mL and shake at 250 rpm, 30°C overnight (*see* **Note 3**).
4. Day 0: 16 h after the inoculation, cell density should be 2–4 × 10^7 cells/mL. Harvest cells by centrifugation at 2,600*g* for 5 min and wash with one vol of pre-warmed SPM. Then resuspend cells in two vol of pre-warmed SPM (*see* **Note 4**).
5. Collect 0 h (premeiotic) sample and return the flask to the shaker. Shake at 250 rpm, 30°C. Collect meiotic samples at appropriate times after transfer to SPM (*see* **Note 5**). Refer to individual subsections below for appropriate volumes and methods for harvesting cells.

3.2. Chromosome-Level DSB Mapping

This method is suitable for observing whole chromosomes and DSB fragments of 0.2–2.0 Mbp after separation by pulsed-field gel electrophoresis. It is necessary to prepare high molecular weight DNA under conditions of gentle cell lysis to prevent any mechanical DNA breakage. This is achieved by embedding cells in low-melting point agarose as soon as they are collected and performing all enzymatic reactions (preparation of spheroplasts, lysis and DNA purification) in the resulting DNA "plugs." Enzymatic degradation of chromosomal DNA is prevented by maintaining high concentrations of EDTA, which inhibits nucleases, at all steps of DNA plug preparation. For indirect end-labeling on Southern blots, a good choice for probe is often the open reading frame closest to one extremity of the chromosome of interest. A unique, unrepeated gene should be chosen.

3.2.1. Preparation of Agarose-Embedded DNA ("plugs")

This method has been described in *(14)*.

1. Collect 15 mL of meiotic culture (30 mL if at times before replication, i.e., 0–2 h after meiotic induction). (This is equivalent to 3×10^8 cells (6×10^8 before replication), and represents [$(13 \times 10^6$ bp$) \times (650 \times 10^6) \times 2$ {diploid} $\times$ 2 {after S phase} $\times$ $(3 \times 10^8$ cells$)/(6.022 \times 10^{23})] = 17$ μg of DNA. This is sufficient for six loadings and PFGE runs.
2. Centrifuge in 50 mL conical tubes 4 min in a table-top centrifuge at 3,200*g*.
3. Resuspend in 5 mL of 50 mM EDTA, pH 7.5. Vortex until completely resuspended.
4. Repeat **Steps 2** and **3**.
5. Repeat **Step 2**.
6. While washing, make the following mix: for five samples, mix 0.83 mL of 1% LMP agarose mix with 0.17 mL of Solution 1 (adjust quantities according to the number of samples you are processing). Vortex and keep at 40°C (*see* **Note 6**).
7. After the last wash, eliminate supernatant and resuspend the cell pellet in 100 μL of 50 mM EDTA pH 7.5. Transfer to a microcentrifuge tube. Equilibrate 30 s at 40°C.
8. Add 200 μL of pre-warmed LMP agarose/Solution 1 mix. Vortex briefly. Pipette into plug mold. 90 μL fills each block. Work quickly to fill the plugs before the solution begins to harden.
9. Chill blocks for 10 min at 4°C.
10. Using a thin spatula, express blocks into 3 mL of Solution 2. Invert gently several times to mix. Incubate 1 h at 37°C.
11. Carefully pour off Solution 2 and replace with 3 mL of Solution 3. Mix by gentle inversion. Incubate overnight at 50°C.
12. Carefully pour off Solution 3. Wash twice with 3 mL of 50 mM EDTA for 15 min on a rotating wheel. Equilibrate with 3 mL of storage buffer for 15 min. Pour off and replace with 3 mL of fresh storage buffer. Store plugs in storage buffer at −20°C.

3.2.2. Pulsed-Field Gel Electrophoresis

1. Prepare 2.2 L of 0.5 × TBE.
2. Use 150 mL of 0.5 × TBE to make the 1.3% agarose gel. Microwave and keep at 55°C.

3. Put the remaining buffer in the CHEF apparatus tank and switch on the circulating pump.
4. Switch on the cooler and set the thermostat to 14°C.
5. Using a clean razor blade, cut one third of a plug (this represents about 1.8 μg DNA) and immerse in 1.5 mL of filter-sterilized 0.5 × TBE in a 2 mL Eppendorf tube (buffer mixing and exchange is more efficient in a 2 mL rather than 1.5 mL tube). Rotate 15 min on a rotating wheel.
6. Prepare the gel casting stand and comb.
7. Using a spatula, take the plug out of solution and lay it on a tooth of the comb. Reserve one tooth for size markers. Seal each sample on the comb with one drop of the reserved 1.3% agarose gel. Let the comb stand horizontally for ~5 min to allow the sealing agarose to set.
8. Install the comb vertically in the gel casting stand and slowly pour the agarose into the stand. Let the gel set for 30 min at room temperature.
9. When set, using a razor blade, cut the drop of set agarose on each tooth of the comb that goes above the gel surface. Then slowly remove the comb, so that the DNA sample plugs remain in the gel.
10. Dismantle the gel from the casting stand and install it in the PFGE electrophoresis tank filled with pre-cooled running buffer. Take care when placing the gel on the bottom of the tank so that it will not move during the electrophoresis.
11. Let the gel equilibrate to the buffer temperature for 15 min.
12. Begin electrophoresis: for separating fragments from 425 to 15 kb, use the following conditions: 15.1 s initial switch time; 25.1 s final switch time; 120° switch angle; 6 V/cm (120 V in the CHEF and DRII Biorad apparatuses); 46 h run time (*see* **Note 7**). An example of separation of DSB fragments after such pulsed-field gel electrophoresis conditions is shown in **Fig. 9.2A**.

3.2.3. Southern Blotting

1. After electrophoresis, stain the gel for 20 min in 0.5 μg/mL ethidium bromide. Rinse in water for 5 min and photograph the gel.
2. Place the gel on a tray in a UV crosslinker and UV irradiate with 180 mJ/cm^2. This step is necessary to nick the DNA to allow efficient transfer.
3. Equilibrate the gel in 250 mL of PFGE transfer buffer for 15 min.
4. Cut the nylon membrane and wet first in water, then in PFGE transfer buffer for 1 min. Assemble the membrane, gasket and gel on the blotter.

5. Set up the vacuum blotter according to manufacturer's instructions. Place the gel on the blotter.
6. Pour PFGE transfer solution directly onto the gel.
7. Vacuum transfer at 55 cm Hg for 2 h, watching that the gel does not become dry. If it does, add more transfer solution.
8. Rinse the membrane with 2 × SSC. Dry the membrane on Whatman 3 MM paper for 1 h. Dry the membrane completely if storage is required.

3.2.4. Hybridization and Wash

1. Prepare 50 mL of hybridization solution.
2. Place the blot into a hybridization bottle and rinse blot with 50 mL of 0.5 M phosphate buffer.
3. Prehybridization: Pour 25 mL of hybridization solution into the bottle and rotate at 65°C for 1 h in a hybridization oven.
4. Place remaining 25 mL of hybridization solution at 65°C.
5. During prehybridization, prepare radioactively labeled probe.
6. Purify the radiolabeled probe using a gel filtration spin column. Denature the probe at 100°C for 5 min. Place on ice.
7. Discard the hybridization solution in the bottle. Add denatured probe to the fresh hybridization solution that has been equilibrated to 65°C (**Step 4**). Pour hybridization solution containing the probe into the hybridization bottle.
8. Rotate hybridization bottle at 65°C overnight (16–48 h).
9. Warm the wash buffer to 65°C.
10. Discard the hybridization solution.
11. Pour approximately 50 mL of wash buffer into the bottle. Wash at 65°C for 15 min. Discard wash buffer. Repeat three times.
12. Remove blot from the hybridization bottle. (Further washing in a tray in a 65°C water bath may be conducted if necessary to remove background.)
13. Wrap the blot with plastic wrap (*see* **Note 8**).
14. Expose to phosphor screen or film.

3.3. Medium-Resolution DSB Mapping

DSB mapping using separation by conventional agarose gel electrophoresis is suitable for observing DSBs across ~2–15 kb regions, with a resolution of ~50–100 bp. For accurate quantification, parental fragments of less than 10 kb are recommended so that transfer during Southern blotting is efficient. Restriction enzymes can be chosen as appropriate to give good spatial separation between the DSB fragments and the parental fragment (*see* **Fig. 9.3A**). We generally use 0.5–1 kb DNA fragments as probes, which can be efficiently labeled by random priming. Ideally, the probe should hybridize to sequence close to one end of the parental size restriction fragment (**Fig. 9.3A**).

3.3.1. Genomic DNA Extraction

This section is based on *(7)*. (*See* **Note 9**)

1. Collect samples from synchronous meiotic culture (*see* **Section 3.1**). The amount of DNA per mL of culture increases according to the progression of meiosis as the yeast cells undergo premeiotic DNA replication. To adjust for this increase, we collect cells from the following volumes of SPM culture: 0 h, 34 mL; 2 h, 31 mL; 3 h, 27 mL; 4 h, 24 mL; after 5 h, 20 mL. Harvest cells by centrifugation 5 min at 3,200*g* in a tabletop centrifuge. Resuspend cell pellet in 1 mL sterile distilled water, transfer to a 2.0 mL Eppendorf tube, and pellet cells with a brief (30 s) spin in a microcentrifuge. Store pellets at −20°C.
2. Add 200 μL of 0.5 M EDTA pH 8.0, 800 μL of 1.2 M sorbitol, 10 μL of β-mercaptoethanol, and 10 μL of zymolyase solution to frozen yeast cells. Suspend yeast cells by vortexing, making sure there are no clumps. Incubate at 37°C for 45 min.
3. Centrifuge 4,500*g* in a microcentrifuge for 5 min. Gently remove supernatant by pipetting and discard.
4. Add 500 μL of extraction solution and 10 μL of proteinase K. Gently suspend the spheroplasts by pipetting. Incubate at 65°C for 1 h.
5. Add 200 μL of 5 M potassium acetate, and mix immediately. Chill tubes on ice for 10 min.
6. Centrifuge 16,000*g* in a microcentrifuge for 30 min at 4°C. Transfer supernatant to new tube. Add 500 μL of isopropanol and invert tube 4–5 times to mix.
7. Centrifuge 16,000*g*, 5 min. Discard supernatant.
8. Add 1 mL of 70% ethanol. Centrifuge 16,000*g* for 5 min at 4°C. Discard supernatant. Centrifuge for 1 min at 4°C. Remove residual ethanol by pipetting.
9. Add 300 μL of TE and 3 μL of RNase A. Do not try to suspend the pellet at this stage, because it causes random DNA shearing. Incubate at 37°C for 1 h.
10. Add 30 μL of 3 M sodium acetate pH 5.2 and 750 μL of 100% ethanol. Mix by inversion. Chill sample at −20°C for 1 h. Centrifuge 16,000*g* for 20 min at 4°C. Discard supernatant.
11. Add 1 mL of 70% ethanol. Centrifuge for 5 min at 4°C. Discard supernatant. Centrifuge for 1 min at 4°C. Remove residual ethanol by pipetting.

12. *Briefly* dry the pellet at 37°C for 20 min. Add 100 μL of TE. Store overnight at 4°C: the DNA will gently dissolve during this step (*see* **Note 10**).
13. Store dissolved DNA at −20°C.

3.3.2. Restriction Enzyme Digestion and Electrophoresis

1. Mix approximately 1 μg of genomic DNA with restriction enzyme buffer, water, and restriction enzyme (10–20 units), in 20 μL total volume (*see* **Note 11**).
2. Digest DNA for 3 h. Heat inactivate the restriction enzyme.
3. Prepare an appropriate concentration agarose gel with 1 × TBE.
4. Add 4 μL of 6 × loading buffer to samples. Mix.
5. Load samples on gel along with appropriate size standards (*see* **Note 12**).
6. Run samples in 1 × TBE at >70 V overnight until dyes migrate to appropriate position (*see* **Note 13**)
7. Place the gel in 1 × TBE containing 0.5 μg/mL of ethidium bromide. Agitate for 20 min. Take a photograph of the gel (*see* **Note 14**).

*3.3.3. Southern Blotting (see **Note 15**)*

1. Cut the nylon membrane to the appropriate size for the gel. Set up the vacuum blotter according to manufacturer's instructions. Place the gel on the blotter.
2. Pour depurination solution directly onto the gel. Apply vacuum at 55 cm Hg for approximately 15 min, taking care that the gel does not dry out. The bromophenol blue dye will turn yellow (*see* **Note 16**).
3. Drain residual depurination buffer and pour denaturation solution directly onto the gel. Denature for approximately 45 min. During this step, the bromophenol blue dye will become blue again.
4. Pour transfer solution directly onto the gel. Transfer for 2 h.
5. Take the membrane out. Fix DNA to the membrane by soaking in freshly prepared 0.4 N NaOH for 5 min.
6. Rinse the membrane with 2 × SSC. Dry the membrane on Whatman 3 MM paper for 1 h. The membrane is ready to use for hybridization as described in **Section 3.2.4**, or it can be dried completely at this stage if storage is required.

3.3.4. Quantification of DSB Frequency

Parental and DSB bands can be visualized using a Phosphoimager and quantified using ImageQuant (Molecular Dynamics) as described in *(7)*. The frequency of DSBs is calculated as the percent of radioactivity in DSB fragments relative to the total radioactivity in the lane (i.e., parental and DSB fragments). Appropriate exposure times can be selected to allow accurate determination of the intensity of the parental band (i.e., within the linear response range of the phosphor screen). Background signals are often present in the lane because of random shearing of genomic DNA during preparation of samples, etc; these background signals should be subtracted.

3.4. High-Resolution DSB Mapping

DSBs can be detected at higher resolution (±10~20 bp) using polyacrylamide instead of agarose gel electrophoresis (*see* **Fig. 9.3B**). We generally obtain better resolution and signal strength using denaturing (6% polyacrylamide containing 8 M urea) rather than nondenaturing gels. Good resolution is achieved when DSB fragments are 150–300 bp long. To provide sufficient spatial separation from DSB fragments of this size, parental fragments should be 500–1,000 bp. Restriction sites should be chosen with these size ranges in mind. The size of the DNA probe fragment should be 100–200 bp, which is often difficult to label to sufficient specific activity by random priming. Therefore, we use PCR to amplify the probe fragment in the presence of [α-^{32}P] dCTP.

3.4.1. Denaturing PAGE Gel Electrophoresis and Semi-dry Transfer

1. Prepare restriction enzyme digested premeiotic and meiotic samples (**Section 3.3.2**).
2. Ethanol precipitate all samples. Rinse with 70% ethanol. Dry pellets. Dissolve each pellet in 6 μL of TE and 3 μL of 3 × loading buffer.
3. For each gel, prepare 10 mL of 6% acrylamide gel solution: 4.2 g urea, 1.5 mL 40% acrylamide (19:1), 1 mL 10 × TBE, 4 mL H_2O.
4. Add 10 μL of TEMED and 100 μL of 10% APS. Mix.
5. Pour the gel immediately into gel casting plates. Insert the comb (12 well) into the gel solution. Normally, the gel will be completely polymerized within 30 min.
6. Assemble the gel into the electrophoresis apparatus with 1 × TBE as the running buffer. Flush urea from the wells using a pipette. Load each well with 9 μL of 1 × loading buffer (diluted from 3 × stock with TE). Pre-run at 240 V for 30 min (*see* **Note 17**).
7. Denature samples at 100°C for 5 min. Place samples on ice.
8. Flush urea from the wells using a pipette. Load samples. Start electrophoresis at 240 V.
9. During electrophoresis, cut the uncharged nylon membrane and five pieces of Whatman 3 MM paper to the size of the gel.
10. Stop electrophoresis when the xylene cyanol dye has migrated close to the end of the gel. Under these conditions, xylene cyanol comigrates with single-stranded DNA of approximately 100 bases.

11. Remove the gel from the gel plates. Equilibrate the gel, membrane, and Whatman 3 MM in 0.5 × TBE for 15 min.
12. Assemble the sandwich for the semi-dry transfer apparatus (+ anodic side: two Whatman 3 MM sheets, membrane, gel, three Whatman 3 MM sheets: – cathodic side). Remove all air bubbles by rolling a pipette over the surface of each layer.
13. Run the transfer at 400 mA for 1 h.
14. Remove the membrane and fix the DNA to the membrane by UV crosslinking at 120 mJ/cm^2. Rinse the membrane once with 2 × SSC. Dry the membrane completely if storage is required.

3.4.2. PCR-Mediated Probe Labeling

1. Prepare labeling mix on ice as following:

333 μM dNTP mix, without dCTP	0.5 μL
[α-^{32}P] dCTP (400 Ci/mmol)	5 μL
20 μM Forward Primer*	0.625 μL
20 μM Backward Primer*	0.625 μL
10 × Buffer	1.25 μL
50 mM $MgCl_2$	0.625 μL
Taq polymerase (5 U/μL)	0.25 μL
Template DNA fragment	5 ng
Water	to 12.5 μL

*These are the same primers used originally to amplify the template fragment from genomic DNA.

2. PCR conditions: 25 cycles of 94°C for 1 min, X°C for 30 s, 72°C for 1 min (X = primer-specific Tm).
3. Purify and denature the labeled probe as in **Section 3.2.4, Step 6**.
4. Perform hybridization and wash as described in **Section 3.2.4**, except at 57°C instead of 65°C.
5. Expose the blot to phosphorimager screen.

3.5. Nucleotide-Resolution DSB Mapping

In order to detect DSB sites at single nucleotide resolution, DSB fragments must be separated on a sequencing gel. Before carrying out this type of analysis, it is recommended to first determine the position and frequency of DSBs at the locus of interest by performing high-resolution mapping (**Section 3.4**). On the basis of the result of this mapping, a restriction enzyme that cuts 150–200 bp from the DSB site can be chosen, along with primer sets to amplify probe and template for sequence standards (*see* **Fig. 9.4**).

Meiotic DSBs made by Spo11 have 2-nucleotide 5′ overhangs (*see* **Note 18**). In *rad50S* or *sae2Δ* mutants, Spo11 remains covalently attached to the 5′ ends. Even with extensive proteinase K treatment, oligo-peptides of Spo11 still remain covalently attached to 5′ DSB ends, which causes the 5′-terminal DNA strand to migrate slower on sequencing gels and thus makes it difficult to map the 5′ ends accurately *(3–6)*. To circumvent this problem, 5′ ends are mapped indirectly by filling in the 3′ ends with a DNA polymerase that does not add extra nucleotides. After the fill-in reaction, the ends of the DSB fragments will be blunt and the filled-in 3′ ends will match the original 5′ ends (*see* **Fig. 9.5**). The filled-in 3′ ends and untreated (original) 3′ ends can then be detected side-by-side on the same gel by probing a Southern blot with a strand-specific probe (**Fig. 9.5**). Sequence standards are prepared by linear amplification in the presence of dideoxy nucleotide triphosphates (ddNTPs) with a primer that corresponds to the end of the sequence cleaved by the restriction enzyme (**Fig. 9.4**, primer a). This method is based on *(5, 15)*.

After defining DSB sites at single-nucleotide resolution by this method, it is advisable to confirm the accuracy by mapping the DSB ends on the other side of the hotspot. This is accomplished by repeating the procedure using an appropriate restriction digestion and probe, as diagrammed in **Fig. 9.4** (probe B, primer c, primer d, and RE2).

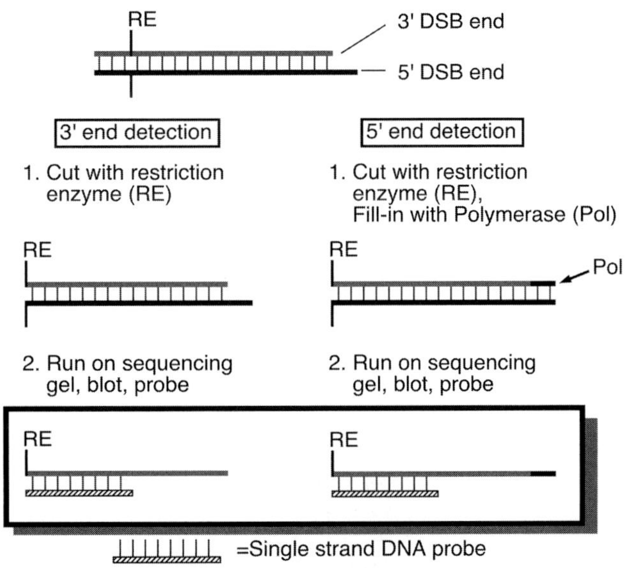

Fig. 9.5. Overview of procedure for nucleotide-level resolution mapping of 5′ and 3′ DSB ends. Genomic DNA is purified from the meiotic cell and digested with an appropriate restriction enzyme (RE). After the separation on a sequencing gel and Southern blotting, 3′ DSB ends are detected with a strand-specific DNA probe. For the indirect detection and mapping of 5′ ends, 3′ ends are filled-in with DNA polymerase to match the size of the 5′ ends.

3.5.1. Preparation of Probes, Sequence Standards, and Samples

1. On the basis of the high-resolution DSB mapping performed as described in **Section 3.4**, choose two restriction enzymes, which cleave 150–200 bp on either side of the DSB site (RE1 and RE2 in **Fig. 9.4**).
2. Design DNA primers a, b, c, d as described in **Fig. 9.4**. Amplify Probe A, Probe B, and template DNA for sequencing standard by PCR. Gel-purify and extract all fragments. Quantify the DNA concentrations.
3. Using the template DNA fragment with primer a or primer d, prepare two sets of G, A, T, and C sequence standards using a thermocycle sequencing kit (e.g., SequiTherm EXEL™ II DNA Sequencing Kit (Epicentre), or similar). These sequence standards need to be appropriately diluted with TE to approximately match the intensity of the DSB bands. For a DSB hotspot where ~20% of DNA molecules are broken, an 80-fold dilution is appropriate. For a DSB hotspot where ~8% of DNA molecules are broken, a 200-fold dilution is appropriate.
4. Prepare two separate restriction enzyme digests (RE1 in **Fig. 9.4**), each of which contains approximately 1 μg of genomic DNA from a meiotic sample (*see* **Note 19**). One sample will be for 5′ end mapping and the other for 3′ end mapping. Also prepare four digests of sample from a 0 h culture (one each for the G, A, T, and C sequence standards) (*see* **Note 20**). Digest for 3 h, then heat inactivate the restriction enzyme.
5. Purify all of the samples using a commercial extraction kit (e.g., NucleoSpin Extract II kit (Macherey-Nagel), or similar). Elute DNA with 39.1 μL (5′ end sample) or 20 μL (the rest of the samples) of 1 mM Tris-HCl, pH 8.5 (this is a five-fold dilution of the elution buffer supplied with the extraction kit).
6. For the 5′ end sample, prepare 50 μL of a fill-in reaction mixture using Phusion™ High-Fidelity DNA Polymerase (FINNZYMES) as follows:

DNA	39.1 μL
5 × buffer	10 μL
25 mM dNTP mix	0.4 μL
Polymerase	0.5 μL
Total volume	50 μL

6. Incubate at 72°C for 5 min. Purify DNA as in **Step 2**, eluting DNA in 20 μL.
7. Dry all samples by speed-vac (takes about 1–2 h).
8. Dissolve the 5′ and 3′ samples in 4 μL of TE.
9. Dissolve the 0 h DNA samples with 4 μL of either G, A, T or C sequence standard, which has been appropriately diluted (*see* **Step 3** of **Section 3.5.1**).
10. Add 2 μL of 3 × loading buffer to each sample. Mix. Samples may be stored at −20°C.

3.5.2. Denaturing Polyacrylamide Gel Electrophoresis

1. Wash a pair of 30 × 40 cm front and back gel plates with soap and water. Wipe well with 100% ethanol and dry.
2. Siliconize one side of each plate by wetting a Kimwipe with Sigmacote solution and wiping the whole plate. After the solution dries, wipe with water and then with 100% ethanol.
3. Assemble gel plates with 0.4 mm spacers. Carefully seal the side and bottom of plates with sticky tape.
4. Prepare 60 mL of 6% acrylamide gel solution as follows: 25.2 g urea, 9 mL of 40% acrylamide (19:1); 6 mL of 10 × TBE; 24 mL of H_2O.
5. Add 60 μL of TEMED and 600 μL of 10% APS. Mix.
6. Pour the gel immediately using a large syringe. With the short plate on top, raise the top of the gel sandwich to about a 30° angle from the bench top and carefully pour the acrylamide between the plates along one side. Insert the flat side of a 0.4 mm shark tooth comb at the top of the gel. Avoid introducing bubbles in the gel sandwich.
7. The acrylamide should polymerize within 2 h. The polymerized gel may be used immediately, or may be wrapped in plastic film and stored at room temperature over night.
8. Remove sticky tape from the gel sandwich. Place the gel sandwich in a sequencing electrophoresis apparatus.
9. Pour 1 × TBE into the apparatus. Remove the shark tooth comb. Wash the top and bottom of the gel using a syringe with needle to remove acrylamide fragments, urea, and air bubbles.
10. Insert the teeth side of the shark tooth comb into the gel sandwich. Flush wells using a syringe with needle. Load 1 × loading buffer (diluted from stock with TE) to all wells. Pre-run at 1,700 V, 70 W for 30 min (*see* **Note 17**).

11. Prior to loading, denature all samples at 100°C for 5 min and chill on ice. Flush wells using a syringe with needle. Load samples gently to avoid spillover between lanes.
12. Begin running the gel at 1,700 V, 70 W.
13. Observe the migration of the xylene cyanol dye, which comigrates with DNA strands of approximately 100 nucleotides. Stop electrophoresis at an appropriate time after the xylene cyanol dye has run out of the gel (*see* **Note 21**).
14. Remove the gel sandwich from the electrophoresis apparatus. Place the sandwich in a cold room (4°C) until the plates are cool. This step makes it easier to handle the gel during the transfer step.

3.5.3. Southern Blotting by Electro-Transfer

This step assumes the use of TE 90 GeneSweep™ Sequencing Gel Transfer Unit (Hoefer Scientific Instruments).

1. Prepare blotting paper and membrane: Cut the GeneScreen™ uncharged nylon membranes to the size of the gel plus 2 cm. Prepare two sheets of 35 × 45 cm blotting paper. Prepare a tray containing 1 × TBE to soak the membrane and blotting paper.
2. Place the gel sandwich on the bench with the short plate on top. Very slowly disassemble the gel sandwich so that the gel remains on the bottom plate. Take out the side spacers. If wrinkles or air bubbles developed between the gel and the plate, pour a small amount of 1 × TBE onto the gel and gently roll the area smooth with a pipette.
3. Wet a sheet of blotting paper with 1 × TBE. Drain off excess buffer with Whatman 3 MM paper. Lay the blotting paper on the gel, starting from the top of the gel. Carefully roll out wrinkles and air bubbles with a pipette.
4. To lift the gel from the glass plate onto the blotting paper, rapidly peel off the blotting paper, taking care not to damage the gel.
5. Place the blotting paper with gel side up onto the GeneSweep platform.
6. Wet the nylon membrane with 1 × TBE. Blot off excess buffer with Whatman 3 MM paper. Lay the membrane onto the gel. Do not move the membrane after it makes contact with the gel. Gently roll out wrinkles and air bubbles with a pipette.
7. Wet another sheet of blotting paper with 1 × TBE. Drain excess buffer with Whatman 3 MM paper. Lay the paper on top of the nylon membrane. Gently roll out wrinkles and air bubbles with a pipette.

8. Lower the GeneSweep arm onto the left edge of the sandwich. Start the transfer. As the arm slides across the sandwich, the display should read between 1.8 and 2 A. If the current is less than 1 A, the sandwich may be too dry. If so, stop the run, lift the arm, and pour 1 × TBE onto the sandwich. Gently roll out excess buffer with a pipette and start the run again.
9. When the arm reaches the right end of the sandwich, stop the run and lift the arm.
10. Carefully take off the upper blotting paper and the nylon membrane. Leave the membrane on the blotting paper (to prevent the membrane from drying out) and immediately fix the DNA to the membrane by UV crosslinking at 120 mJ/cm^2. The membrane can then be removed from the blotting paper.
11. Rinse the membrane once with 2 × SSC. The membrane is now ready for hybridization (next section), or can be dried if storage is required.

3.5.4. Strand-Specific Probe Preparation and Hybridization

1. Prepare labeling mix on ice as follows (*see* **Note 22**):

333 μM dNTP mix without dCTP	0.5 μL
[α-^{32}P] dCTP (6,000 Ci/mmol.)	5 μL (50 μCi)
20 μM Primer*	0.625 μL
10 × Buffer	1.25 μL
50 mM $MgCl_2$	0.625 μL
Taq polymerase (5 U/μL)	0.25 μL
Template DNA fragment**	5 ng
Water	to 12.5 μL

*Primer b, **Probe A in **Fig. 9.4**

2. PCR conditions: 25 cycles of 94°C for 1 min, X°C for 30 s, 72°C for 1 min (X = primer specific Tm).
3. Purify and denature labeled probe as described in **Step 6** of **Section 3.2.4**.
4. Perform hybridization and wash as described in **Section 3.2.4**, except at 57°C instead of 65°C.
5. Expose the blot to the phosphorimager screen. Compare migration of the DSB bands to the sequencing standards to determine the positions of 5′ and 3′ DSB ends (*see* **Fig. 9.6** for an example).

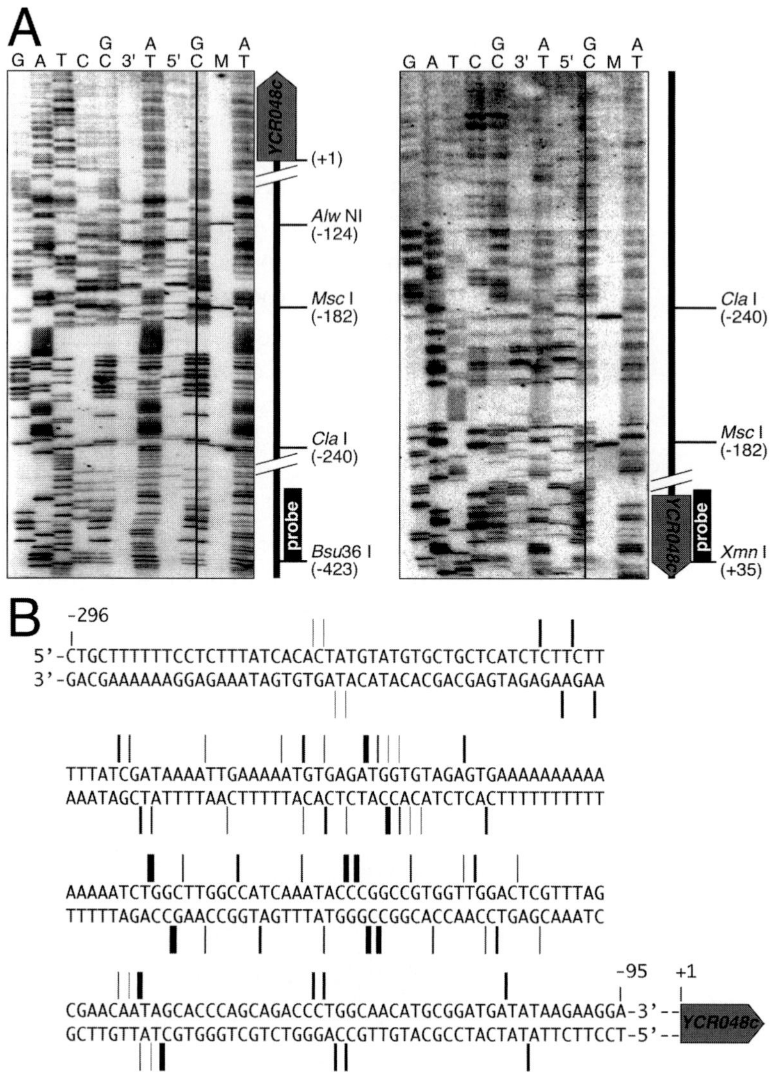

Fig. 9.6. Nucleotide-level resolution mapping of DSB at the *YCR048w* promoter region. (**A**) Examples of sequencing gel mapping of DSBs. Images of Southern blots probed with two different probes are shown. Samples were digested with *Bsu36* I (*left*) or *Xmn* I (*right*) and probed with single-stranded DNA in the vicinity of each restriction enzyme site (illustrated on the right of each panel; *see* also **Fig. 9.4**). Lanes marked 5′ and 3′ contain separate DSB end-mapping samples, prepared using genomic DNA purified from cells 9 h after transfer to SPM. Lanes marked G, A, T, and C are nucleotide sequence standards. Lanes marked GC and AT are sequencing standard containing pooled G+C or A+T standards, respectively. All of the sequence standards contain additional restriction-digested genomic DNA (from a 0 h culture) so that the amount of total genomic DNA is the same as in the DSB sample lanes. Lanes marked M contain DNA from 0 h cells, digested with either *Bsu36* I (*left*) or *Xmn* I (*right*), plus 1% of DNA subjected to secondary digest with either *AlwN* I, *Msc* I, or *Cla* I (*see* **Note 12**). Numbers correspond to positions relative to the translation start of *YCR048w*. (**B**) Location of DSBs in the *YCR048w* promoter region. The vertical bars represent DSB locations, defined by mapping of DSB 5′ ends using probes on both sides of the DSB hotspot. Thickness of the bars provides a semi-quantitative representation of the signal strength of each DSB band on the Southern blots.

4. Notes

1. These instructions assume the use of diploids of the SK1 strain background, with either *rad50S* or *sae2Δ* mutation. If other strain backgrounds are used, adjustments to the culture media and/or times for sample collection may be necessary.
2. Be careful that the agarose solution is not too hot when you pour it because this may partially melt the DNA agarose plugs and detach them from the comb.
3. The volume of SPS should be less than 20% of the flask volume to ensure good aeration. Use a flask of ≥250 mL for 50 mL of SPS culture.
4. The volume of SPM should be less than 10% of the flask volume to ensure good aeration, for example, use a 1 L flask for 100 mL of SPM culture.
5. Appropriate time points to collect cells are dependent on the purpose of the experiment. Usually, the DSB frequency reaches a maximum by 5–6 h after transfer to SPM in the SK1 background when using *rad50S* or *sae2Δ* mutants.
6. The mixture of Solution 1 and the LMP agarose mix should be kept at 40°C as briefly as possible (maximum 3–4 min) in order not to inactivate the zymolyase enzymatic activity. Therefore, it is better not to make an agarose mix for more than six samples at a time.
7. If you have more than 15 samples to run on the same gel, you may run a 14 cm long by 21 cm wide gel using the same gel casting stand, but with a 21 cm wide comb. The running conditions must be adjusted such that total run time is 30.5 h but all the other conditions are unchanged. An example of such a run is shown in **Fig. 9.2B**.
8. Try to avoid leaks to protect the phosphor screen and to prevent the blot from drying. If the blot is kept moist, the probe can be stripped by washing twice with 1% SDS for 15 min. However, once the blot is dried with the hybridized probe, it becomes difficult to strip the probe.
9. If the region to be analyzed is larger than 10 kb, it is recommended to extract and digest genomic DNA in low melting point agarose as described in **Section 3.2.1**.
10. If a residual pellet exists after overnight incubation at 4°C, mix the DNA by gentle tapping. Do not vortex. Often, meiotic samples are cloudy, presumably because of polysaccharide or other components from the ascus or spore walls. However, this does not affect later steps (restriction enzyme digestion, etc.).

11. Usually 10 μL of sample corresponds to approximately 1 μg of DNA. For more precision to ensure that the quantity of DNA is similar from sample to sample, the DNA concentration can be quantified by Hoechst dye fluorescence using a fluorometer.
12. For accurate DSB mapping, it is important to include appropriate DNA size standards on the same gel, and it is also important that they appear in the autoradiograph of the Southern blot after probing. There are two convenient ways to achieve this goal. The first is to use a commercially available molecular weight marker, which is run in an adjacent lane in the gel (e.g., lambda DNA digested with BstE II; *see* **Fig. 9.3A, lane M**). To visualize the marker on the autoradiograph, marker DNA is also added to the random primed labeling reaction at a probe-to-marker ratio of 1,000:1. Alternatively, molecular weight markers can be prepared from yeast genomic DNA. To do so, approximately 1 μg of 0 h sample (i.e., DNA from a premeiotic culture) is digested with the same restriction enzyme used to digest the meiotic samples. Then, an aliquot of the digested DNA is digested with an appropriate second enzyme which cuts within the region of interest. A set of such double digests is then pooled with the undigested DNA after heat inactivation of the restriction enzymes (*see* **Fig. 9.3B, lane M**). The amount of DNA that is subjected to double digestion should be adjusted dependent on the expected DSB frequency, such that the marker fragments are not stronger than the DSB signal. A typical starting point would be to perform the secondary digestions on aliquots of 1% of the first digest (i.e., ~0.01 μg of DNA). Either method works well for providing size standards, but the second method provides somewhat more accurate size information because it allows one to control for DNA sequence composition and for the amount of DNA loaded in the lane.
13. 1 × TAE may be used instead of TBE. If so, it is essential to circulate the buffer during electrophoresis.
14. If the restriction digest worked well, each sample will show the same pattern on ethidium stained gel.
15. We obtain good results using a vacuum blotter to transfer DNA from the gel to a hybridization membrane. Alternative methods using capillary transfer under neutral or denaturing conditions may also be satisfactory.
16. Depurination is not necessary for DNA fragments less than 5 kb, but we observe that transfer is partial for DNA fragments larger than this. Thus, depurination (and subsequent nicking of apurinic sites) is critical to obtain accurate estimation of DSB frequencies. The depurination step can be replaced by UV treatment at 120 mJ/cm^2, which produces alkali-labile photoproducts in the DNA.

17. This step is to check the quality of the gel. If there is a problem (e.g., leaks or air bubbles that cannot be removed, etc.), discard the gel and prepare it again.
18. We have mapped breaks at single nucleotide resolution at several DSB hotspots (Murakami et al., manuscript in preparation). When the 5′ ends were mapped independently from both sides of DSB hotspots using separate restriction digests and probes, we confirmed that DSB ends always have 2-nucleotide 5′ overhangs, as reported previously *(4, 5)*. However, it is important to note that some 3′ DSB ends were not identical to what was expected based on the mapping of 5′ ends from the other side of the DSB site. Specifically, 3′ ends were often 1–2 nucleotides longer than expected. Therefore, it appears that 3′ ends may sometimes be filled in by DNA polymerase, either in vivo or during the preparation of genomic DNA (B. de Massy, personal communication).
19. If the DSB signals are too weak, more genomic DNA can be used. We have confirmed that up to 5 μg of DNA can be loaded in a single lane.
20. Mixing the sequence standard with genomic DNA from the 0 h sample allows for equal amounts of total DNA to be loaded in each lane. This controls for the effects of genomic DNA on the migration pattern on the sequencing gel, and is essential for accurate DSB mapping.
21. The electrophoresis time depends on the length of the DSB fragment to be resolved. The following are optimal total times using the conditions described in this protocol:

150 nucleotides	190 min
180 nucleotides	250 min
220 nucleotides	290 min
240 nucleotides	320 min

22. If the DSB hybridization signals are too weak, increase the total volume with the same concentrations of all reagents. We have used up to 50 μL of total reaction volume successfully.

References

1. Petes, T. D. (2001) Meiotic recombination hot spots and cold spots. *Nat. Rev. Genet.* **2**, 360–369.
2. Baudat, F. and Nicolas, A. (1997) Clustering of meiotic double-strand breaks on yeast chromosome III. *Proc. Natl. Acad. Sci. U. S. A.* **94**, 5213–5218.
3. Xu, F. and Petes, T. D. (1996) Fine-structure mapping of meiosis-specific double-strand DNA breaks at a recombination hotspot

associated with an insertion of telomeric sequences upstream of the *HIS4* locus in yeast. *Genetics* **143**, 1115–1125.

4. Xu, L. and Kleckner, N. (1995) Sequence non-specific double-strand breaks and interhomolog interactions prior to double-strand break formation at a meiotic recombination hot spot in yeast. *EMBO J.* **14**, 5115–5128.
5. Liu, J., Wu, T. C., and Lichten, M. (1995) The location and structure of double-strand DNA breaks induced during yeast meiosis: evidence for a covalently linked DNA-protein intermediate. *EMBO J.* **14**, 4599–4608.
6. de Massy, B., Rocco, V., and Nicolas, A. (1995) The nucleotide mapping of DNA double-strand breaks at the CYS3 initiation site of meiotic recombination in *Saccharomyces cerevisiae. EMBO J.* **14**, 4589–4598.
7. Vedel, M. and Nicolas, A. (1999) *CYS3*, a hotspot of meiotic recombination in *Saccharomyces cerevisiae.* Effects of heterozygosity and mismatch repair functions on gene conversion and recombination intermediates. *Genetics* **151**, 1245–1259.
8. Borde, V., Goldman, A. S., and Lichten, M. (2000) Direct coupling between meiotic DNA replication and recombination initiation. *Science* **290**, 806–809.
9. Neale, M. J., Pan, J., and Keeney, S. (2005) Endonucleolytic processing of covalent protein-linked DNA double-strand breaks. *Nature* **436**, 1053–1057.
10. Keeney, S. and Kleckner, N. (1995) Covalent protein-DNA complexes at the 5′ strand termini of meiosis-specific double-strand breaks in yeast. *Proc. Natl. Acad. Sci. U. S. A.* **92**, 11274–11278.
11. Blitzblau, H. G., Bell, G. W., Rodriguez, J., Bell, S. P., and Hochwagen, A. (2007) Mapping of meiotic single-stranded DNA reveals double-strand-break hotspots near centromeres and telomeres. *Curr. Biol.* **17**, 2003–2012
12. Buhler, C., Borde, V., and Lichten, M. (2007) Mapping meiotic single-strand DNA reveals a new landscape of DNA double-strand breaks in *Saccharomyces cerevisiae. PLoS Biol.* **5**, e324.
13. Goyon, C. and Lichten, M. (1993) Timing of molecular events in meiosis in *Saccharomyces cerevisiae*: stable heteroduplex DNA is formed late in meiotic prophase. *Mol. Cell. Biol.* **13**, 373–382.
14. Borde, V., Wu, T. C., and Lichten, M. (1999) Use of a recombination reporter insert to define meiotic recombination domains on chromosome III of *Saccharomyces cerevisiae. Mol. Cell. Biol.* **19**, 4832–4842.
15. Buhler, C., Lebbink, J. H., Bocs, C., Ladenstein, R., and Forterre, P. (2001) DNA topoisomerase VI generates ATP-dependent double-strand breaks with two-nucleotide overhangs. *J. Biol. Chem.* **276**, 37215–37222.

Chapter 10

Genome-Wide Mapping of Meiotic DNA Double-Strand Breaks in *Saccharomyces cerevisiae*

Cyril Buhler, Robert Shroff, and Michael Lichten

Abstract

DNA double-strand breaks (DSBs) initiate meiotic recombination in eukaryotes. We describe two strategies that use microarrays to determine the genome-wide distribution of meiotic DSBs in the yeast *Saccharomyces cerevisiae*. The first is a chromatin immunoprecipitation (ChIP) approach that targets the Spo11 protein, which remains covalently attached to DSB ends in certain mutant backgrounds. The second approach involves BND cellulose enrichment of the single-strand DNA (ssDNA) recombination intermediate formed by end-resection at DSB sites following Spo11 removal.

Key words: Double-strand break, Spo11, *DMC1*, *rad50S*, chromatin immunoprecipitation, single-strand DNA, BND cellulose, microarray, background normalization, quantitative PCR.

1. Introduction

Meiotic recombination is initiated by a program of transient DNA DSBs, which are repaired by homolog-directed recombination *(1)*. Recombinant products include both crossovers (COs) and noncrossovers (NCOs), which differ by the presence or absence of exchange of flanking parental sequences, respectively. COs ensure that homologous chromosomes are physically linked at the first meiotic division, and this is essential for accurate segregation of homologs at the first meiotic division *(2)*. The mechanism of DSB formation and the control of DSB distribution across the genome are still under investigation.

While mapping meiotic COs genome-wide is of considerable interest in studies of a variety of organisms, doing so requires both a large number of heterozygous markers and the ability to

Scott Keeney (ed.), *Meiosis, Volume 1, Molecular and Genetic Methods, vol. 557*

DOI 10.1007/978-1-59745-527-5_10 Springerprotocols.com

efficiently distinguish parental markers in progeny. While productive efforts towards this end have been initiated *(3–5)*, increases in resolution require dauntingly commensurate increases in marker density and detection. These challenges, as well as an intrinsic interest in the mechanism and control of meiotic recombination initiation, have led to the development of whole-genome DSB mapping using microarrays *(6)*, as this can be done independent of genetic marker density. Furthermore, increases in resolution can, in theory, be accomplished by increasing the density of array elements, at least until probe size becomes limiting for resolution.

The identification of budding yeast mutants that accumulate unrepaired DSBs has been particularly useful in the characterization of the early steps of meiotic recombination. The discovery of Spo11 as the protein directly responsible for DSB formation was made possible by the identification of mutants (*rad50S*, *mre11S*, and *sae2Δ*, hereafter referred to as *rad50S*-like) that accumulate DSBs with unprocessed ends *(7–9)*. In *rad50S*-like mutants, Spo11 remains covalently bound to each 5′ end of meiotic DSBs by the tyrosine phosphoester linkage formed during the cleavage reaction *(10)*. This stable covalent linkage has allowed development of the standard approach to measure and localize meiotic DSBs, by immunoprecipitating Spo11 from meiotic extracts of *rad50S*-like mutants and analyzing the accompanying DNA sequences *(6, 11–13)*.

After DSB formation, Spo11 is removed from DSB ends, which are then processed to form ssDNA tails with free 3′ ends *(14)*. This ssDNA recruits Rad51 and Dmc1, which catalyze the invasion of homologous duplex DNA and initiate homologous recombination *(15)*. Unrepaired DSBs associated with ssDNA accumulate in mutants lacking these strand transferases, as well as in mutants lacking cofactors essential for their function. This ssDNA offers an alternative method for localizing Spo11-induced DSBs, namely the isolation of break-associated ssDNA by binding to benzoylated naphthoylated DEAE (BND) cellulose *(16, 17)*. This method can be used to map both the transient DSBs that form in wild-type, and the unrepaired DSBs that accumulate in *dmc1* or *dmc1 rad51* mutants.

In this chapter, we describe the use of both methods (Spo11 immunoprecipitation from *rad50S*-like mutants; BND cellulose enrichment of ssDNA from *dmc1* and *dmc1 rad51* mutants) to isolate break-associated DNA sequences and analyze their distribution on whole-genome microarrays (**Fig. 10.1**). We also describe the method of probe preparation, microarray hybridization, and data analysis currently in use in our laboratory. In addition, we describe the implementation of a novel normalization method, based on biological criteria, that allows the comparison of array datasets derived from different enrichment procedures. While these protocols are designed for the analysis of DSB patterns in budding yeast meiosis, suitable modification should allow their application to DSB mapping in other organisms.

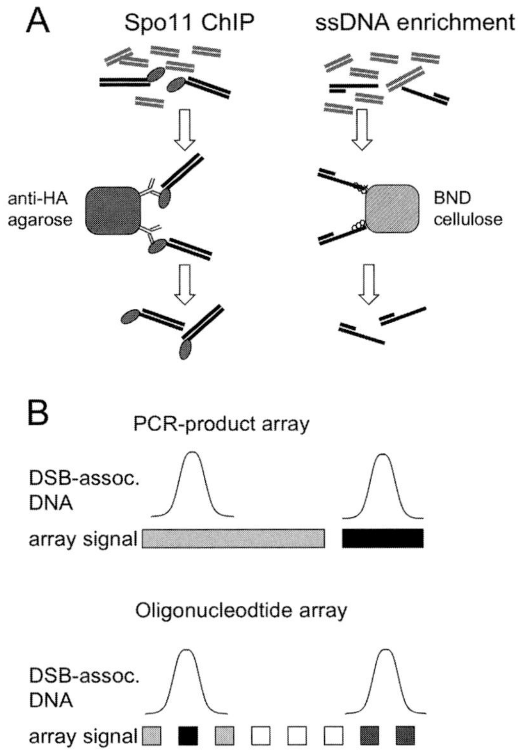

Fig. 10.1. Probe preparation and microarray choice. **A.** Illustration of methods for probe preparation. *Left*: DSB-adjacent DNA isolated from *rad50S*-like mutants by immunoprecipitation with antibody directed against Spo11 protein (in this case, against the influenza hemaglutinin antigen in a Spo11-HA fusion protein). *Right*: DSB-adjacent DNA isolated from *dmc1*Δ mutants by binding ssDNA to BND cellulose. **B.** Illustration of advantages of closely spaced oligonucleotide arrays over PCR-amplified ORF/IGR arrays. Varied array element size in ORF/IGR arrays can lead to different element signal levels for DSBs of similar intensity. Bell curves represent the distribution of DSB-associated DNA fragments.

In addition to specific technical considerations that are discussed within each protocol section, two overall concerns need to be considered. The first involves the choice of mutant background to analyze, as both *rad50S*-like and *dmc1* (or *dmc1 rad51*) mutants have specific advantages and disadvantages. *rad50S*-like mutants block DSB formation before resection, and the target used to enrich for break associated sequences, Spo11p, is located at the DSB end itself. Because DSBs accumulate unprocessed with Spo11 at break ends, *rad50S*-like mutants are the only background that can be used in strains where meiosis is slow and asynchronous, and where the continued resection that occurs in *dmc1* mutants would seriously hamper break recovery and resolution. However, several recent studies indicate that wild-type cells undergo DSB formation and meiotic recombination at significant levels in regions that are "DSB-cold" in *rad50s*-like mutants [*(16–19)*

A. Barton and D. Kaback, personal communication; S. Chen and J. Fung, personal communication], and the degree to which *rad50S*-like mutants under-measure true recombination activity in different parts of the genome remains to be clearly determined. On the other hand, DSBs appear to be more evenly distributed in *dmc1* and *dmc1 rad51* mutants, but the continuous resection that occurs in these mutants, and the random-priming method used to amplify break-associated DNA, tends to create the DSB signal several hundred nucleotides away from the break itself. This limits resolution and, in some cases, can lead to a diminished signal at the DSB site itself. Alternatives to random priming, that will produce a signal closer to the DSB itself, are currently under development in our group, but at the current time the limited resolution, and the need to have synchronous meiosis, remains a liability of *dmc1* mutants.

A second concern applies to the analysis of both *rad50S*-like and *dmc1* mutants. Because both protocols involve substantial amplification of limited input material, and because the uniform and quantitative amplification of background is as important to success as is the amplification of peaks, it is critical to monitor the relative recovery of both hotspots and background at multiple steps during the procedure. We highly recommend the early identification of DSB hot- and cold-spots on Southern blots, and the monitoring of DSB accumulation in meiotic cultures by Southern blot analysis of selected DSB hot-spots before purification for microarray analysis. Furthermore, monitoring recovery of DSB hot-spots relative to DSB cold-spots by quantitative PCR (qPCR), and ensuring that the same recovery ratio is maintained throughout amplification, is essential to successful reproducibility of the analysis. Although we recommend two DSB hot-spots for monitoring, strain-to-strain variation in DSB hot-spots has been documented *(6)*. Thus, it is important to confirm that a given DSB hot-spot is active in the particular strain to be used before proceeding. If these methods are applied to species other than *S. cerevisiae*, a bootstrap procedure of initial hot-spot and cold-spot identification on pilot microarrays, followed by qPCR and/or Southern blot verification, may be necessary.

2. Materials

2.1. Meiotic Cultures

Supplier of materials is given only when deemed important; in absence of an indication, any high-quality source will do.

1. YEPD broth: 1% (w/v) Bacto^{TM} yeast extract (BD, Franklin Lakes, NJ), 2% (w/v) Bacto^{TM} peptone (BD), 2% D-glucose, 0.004% adenine. Adjust pH to 5.5 with 1 N HCl. Broth

without glucose is prepared as a 1.1 × solution and autoclaved; this is mixed with sterile 20% (w/v) glucose in a 9:1 ratio before use. YEPD agar is identical but contains 2% Bacto agar (BD), and is prepared by autoclaving all components together.

2. Pre-sporulation medium (SPS): 1% (w/v) Bacto™ peptone (BD), 0.5% (w/v) Bacto™ yeast extract (BD), 1% (w/v) potassium acetate, 0.17% (w/v) Difco™ yeast nitrogen base w/o amino acids (BD), 1% (w/v) ammonium sulfate, 0.5% (w/v) potassium hydrogen phtalate (Sigma). Adjust pH to 5.5 with 10 N KOH.
3. Sporulation medium (KAc): 1% (w/v) potassium acetate supplemented with nutrients according to the auxotrophic requirements at 1/5 levels used in vegetative growth medium (**Table 10.1**) and with 0.001% polypropylene glycol 2000 (Aldrich Chemicals, Milwaukee, WI) as an anti-clumping agent.
4. Spheroplast storage buffer (SSB): 50 mM potassium phosphate, 10 mM ethylenediamine tetraacetic acid (EDTA), 1.2 M sorbitol (Sigma), 20% (v/v) glycerol pH 7.5.
5. Tris-buffered saline (TBS): 10X solution, 1.5 M NaCl, 1 M Tris-HCl pH 7.5.
6. 2.8 L triple-baffled Fernbach flask (Bellco Glass, Vineland, NJ).

Table 10.1
Supplements for sporulation media

Nutrient	Stock concentration	mL/L
adenine	0.5% in 0.05 M HCl	1.6
arginine	2.0%	0.2
histidine	2.0%	0.2
isoleucine	1.0%	0.6
leucine	1.0%	1.2
lysine	1.5%	0.4
methionine	2.0%	0.2
threonine	6.0%	1.0
tryptophan	1.0%	0.4
tyrosine	0.25%	2.0
valine	3.0%	1.0
uracil	0.2% in 1% Na_2CO_3	1.0

2.2. Spo11 ChIP in Break Processing Defective Mutants

1. Bead Beater, either Mini or Maxi (Biospec Products, Bartlesville, OK).
2. Misonia XL2020 Sonicator and CL4 microprobe (Misonix, Farmingdale, NY) or equivalent.
3. Acid-washed glass beads, 425–600 μm (Sigma).
4. Anti-HA antibody $IgG_{2b}\kappa$ mouse monoclonal antibody 12CA5 (Roche Diagnostics Indianapolis, IN). 200 μg lyophilized antibody is dissolved in 200 μL of 1X phosphate buffered saline (137 mM NaCl, 10 mM phosphate, 2.7 mM KCl, pH 7.4) and 10 μL aliquots are stored at –20°C.
5. Protease inhibitor cocktail III (PIC3, Calbiochem, EMD, San Diego, CA) stored in 50 μL aliquots at –20°C.
6. Ultralink Immobilized Protein G Plus (Pierce, Rockford, IL).
7. 1 M NaCl lysis buffer: 50 mM HEPES/KOH pH 7.5, 1 M NaCl, 5 mM EDTA, 1% Triton X-100, 0.1% (w/v) sodium deoxycholate.
8. 0.5 M NaCl lysis buffer: 50 mM HEPES/KOH pH 7.5, 0.5 M NaCl, 5 mM EDTA, 1% Triton X-100, 0.1% sodium deoxycholate.
9. Wash solution 3: 10 mM Tris-HCl pH 8.0, 0.25 M LiCl, 0.5% NP-40, 0.5% sodium deoxycholate, 5 mM EDTA.
10. Proteinase K: 20 mg/mL in 10 mM Tris-HCl pH 7.5, 20 mM calcium chloride, and 50% (v/v) glycerol. Store aliquots at –20°C.

2.3. Meiotic DNA Preparation by Phenol Chloroform Extraction

1. Spheroplast buffer: 50 mM potassium phosphate, 10 mM EDTA, 1.2 M sorbitol pH 7.5, 0.25 mg/mL (w/v) Zymolyase 100T (MP Biomedical, Solon, OH), 1% 2-mercaptoethanol.
2. Spheroplast stop solution: 100 mM NaCl, 50 mM Tris-HCl pH 8.0, 50 mM EDTA.
3. Phenol:Chloroform:Isoamyl alcohol (25:24:1). Use buffer-equilibrated phenol.
4. Chloroform:Isoamyl alcohol (24:1).
5. DNase-free RNase (Roche Diagnostics).
6. NanoDrop *ND*-1000 Spectrophotometer (Wilmington, DE).
7. Screw-capped polypropylene microfuge tubes are preferred to avoid leakage during extraction.

2.4. Batch Enrichment of ssDNA on BND Cellulose

1. *Hind*II and *Ssp*I restriction enzymes (Roche Diagnostics).
2. 1 M TEN: 10 mM Tris-HCl, 1 mM EDTA, 1 M NaCl, pH 8.0.
3. 1 M TENC: 10 mM Tris-HCl, 1 mM EDTA, 1 M NaCl, 1.7% (w/v) caffeine (Sigma). Prepare just before use to avoid caffeine crystallization. Mild heat and extended mixing will help dissolve caffeine. Care should be used in handling caffeine, as it is toxic.

4. Benzoylated naphthoylated DEAE cellulose, medium mesh (Sigma), prepared as follows, with all steps performed at room temperature, and all centrifugation in a swinging-bucket rotor:
   a. Resuspend 2 g of BND cellulose in 50 mL of 5 M NaCl, using a spatula to break up clumps.
   b. Remove fines by centrifugation, 3,200*g*, 1 min and discard supernatant.
   c. Resuspend pellet in 5 M NaCl and repeat three more times, or until the remaining BND cellulose settles by gravity. Remove excess 5 M NaCl.
   d. Add 20 mL ultrapure H_2O to packed BND cellulose. Mix by vortexing and allow slurry to settle for 5 min, then remove 5 mL supernatant.
   e. Mix slurry and transfer to a 15 mL conical tube. Centrifuge 3,200*g*, 1 min, remove supernatant. The packed BND cellulose should occupy a volume of about 3 mL.
   f. Add 1 M TEN to the BND cellulose pellet to a final volume of 15 mL. Resuspend slurry by vortexing. Centrifuge, 3,200*g*, 1 min and remove supernatant.
   g. Repeat 1 M TEN wash, as above, three times.
   h. Resuspend pellet in 1 M TEN to 15 mL final volume and store at 4°C. After ~2 h, remove excess buffer so that the slurry is about one half total volume. BND cellulose can be stored in this form for up to 1 year at 4°C. Binding capacity and specificity can be checked by applying a mixture of linearized plasmid DNA and øX174 ssDNA, and proceeding as in **Section 3.4**.
5. Empty macro spin columns (Harvard Apparatus, Holliston, MA).
6. Microcon YM-50 centrifugal filters (Millipore, Bedford, MA).

2.5. Quantitative PCR of DSB-Associated Sequences

1. Ultra Pure UV-treated deionized H_2O; 1 mL aliquots stored at –20°C.
2. icycler IQ thermal cycler or equivalent (Bio-Rad Laboratories, Hercules, CA)
3. iQ SYBR Green supermix (Bio-Rad).
4. Primers (MWG Biotech, High Point, NC). 200 μM stock prepared in 1X TE (10 mM Tris-HCl, 1 mM EDTA pH 8.0) and stored at –20°C.
   rDNAup: 5′-CTGATGTCTTCGGATGGATTTGAG-3′;
   rDNAdw 5′-TTTCCTCTGGCTTCACCCTATTC-3′;
   qBUD23up 5′-TATGTCGTCCACCTGGTCGTCG-3′;
   qBUD23dw 5′-TCCTAAACAGCGGTTGATGAGG-3′;
   qERG1up 5′-CAGTCATACCACCACCAGTCAATG-3′;
   qERG1dw 5′-GCCAAACTCCTACTTGCCAGC-3′;

2.6. High-Density Microarray Analyses of DSB-Associated Sequences

1. Sequenase^{TM} Version 2.0 (USB, Cleveland, OH) and *Taq* DNA polymerase (Promega, Madison, WI).
2. Primer A: 5′-GTTTCCCAGTCACGATCNNNNNNNNN-3′ and primer B: 5′-GTTTCCCAGTCACGATC-3′ (MWG Biotech). "N" is an equal mixture of A, T, G, or C. The first 17 nucleotide sequence of primer A is absent from the *S. cerevisiae* genome sequence.
3. SYBR® Gold nucleic acid gel stain (Molecular Probes, Invitrogen, Carlsbad, CA).
4. Sodium-aminoallyl-dUTP (aa-dUTP 86% pure), 1 mg vial (Sigma). Prepare a 10 mM stock solution by adding 160 μL of 0.1 M potassium phosphate, pH 7.5 directly into the vial.
5. 20X aa-dUTP/dNTP mix. Mix 25 μL each of 100 mM stocks of dATP, dGTP, dCTP (USB), 10 μL of dTTP (USB), 150 μL aa-dUTP (10 mM), dH_2O to a total volume of 500 μL. Store in 50 μL aliquots at –20°C. The 3/2 molar ratio of aa-dUTP/dTTP may not be optimal for aa-dUTP incorporation.
6. 4 M hydroxylamine-HCl (Sigma).
7. Cy^{TM}3 Monoreactive dye pack (GE Healthcare/ Pharmacia). Resuspend by adding 20 μL ultrapure DMSO (MP Biomedical), flicking vigorously, followed by brief centrifugation. Repeat resuspension and centrifugation twice. All manipulations should be performed in semi-darkness (*see* **Note 1**).
8. Cy^{TM}5 Monoreactive dye pack (GE Healthcare/ Pharmacia), resuspended as in **Step 7**.
9. 100 mM Na_2CO_3. Adjust pH to 8.9 with 10 N HCl. Dilute to 10 mM for exchange buffer; dilute to 80 mM for coupling buffer.
10. Microcon YM-30 and YM-100 centrifugal filters (Millipore). Residual particles from manufacture are removed as follows: add 500 μL relevant buffer to upper chamber. Spin briefly (~1 min, 14,000*g* for YM-30, 500*g* for YM-100) to wet filter. Invert upper chamber in microfuge tube, spin as above to remove wash solution. Use immediately.
11. *S. cerevisiae* oligonucleotide microarrays and the following reagents are from Agilent Technologies, Santa Clara, CA. 4 × 44 k yeast whole genome oligonucleotide array (Cat #G4493A), DNA microarray hybridization chambers, gasket slides; stabilizing and drying solutions, 2X hybridization buffer.
12. 20X SSPE: 3 M NaCl, 0.2 M NaH_2PO_4, 20 mM EDTA, pH 7.4.
13. Microarray wash solution I: 6X SSPE, 0.05% (v/v) N-lauroylsarcosine (Sigma), the latter diluted from a 20% (v/v) stock solution stored at room temperature.
14. Microarray wash solution II: 0.06X SSPE.

15. Rotisserie oven for slide hybridization (Agilent or equivalent).
16. Axon Scanner 4000B, GenePix Pro 6.0 software (Molecular Devices, Sunnyvale, CA), or any microarray scanner capable of 5 μm resolution.
17. Glass slide washing rack and reservoir, rectangular, eight-slide capacity, ~500 mL reservoir. At least three sets.

3. Methods

3.1. Meiotic Cultures

The following protocol is designed for efficient and relatively synchronous sporulation of budding yeast strains of the SK1 strain background.

1. Use a freshly isolated single colony grown on a YEPD agar plate at 30°C for ≤48 h to inoculate a 5 mL culture in YEPD broth. Incubate at 30°C with aeration overnight.
2. About 18 h before the desired time of sporulation initiation, inoculate presporulation cultures at dilutions in 2-fold increments ranging from 1/750 to 1/3,000 into 200 mL of SPS (2 L Erlenmeyer flasks) and incubate overnight at 30°C with agitation at 300 rpm.
3. When the OD_{600} of the culture is 1.35–1.4 (~2 × 10^7 cells/mL, *see* **Note 2**), harvest cells by centrifugation (3,200*g*, 3 min, room temperature), and resuspend in 200 mL of prewarmed (30°C) KAc without supplements. Harvest cells by centrifugation as above, and resuspend in 400 mL of prewarmed supplemented KAc, transfer to a 2.8 L baffled Fernbach flask, and incubate with vigorous agitation (at least 350 rpm, preferably greater).
4. For Spo11 ChIP, harvest a 50 mL sample by centrifugation (2 min, 4,000*g*, 4°C). Resuspend cell pellet in 20 mL 1X TBS (4°C) and centrifuge as before. Resuspend cell pellet in 1 mL 1X TBS (4°C), transfer to a 1.5 mL screw cap tube, and centrifuge at maximum speed in a refrigerated tabletop centrifuge for 1 min. Remove supernatant, freeze cell pellet on dry ice, and store at –80°C.
5. For phenol chloroform genomic DNA extraction, harvest 25 mL of culture by centrifugation (2 min, 4,000*g*, 4°C). Resuspend cell pellet in 20 mL of SSB (4°C) and centrifuge as before (2 min, 4,000*g*, 4°C). Resuspend cell pellet in 1 mL SSB (4°C), transfer to a 1.5 mL screw cap tube, centrifuge at 14,000*g* for 10 s, remove supernatant, freeze cell pellet on dry ice and store at –80°C.

3.2. Spo11 ChIP in Break Processing Defective Mutants

In *rad50S*, *mre11S*, or *sae2Δ* mutants, Spo11p remains covalently attached to break ends, and therefore no crosslinking is required to enrich for DSB-adjacent DNA by Spo11-immunoprecipitation. The following protocol, adapted from Strahl-Bolsinger et al. *(20)*, is designed to immunoprecipitate DSB-adjacent sequences from strains where Spo11 is tagged with the influenza hemaglutinin antigen [HA, *(21)*]. High salt concentrations are used during anti-HA immunoprecipitation to reduce nonspecific background, although this results in reduced recovery. Enrichment of DSB-associated DNA is estimated by measuring genome equivalents of target sequences relative to genome equivalents of ribosomal DNA (rDNA), which represents background sequences not associated with meiotic DSBs. Primers close to the *BUD23* and *ERG1* promoter regions are used to measure recovery of DSB-associated sequences. Dilutions of genomic DNA are used to establish standard curves, and enrichment is calculated as the ratio of target locus genome equivalents/rDNA genome equivalents in samples.

1. Thaw frozen cell pellets on ice and resuspend in 800 μL of 1 M NaCl lysis buffer + 1:100 PIC3.
2. Add an equal volume of glass beads. Agitate tubes in the bead beater for 3 min on the highest speed at 4°C. Cool sample for 1 min, and then agitate for a further 3 min.
3. Collect lysate by punching a hole in the bottom of the microfuge tube with a 20 gauge needle. Place the tube in a 5 mL round-bottom polypropylene tube and centrifuge at 200*g*, 10 s. Transfer lysate to a 1.5 mL microfuge tube, centrifuge the collected lysate in a microfuge (14,000*g*, 10 min, 4°C) and transfer the supernatant to a new 5 mL round-bottom tube.
4. Resuspend lysate pellet in 400 μL of 1 M NaCl lysis buffer + 1:100 PIC3, and perform a second round of glass bead lysis as described in **Steps 3** and **4**. Supernatants are then combined (*see* **Note 3**).
5. Fragment genomic DNA by sonication to an average size of approximately 1 kb using eight cycles of 10 s pulse followed by 30 s hold on ice. Use a low power output (5%) to avoid foaming. The extent of sonication is determined by gel electrophoresis (*see* **Note 4**). Centrifuge lysate 5 min, 14,000*g*, 4°C. Save 10 μL of the lysate as input material and store at –20°C
6. Immunoprecipitate Spo11-3HA by adding 10 μg of anti-HA antibody and 1:100 PIC3. Incubate lysate for 2 h on a mini-rotator wheel, 30 rpm, room temperature.
7. Add 25 μL of Protein G plus bead suspension and incubate sample overnight at 4°C (*see* **Note 5**).
8. Pellet chromatin immunoprecipitates by centrifugation (14,000*g* for 10 s). Carefully decant lysate.

9. All the following washes are done by 5 min incubation at room temperature on a mini-rotator at 30 rpm. Wash five times, 1 mL of 1 M NaCl lysis buffer; three times, 1 mL of 0.5 M NaCl lysis buffer; once, 1 mL Wash Solution 3; once, 1 mL of TE.
10. Remove all traces of wash solution from the Protein G beads (*see* **Note 6**) and resuspend beads in 50 μL of TE+0.2 μg/mL proteinase K. Incubate 1 h, 60°C with occasional vortexing. Extract once with phenol:chloroform:isoamyl alcohol (25:24:1) pH 7.5 and once with chloroform:isoamyl alcohol (24:1). Extractions are performed by adding an equal volume of organic mixture, vortexing vigorously, and centrifuging (14,000*g*, 3 min, room temperature) to separate phases. The aqueous (top) phase is carefully removed and placed in a fresh tube.
11. Remove residual phenol and chloroform by ethanol precipitation. Add linear acrylamide (Ambion, Austin TX) to 10 μg/mL; add 0.1 vol 3 M sodium acetate. Add 2 vol ethanol, incubate at –20°C 1 h. Spin in microfuge (14,000*g*, 20 min, 4°C); remove supernatant; wash with 70% ethanol (v/v); remove all traces of wash and air dry. Resuspend in 50 μL of TE.
12. Measure immunoprecipitation efficiency (%IP) and relative enrichment of Spo11-associated sequences by qPCR. 1 μL of ChIP and 1 μL of a 1/1,000 dilution of input are most often used, and are compared to a dilution series of a known amount of genomic DNA (~13 fg DNA/genome). The following PCR program is used with the primers recommended above: 95°C for 10 s; 57°C for 10 s; 72°C for 10 s; 40 cycles. Only PCR reactions with amplification efficiencies between 90% and 100% are used.

3.3. Meiotic DNA Preparation by Phenol Chloroform Extraction

The following protocol isolates DNA from a cell pellet from 25 mL of a meiotic culture stored in spheroplast buffer + 20% glycerol (*see* **Section 3.1**). It is designed to minimize artifactual ssDNA formation due to shearing and to nuclease activity during preparation (*see* **Note 7**).

1. Thaw cell pellets on ice and resuspend in 0.5 mL chilled spheroplast buffer + Zymolyase on ice.
2. When all the samples are resuspended, spheroplast cells by incubating at room temperature (23°C) for 15–20 min (*see* **Note 8**). Gently pellet spheroplasts in a microfuge (1,000*g*, 20 s, room temperature).
3. Remove supernatant. Resuspend pelleted spheroplasts in 500 μL of spheroplast stop solution by gentle up and down pipetting.

4. Add 25 μL of a 20% SDS solution, then 1 vol of phenol:-chloroform:isoamyl alcohol. Extract gently, 10–15 min on a rotary wheel, 25 rpm, room temperature. Separate organic and aqueous phases by centrifugation, 14,000*g*, 5 min, room temperature.
5. Remove aqueous phase to fresh tube; extract with chloroform: isoamyl alcohol as in **Step 4**.
6. Remove aqueous phase to fresh tube, add 1/10 volume of 5 M NaCl. Slowly invert tube to mix. Add 1 vol isopropanol and gently invert tube to precipitate nucleic acids.
7. Let the pellet settle to bottom of the tube. Centrifugation is not recommended, as the resulting nucleic acid pellet will be difficult to dissolve, resulting in low yields. Carefully remove supernatant using a transfer pipette and allow pellet to dry briefly.
8. Resuspend pellet in 300 μL of TE + 0.05 μg/mL DNase-free RNase. Incubate at 37°C for 30 min with occasional mixing by inversion.
9. Add 30 μL 3 M sodium acetate, then 200 μL isopropanol. Mix by inverting gently to precipitate DNA. Remove supernatant as above and wash pellet with 70% ethanol.
10. Air dry pellet and resuspend in 100 μL of 1X TE pH 8.0.
11. Check concentration and purity of DNA with NanoDrop spectrophotometer. Typical yields are 0.5–1 μg DNA per mL of meiotic culture.

3.4. Batch Enrichment of ssDNA on BND Cellulose

The single-strand DNA enrichment procedure described below is adapted from Gamper et al. *(22)* and from the Huberman lab protocol (http://asajj.roswellpark.org/huberman/BNDCellulose.html). Genomic DNA size is reduced by restriction enzyme digestion rather than by sonication to prevent ssDNA formation. The combination of 4-base blunt-end cutters *Hin*dII and *Ssp*I results in a predicted median fragment size of 407 bp. All centrifugations (3,200*g* for 1 min at room temperature) are performed using a swinging-bucket or horizontal rotor to allow pelleting of the BND cellulose slurry to the bottom of the tube.

1. Digest 10 μg of meiotic DNA, prepared as in **Section 3.3** (in supplier's recommended restriction enzyme buffer), with 20 units each of *Hin*dII and *Ssp*I restriction enzymes for 4 h, 37°C in 400 μL. Add 5 M NaCl solution to give a final concentration of 1 M NaCl.
2. While DNA is digesting, place 100 μL of resuspended BND cellulose in a microfuge tube (~50 μL of packed BND cellulose). Add 400 μL 1 M TEN; mix with a heat-sealed Pasteur pipette or sealed pipette tip. Let it sit, 5 min and pellet briefly

in microfuge. Remove supernatant and resuspend pellet in 450 μL of 1 M TEN. Repeat two more times, and remove supernatant.

3. Add digested sample to BND cellulose pellet and mix with sealed Pasteur pipette. Let sit, 5 min at room temperature. Repeat mixing and incubation twice.
4. Centrifuge (3,200*g*, 30 s, room temperature). Remove supernatant and save; this is the unbound fraction. Wash pellet five times with 450 μL 1 M TEN as follows: resuspend in 450 μL 1 M TEN with sealed Pasteur pipette; let sit, 5 min, centrifuge (3,200*g*, 30 s, room temperature), remove supernatant. Save wash supernatants and combine with the unbound fraction.
5. Elute ssDNA from the matrix with four washes of 450 μL of 1 M TENC, performed as above.
6. Combine caffeine elution fractions and remove traces of BND cellulose by centrifuging (3,200*g*, 10 s, room temperature) through glass filter spin columns.
7. Exchange buffer by successive washes in a Microcon YM-30 spin filter. Load sample into spin filter, spin in microfuge until all liquid has flowed through (14,000*g*, ~10 min, room temperature) and remove filtrate. Add 500 μL TE to upper chamber and spin as above. Monitor caffeine concentration in filtrate by measuring absorption at 273 nm and repeat wash cycles until no further decrease is detected; about six wash cycles usually are needed. After the last TE wash, add 50 μL TE to the filter chamber, let sit for 3 min at room temperature, and recover eluate by inverting the filter chamber into a clean microfuge tube and centrifuging at 1,000*g*, 3 min as indicated in the manufacturer's instructions.
8. Measure recovery and DSB hot-spot enrichment by qPCR, as in **Section 3.2 Step 12**.

3.5. Amplification and Labeling of ChIP-Associated Sequences and ssDNA Enrichment Material by Random Primer Extension

Input and ssDNA-enriched material from BND cellulose fractionation and Spo11-ChIP fractions and whole cell extracts from Spo11 immunoprecipitates are amplified using a previously described random priming amplification procedure with minor modifications [*(6)*; deRisi lab protocols at http://cat.ucsf.edu/resources/index.html]. Two successive primer extensions (Round A) are performed in a thermocycler that allows temperature ramping times to be adjusted. For ssDNA-enriched material, the first extension is performed without prior denaturation to select for the single strand templates; for ChIP samples both extensions are preceded by denaturation. The protocol below is written for single-strand DNA amplification; to amplify ChIP, input, and genomic DNA samples, an initial 30 s, 95°C step should be added.

1. Combine in a 200 μL PCR tube, on ice: input sample (~10^5 genome equivalents of rDNA, *see* **Note 9**); 1X Sequenase buffer; 0.3 mM each dNTP; 5 mM DTT; 40 pmol primer A; four units of Sequenase; 15 μL final volume. Place in thermal cycler set at 10°C.
2. Raise temperature from 10°C to 37°C over 8 min. Hold at 37°C for 8 min. Rapidly ramp to 94°C and hold for 2 min. Rapidly cool to 10°C and hold for 5 min. During the 5 min hold period, remove tubes to an aluminum block on ice, add four units Sequenase, mix by flicking the tube, pulse-spin in a microfuge, and return to the thermocycler. Ramp from 10°C to 37°C over 8 min, hold at 37°C for 8 min, rapidly ramp down to 23°C and hold for 10 min. Return to ice and add 45 μL of dH_2O.
3. Split round A amplification into six 10-μL aliquots, each to be used as input for a round B amplification using *Taq* DNA polymerase and the following conditions: 0.2 mM each dNTP; 20 pmol of primer B; 2.5 mM $MgCl_2$; 1X Promega buffer; 100 μL total; with an amplification program of 95°C for 30 s; 47°C for 30 s; 72°C for 2 min; for 0, 15, 18, 21, 24, or 27 cycles, removing tubes at the end of the 72°C extension step.
4. Measure the extent of amplification by displaying 5% of each reaction on a 1.5% agarose gel, run in 1X TBE. Stain gel with SYBR gold at supplier's recommended concentration in deionized H_2O for 15 min, rinsed twice in deionized H_2O for 20 min, and image gel by epifluoresence at 460 nm. Select the sample with the lowest number of amplification cycles that shows detectable product (about 1 ng/μL).
5. Remove buffer and primers by two successive filtration/wash cycles using 500 μL of TE and a Microcon YM-100. Add sample to upper filter chamber and centrifuge at 500*g*, ~30 min or until all liquid has passed through the filter. Add 500 μL TE and centrifuge as above. Repeat wash, add 100 μL TE to filter chamber, let sit 3 min at room temperature, invert filter chamber into clean microfuge tube and recover eluate by centrifuging at 1,000*g*, 3 min.
6. Incorporate aa-dUTP (Round C) using the same program and reactions as in Round B (95°C for 30 s; 47°C for 30 s; 72°C for 2 min for 20 cycles), but with the following reaction: 5 μL Round B as input; 1X aa-dUTP/dNTP mix, 2.5 mM $MgCl_2$, 1X Promega buffer and 15 units of *Taq* DNA polymerase per reaction in 100 μL.
7. Combine two Round C reactions (200 μL) and purify by three successive filtration/wash cycles using Microcon YM-100 spin filters and 500 μL of 10 mM Na_2CO_3 pH 8.9 as described in **Step 5**.

8. Add 20 μL of 10 mM Na_2CO_3 pH 8.9, wait 1 min, invert filter chamber over a fresh microfuge tube and spin 3 min at 1,000*g* to recover DNA
9. Measure final DNA concentration with a NanoDrop spectrophotometer (*see* **Note 10**).
10. Dry 3 μg amplified sample for each dye in a Speed-Vac. To avoid over-drying, dry for ~2 min/μL at medium heat, monitoring drying visually every 5 min. Resuspend in 5 μL of coupling buffer. Add 5 μL monoreactive dye. Mix thoroughly by flicking the tube, then incubate 1 h, room temperature, flicking tubes every 15 min. Avoid exposure to light (*see* **Note 1**).
11. Add 4.5 μL of 4 M hydroxylamine to quench the reaction. Incubate 15 min at room temperature. Add 400 μL of 50 mM MES, pH 7.2 to each reaction.
12. Remove uncoupled dye by filtration through a YM-50 Microcon filter as in **Step** 7 with centrifugation at 13,000*g* for 8 min. Repeat buffer exchange two more times with 500 μL 50 mM MES, pH 7.2.
13. Add 20 μL of 50 mM MES, pH 7.2 to recover labeled DNA and recover by inverting filter over new tube, centrifuging at 1,000*g*, 3 min.
14. Use NanoDrop spectrophotometer to determine dye and DNA content (using preset programs) and calculate labeling efficiencies. Successful labelings contain one dye moiety per 10–30 bp. If another spectrophotometer without preset programs for Cy3 and Cy5 is used, measure absorbance at 550 and 650 nm for Cy3 and Cy5 respectively, and calculate concentrations using $c = El/A$ where c = concentration (M), E = extinction coefficient (150,000/M*cm for Cy3, 250,000/M*cm for Cy5), l = light path length (cm) and A = absorbance.

3.6. High-Density Microarray Analyses of DSB-Associated Sequences

The microarray analysis presented here uses high-density microarrays containing oligonucleotides (~60 nt each) corresponding to *S. cerevisiae* reference strain S288c single-copy sequences, with array element locations spaced roughly every ~290 nucleotides. The specific microarrays used are manufactured by Agilent Technologies, and contain about 44,000 individual array elements, synthesized in situ on the slide itself. These arrays have two major advantages over spotted PCR product arrays (average element size ~1 kb), which have also been used in DSB-mapping *(6, 23)*. First, the smaller element size of oligonucleotide arrays appears to confer a lower nonspecific hybridization background. Second, the uniform element size, and the close spacing of array elements, in theory should allow more accurate detection of

DSB peaks, whereas the variation in element size and spacing in PCR-product arrays can result in variation in the detection of DSBs of similar frequency (**Fig. 10.1b**). While we have used microarrays manufactured by Agilent, oligonucleotide-based microarrays from other suppliers (e.g. Nimblegen) should work equally well, as long as the appropriate changes are made to hybridization and washing protocols, as recommended by the manufacturer.

Because the fluorescent dyes used below are light-sensitive, all steps using dye-labeled probe should be performed in the dark or in reduced light (*see* **Note 1**). The following instructions are for a single array hybridization on a 4 × 44 k array.

1. Prewarm rotisserie oven to 60°C. Place 300 µL of 2X Agilent hybridization buffer into 1.5 mL screw cap tube. Incubate at 60°C until probe is ready
2. Mix 100 ng of Cy5-labeled sample with 100 ng of Cy3-labeled sample; add ultrapure H_2O to a final volume of 50 µL.
3. Denature probe at 95–100°C for 3 min; transfer to 60°C for 30 s. While probe is denaturing, place Agilent gasket slide into slide hybridization chamber.
4. Add 50 µL of the 2X hybridization buffer to each denatured sample; mix by gently flicking the tube. Centrifuge at 14,000*g*, 30 s.
5. Place 95 µL in the center of each single gasket, pipetting slowly to avoid bubbles.
6. Slowly place the DNA microarray slide on top of the sample, with the word "Agilent" face down and barcode face up.
7. Tighten the hybridization chamber with the knob and manually rotate the chamber several times to evenly mix the sample.
8. Insert hybridization chamber in rotisserie oven and set the rotation speed to 10 rpm. Incubate 15–17 h at 60°C.
9. Carefully remove slides from the chamber and immerse gasket-microarray sandwich in 500 mL of wash solution I. Slowly separate the immersed gasket-microarray sandwich using plastic tweezers. Quickly transfer microarray slide to a rack previously positioned in another 500 mL of wash solution I, slowly stirring with a magnetic stir bar (about 120 rpm). Microarrays can remain in this wash solution while other slides are prepared.
10. If multiple slides are being used wait until all the slides are in solution I and wash for 5 min. Remove tray, quickly remove excess solution I by gentle shaking, and transfer to solution II.
11. Wash for 5 min in solution II, using a magnetic stir bar as above.

12. Using gloved hands (powder-free gloves), transfer each slide to a slide rack positioned in Agilent stabilizing and drying solution, soak for 2 min (*see* **Note 11**). Carefully and slowly remove each slide to a dry slide rack. Let slides dry, 10s, and place them in a plastic slide box. Protect from light.
13. Scan slides immediately using an Axon 4000B scanner, set at 5 μm resolution with automatic laser PMT adjustment to achieve a maximum fluorescence saturation of 0.005%. Scanner should be turned on at least 15 min before use.
14. Extract fluorescence data with GenePix 6.0 and recognize spots using the irregular feature detection mode. Filter features in each channel according to the following criteria. First, remove all spots with a diameter of less than 33 μm. Then calculate background fluorescence in each channel using the mean fluorescence value of the 685 empty array elements (labeled (-)3xSLv1 and NC2_) incorporated into each array. Remove all array elements with fluorescence less than background + 2 standard deviations, as well as array elements with a signal-to-background ratio [(signal +background)/ background] less than 3. Our experiments use strains of the SK1 strain background, which diverges by about 1% from the standard *S. cerevisiae* sequence, and we typically remove about 2% of all array elements.

3.7. Background Normalization of ChIP-chip and ssDNA-chip Microarray Data

The normalization of microarray datasets where the hybridization probe is derived from ChIP material presents a particular challenge *(24)*, in that overall element signals are expected to have distributions that differ between input and enriched samples, and that may differ between different enrichment methods (e.g. Spo11-ChIP versus ssDNA-enrichment). In particular, ChIP samples usually display skewed distributions, while control samples approximate a normal distribution. As a consequence, normalization methods based on statistical measures (for example, median values) will tend to reduce positive signals and to increase negative signals artifactually (**Fig. 10.2a**). Ideally, array data should be normalized in two steps: first using all elements that represent background in both data sets; and second, using nonbackground elements that are expected (from other experimental measures) to have the same positive signal in both data sets, to adjust the dynamic range. In many cases, identification of both types of elements can be challenging and even impossible.

In the case of comparison of DSB data sets, however, it is possible to identify a subset of probes for which the presence of meiotic single-strand DNA is unlikely, in that Southern blot analyses and fine-structure mapping of DSB hotspots have shown that meiotic DSBs are generally absent from protein coding regions *(19, 25)*. Assuming an average shear size and/or resection tract of about 1 kb, it is likely that the hybridization signal of array elements located at least 2 kb away from the 3′ and 5′ ends

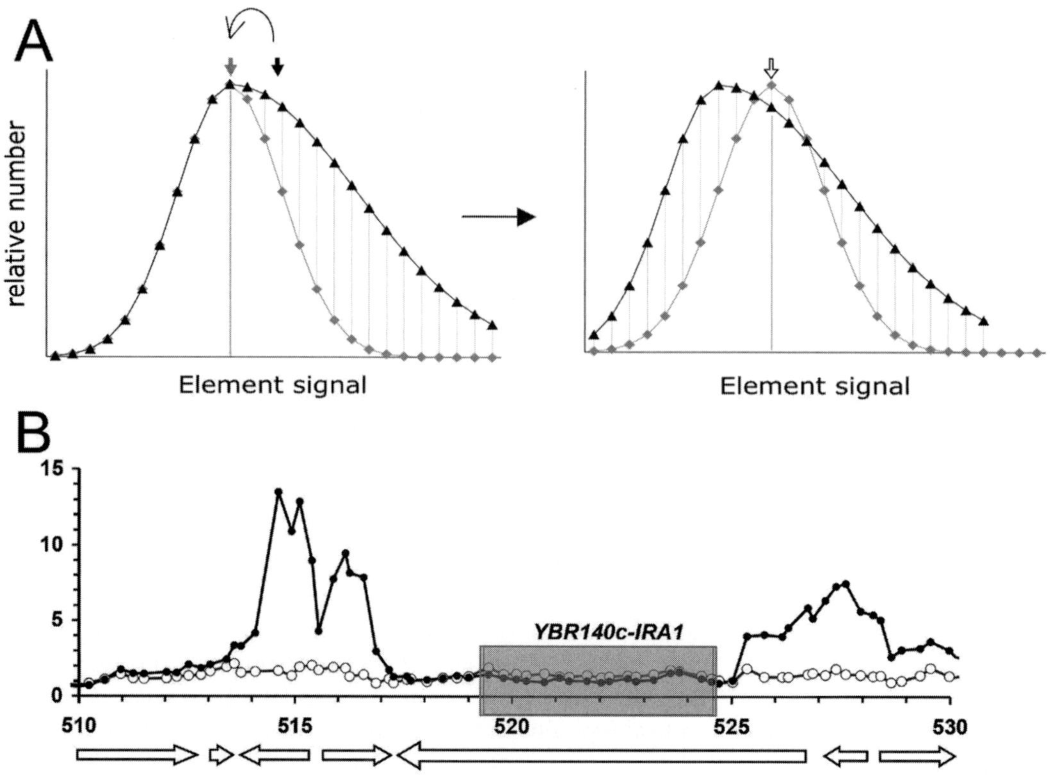

Fig. 10.2. Principle of background normalization. **A.** Median-based normalization procedures can lead to inaccurate assessment of element signals. *Gray diamonds*—background or input signals, which tend to approximate a normal distribution. *Black triangles*—ChIP signals, which tend to be skewed in the positive direction. *Arrows*—medians of the distributions. Median normalization shifts the skewed ChIP signals to the left, which leads to the inclusion of nonbackground elements in the background signal, and an under-counting of elements with positive signal. **B.** An illustration of the choice of intra-ORF elements for background normalization. *Filled circles*—array signal from BND cellulose enriched ssDNA from a *dmc1*Δ strain. *Open circles*—array signal from BND cellulose enriched ssDNA from a DSB-negative *dmc1*Δ *spo11* strain. *Arrows* below the x-axis—gene coding sequences. *Gray box*—array elements selected for background normalization.

of protein-coding sequences are likely to represent background (**Fig. 10.2b**). The median fluorescence intensity of elements meeting this criterion are selected from the largest open reading frames in the budding yeast genome (**Table 10.2**), and this is used to normalize the fluorescence intensity for each element in each single-channel array data set by dividing the signal for each element by the median background value.

We have found that, in practice, a second step of dynamic-range normalization is not necessary, and that simple background normalization is sufficient to bring into near equivalence peak values at common hotspots that have been shown to give the same DSB value under all experimental conditions (in the case of the SK1 strain background, DSB hotspots at *BUD23* and *ERG1*). If this is found not to be the case, then peak-normalization should be performed using background-subtracted datasets.

Table 10.2
Regions used for background normalization

Chr	First element (start)	Last element (end)	Name
2	53074	57163	*YBL088C (TEL1)*
2	229554	233249	*YBL004W (UTP20)*
2	507687	510710	*YBR136W (MEC1)*
2	519653	524587	*YBR140C (IRA1)*
4	759731	761703	*YDR150W (NUM1)*
4	1371894	1377499	*YDR457W (TOM1)*
7	133629	137600	*YGL195W (GCN1)*
8	304720	311776	*YHR099W (TRA1)*
8	431726	434631	*YHR165C (PRP8)*
9	108225	111206	*YIL129C (TAO3)*
10	561447	564432	*YJR066W (TOR1)*
11	58137	61409	*YKL203C (TOR2)*
11	537391	545446	*YKR054C (DYN1)*
12	56219	61500	*YLL040C (VPS13)*
12	308923	313883	*YLR087C (CSF1)*
12	351204	361576	*YLR106C (MDN1)*
12	1046148	1049478	*YLR454W (FMP27)*
15	173348	178370	*YOL081W (IRA2)*
16	762356	765500	*YPR117W*

The locations in each region of the first and last array element used for background normalization are given in chromosome coordinates (nt).

4. Notes

1. All steps involving fluorescent dyes should be performed in semi-darkness to minimize light exposure. We generally perform manipulations as rapidly as possible in a darkened room with minimal indirect light (from a partially opened door), and all incubations are performed in a closed drawer.
2. Because different spectrophotometers have different source-cuvette-detector geometries, this light-scattering measurement may not be an accurate measure of cell density. In general,

spectrophotometers with at least 1 cm distance between the cuvette and the detector will give an OD_{600} of ~1.4 at the desired cell concentration. Other spectrophotometers will require empirical calibration. SK1 cultures grown in SPS to this concentration will be at the transition between exponential growth and stationary phase; the corresponding transition point will need to be empirically determined if other strains and/or other pregrowth regimes are used.

3. Immunoprecipitating in 1 M NaCl buffer results in much less nonspecific binding of DNA to the Protein G matrix, therefore increasing the overall dynamic range of signal over background. However, higher salt concentrations result in reduced yield measured as a percentage of input. When analyzing by microarray we compensated for this reduced yield by using at least 10 times more starting culture (i.e., ten 50 mL culture samples, processed individually, and then pooled before amplification).
4. DNA does not fragment as readily by sonication in higher salt concentrations. We found that five to eight 10 s pulses, intervening 30 s on ice, were required to fragment the DNA such that the majority of fragments were in the 500–1,000 bp size. These conditions will need to be calibrated for different sonicators.
5. The yields are higher with longer antibody and Protein G incubation times. It is useful to do an overnight Protein G incubation assay by microarray to maximize yields.
6. We use more washing steps than most published ChIP protocols. We found that a greater amount of background removal comes from the number of dilutions of the beads rather than the actual length of washing time. With careful removal of the wash solutions this does not affect the overall yield.
7. To avoid high background levels of BND cellulose binding, it is important to handle all samples gently prior to restriction endonuclease digestion, to minimize ssDNA formation through shearing. The inclusion of rapid lysis directly into phenol-chloroform was adopted after empirical tests indicated that even minimal incubation of lysates prior to extraction led to increased background ssDNA levels, presumably due to nonspecific nuclease activity. Other methods of DNA extraction may be used, but the DSB-independent background will be considerably greater.
8. The extent of spheroplasting is monitored every few minutes by mixing 2 μL of spheroplast solution with either 2 μL of 1% SDS or 2 μL of spheroplast buffer without zymolyase on a microscope slide. Optimal spheroplasting is achieved when about 90% of cells are lysed (ghosts) in SDS compared to the spheroplast solution control.

9. The amount of material to be used in Round A is determined by qPCR measurement of the genome-equivalents of rDNA (*see* **Section 3.2 Step 12**). Typically, 10^5 genome-equivalents of rDNA are used. It may be necessary to use up to ten Spo11 ChIP samples to obtain this amount; sporulation cultures are processed in 50 mL aliquots and then combined, as it was found that ChIP using larger volumes resulted in reduced yield. To minimize amplification artifacts, it is important that all samples (both ChIP/enriched and input/control) be amplified from the same genome-equivalent input.
10. qPCR should be performed on final amplification products to confirm that enrichment levels present in the starting material were maintained throughout amplification. If final enrichment levels differ substantially from those in starting material, use qPCR to troubleshoot intermediate steps.
11. Agilent drying solution should be prewarmed at 37°C for 1 h before use to dissolve precipitated components. Because the drying solution contains a proprietary volatile and toxic component, drying should be performed in dedicated slide rack and bath placed in a fume hood.

Acknowledgments

We thank Jennifer Gerton, Hugh Cam and Shiv Grewal for technical advice, Dhruba Chattoraj and Yikang Rong for advice regarding the manuscript, and David Kaback and Jennifer Fung for communicating data in advance of publication.

References

1. Keeney, S. (2001) Mechanism and control of meiotic recombination initiation. *Curr. Top. Dev. Biol.* **52**, 1–53.
2. Koehler, K. E., Hawley, R. S., Sherman, S., and Hassold, T. (1996) Recombination and nondisjunction in humans and flies. *Hum. Mol. Genet.* **5**, 1495–1504.
3. Myers, S., Bottolo, L., Freeman, C., McVean, G., and Donnelly, P. (2005) A fine-scale map of recombination rates and hotspots across the human genome. *Science* **310**, 321–324.
4. Drouaud, J., Camilleri, C., Bourguignon, P. Y., Canaguier, A., Berard, A., Vezon, D., Giancola, S., Brunel, D., Colot, V., Prum, B., Quesneville, H., and Mezard, C. (2006) Variation in crossing-over rates across chromosome 4 of *Arabidopsis thaliana* reveals the presence of meiotic recombination "hot spots". *Genome Res.* **16**, 106–114.
5. Winzeler, E. A., Richards, D. R., Conway, A. R., Goldstein, A. L., Kalman, S., McCullough, M. J., McCusker, J. H., Stevens, D. A., Wodicka, L., Lockhart, D. J., and Davis, R. W. (1998) Direct allelic variation scanning of the yeast genome. *Science* **281**, 1194–1197.
6. Gerton, J. L., DeRisi, J., Shroff, R., Lichten, M., Brown, P. O., and Petes, T. D. (2000) Global mapping of meiotic recombination hotspots and coldspots in the yeast *Saccharomyces cerevisiae*. *Proc. Natl. Acad. Sci. U. S. A.* **97**, 11383–11390.

7. Nairz, K. and Klein, F. (1997) *mre11S*–a yeast mutation that blocks double-strand-break processing and permits nonhomologous synapsis in meiosis. *Genes Dev.* **11**, 2272–2290.
8. Prinz, S., Amon, A., and Klein, F. (1997) Isolation of *COM1*, a new gene required to complete meiotic double- strand break-induced recombination in *Saccharomyces cerevisiae*. *Genetics* **146**, 781–795.
9. Alani, E., Padmore, R., and Kleckner, N. (1990) Analysis of wild-type and *rad50* mutants of yeast suggests an intimate relationship between meiotic chromosome synapsis and recombination. *Cell* **61**, 419–436.
10. Keeney, S., Giroux, C. N., and Kleckner, N. (1997) Meiosis-specific DNA double-strand breaks are catalyzed by Spo11, a member of a widely conserved protein family. *Cell* **88**, 375–384.
11. Prieler, S., Penkner, A., Borde, V., and Klein, F. (2005) The control of Spo11's interaction with meiotic recombination hotspots. *Genes Dev.* **19**, 255–269.
12. Robine, N., Uematsu, N., Amiot, F., Gidrol, X., Barillot, E., Nicolas, A., and Borde, V. (2007) Genome-wide redistribution of meiotic double-strand breaks in *Saccharomyces cerevisiae*. *Mol. Cell. Biol.* **27**, 1868–1880.
13. Cromie, G., Hyppa, R. W., Cam, H., Farah, J. A., Grewal, S., and Smith, G. R. (2007) A discrete class of intergenic DNA dictates meiotic DNA break hotspots in fission yeast. *PLoS Genet.* **3**, e141.
14. Neale, M. J., Pan, J., and Keeney, S. (2005) Endonucleolytic processing of covalent protein-linked DNA double-strand breaks. *Nature* **436**, 1053–1057.
15. Shinohara, A. and Shinohara, M. (2004) Roles of RecA homologues Rad51 and Dmc1 during meiotic recombination. *Cytogenet. Genome Res.* **107**, 201–207.
16. Blitzblau, H. G., Bell, G. W., Rodriguez, J., Bell, S. P., and Hochwagen, A. (2007) Mapping of meiotic single-stranded DNA reveals double-strand break hotspots near telomeres and centromeres. *Curr. Biol.* **17**, 2003–2012.
17. Buhler, C., Borde, V., and Lichten, M. (2007) Mapping meiotic ssDNA reveals a new landscape of DNA double-strand breaks in *Saccharomyces cerevisiae*. *PLoS Biol.* **5**, e324.
18. Borde, V., Wu, T. C., and Lichten, M. (1999) Use of a recombination reporter insert to define meiotic recombination domains on chromosome *III* of *Saccharomyces cerevisiae*. *Mol. Cell. Biol.* **19**, 4832–4842.
19. Baudat, F. and Nicolas, A. (1997) Clustering of meiotic double-strand breaks on yeast chromosome *III*. *Proc. Natl. Acad. Sci. U. S. A.* **94**, 5213–5218.
20. Strahl-Bolsinger, S., Hecht, A., Luo, K., and Grunstein, M. (1997) *SIR2* and *SIR4* interactions differ in core and extended telomeric heterochromatin in yeast. *Genes Dev.* **11**, 83–93.
21. Kee, K. and Keeney, S. (2002) Functional interactions between *SPO11* and *REC102* during initiation of meiotic recombination in *Saccharomyces cerevisiae*. *Genetics* **160**, 111–122.
22. Gamper, H., Piette, J., and Hearst, J. E. (1984) Efficient formation of a crosslinkable HMT monoadduct at the *Kpn* I recognition site. *Photochem. Photobiol.* **40**, 29–34.
23. Borde, V., Lin, W., Novikov, E., Petrini, J. H., Lichten, M., and Nicolas, A. (2004) Association of Mre11p with double-strand break sites during yeast meiosis. *Mol. Cell* **13**, 389–401.
24. Buck, M. J. and Lieb, J. D. (2004) ChIP-chip: considerations for the design, analysis, and application of genome-wide chromatin immunoprecipitation experiments. *Genomics* **83**, 349–360.
25. Liu, J., Wu, T. C., and Lichten, M. (1995) The location and structure of double-strand DNA breaks induced during yeast meiosis: evidence for a covalently linked DNA-protein intermediate. *EMBO J.* **14**, 4599–4608.

Chapter 11

Detection of Meiotic DNA Breaks in Mouse Testicular Germ Cells

Jian Qin, Jaichandar Subramanian, and Norman Arnheim

Abstract

The study of location and intensity of double-strand breaks (DSBs) in mammalian systems is more challenging than in yeast because, unlike yeast, the progression through meiosis is not synchronous and only a small fraction of all testis cells are actually at the stage where DSB formation is initiated. We devised a quantitative approach that is sensitive enough to detect the position of rare DNA strand breaks in mouse germ cell-enriched testicular cell populations. The method can detect DNA breaks at any desired location in the genome but is not specific for DSBs—overhangs, nicks, or gaps with a free 3′ OH group are also detected. The method was successfully used to compare testicular cells from mouse strains that possess or lack an active recombination hot spot at the *H2-Ea* gene. Breaks that were due to meiotic hot spot activity could be distinguished from the background of DNA breaks. This highly sensitive approach could be used to study other biological processes where rare DNA breaks are generated.

Key words: Meiosis, mouse recombination hot spots, DNA double-strand breaks, terminal deoxynucleotidyl transferase.

1. Introduction

DNA double-strand breaks (DSBs) have been found at recombination hot spots during budding and fission yeast meiosis *(1–5)* although the distance between the crossover site and the DSB in fission yeast can, in some cases, be considerable *(6)*. Direct biochemical evidence for meiotic DSBs at a recombination hot spot is more difficult to obtain in mammals, such as mice, than in yeast *(7–10)*. First, any one male meiotic cell undergoes only ~25 crossover events detected as foci of MLH1 protein *(11–13)* although evidence exists that more meiotic DSBs do not resolve as a crossover event (*see 14, 15*).

Scott Keeney (ed.), *Meiosis, Volume 1, Molecular and Genetic Methods, vol. 557*

DOI 10.1007/978-1-59745-527-5_11 Springerprotocols.com

Mouse DSBs *(7, 8)* are distributed across the total haploid genome length of ~3,000 Mb, and on average, a genetic distance (based on crossovers) of 1 cM (centimorgan) corresponds to a physical distance of 2 Mb *(16)*. Per unit of physical distance, mice have ~670 times fewer meiotic exchange events than budding yeast, where a genetic distance of 1 cM corresponds to a physical distance of ~3 kb *(17–19)*. Second, both budding and fission yeast cells can be synchronized to enter meiosis so a relatively large proportion of the cells will have a break at a particular recombination hot spot. In the adult mouse testis, primary spermatocytes account for only ~20% of the germ cell population *(20)*, and only a small proportion of these cells can possibly have an unligated DNA break at any one site at any instance in time. To make it worse, when genomic DNA is extracted from mammalian cells, artificial DNA breaks are unavoidably generated due to mechanical shearing force. These background DNA breaks will outnumber the level of meiotic DNA breaks if the crossover fraction is too low. These factors dramatically reduce the chance that region-specific meiotic DNA strand breaks can be directly detected in adult testis cells.

To address these issues, we developed an assay (outlined in **Figs. 11.1 and 11.2**) to tag naturally occurring DNA breaks in mammalian cells before any artificial breaks are generated by DNA extraction *(9)*. This method is similar to those used previously to detect DNA breaks at high resolution in budding and fission yeast *(6, 21)*, except that modifications were made to enhance the specificity and sensitivity of the assay to allow its application to more complex genomes. The method is not specific for DSBs but will also identify any recombination intermediate with a free 3′ OH group.

In the assay described below we studied DNA breaks in and around a 418-bp region of the *Ea* gene in the mouse Major Histocompatability Complex (H-2). Crossing over in this small interval has been estimated to give rise to a recombination fraction of ~0.4–2% in those mouse strains where this recombination hot spot is active *(22)*. We were able to demonstrate a significantly greater number of breaks and locate their positions at high resolution in the *Ea* region of mouse strains where the hot spot is active compared to mice where it is inactive *(9)*.

2. Materials

2.1. Testis Isolation and Cell Dissociation

1. In mice, only certain combinations of inbred strains exhibit recombination hot spot activity *(23–25)*. Our method was designed to analyze DNA breaks at the *Ea* hot spot. Mice of the B10.S(9R) and B10.F(13R) strains (also referred to as 9R and 13R, respectively) and their F_1 progeny (resulting

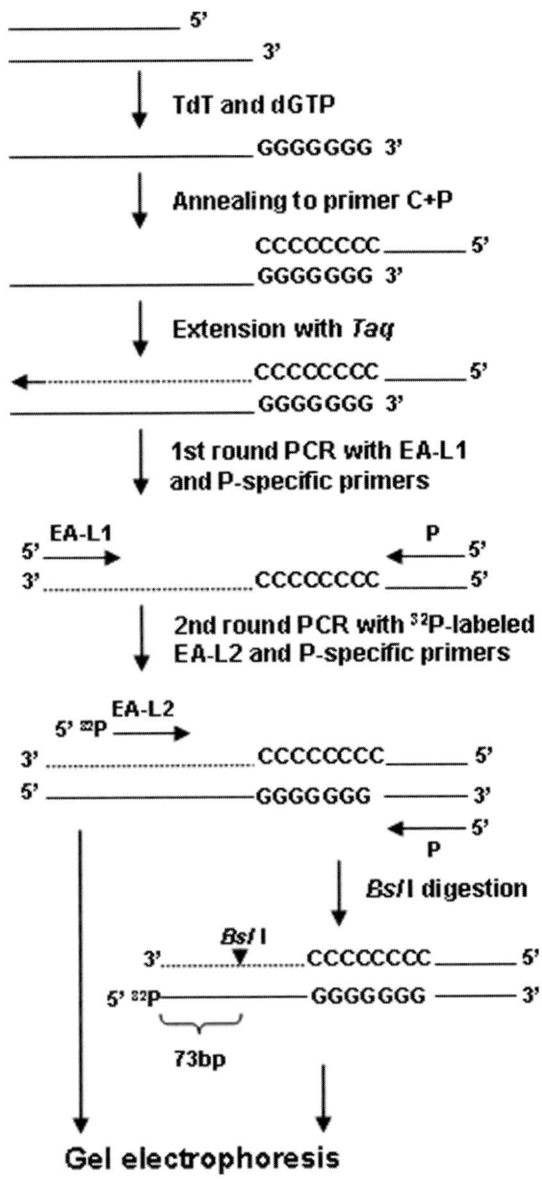

Fig. 11.1. Outline of the method to detect rare DNA breaks. A DNA fragment with a 3′ overhang is used in this example. The method is applicable to any DNA that has an available 3′ OH group. For a detailed explanation see the text and (*9*). Reprinted with the permission of the ASM.

from mating 13R males and 9R females and vice versa) were used for the measurement of strand breaks at the *Ea* recombination hot spot *(9)*. They were originally provided by Drs. Laura L. Richardson and Mary Ann Handel. F_1 mice heterozygous at the *Ea* gene for a *p* haplotype (from P/J females) and a *d* haplotype (from B10 · D2-H2^{d}H2-T18^{c}Hc^o/oSnJ

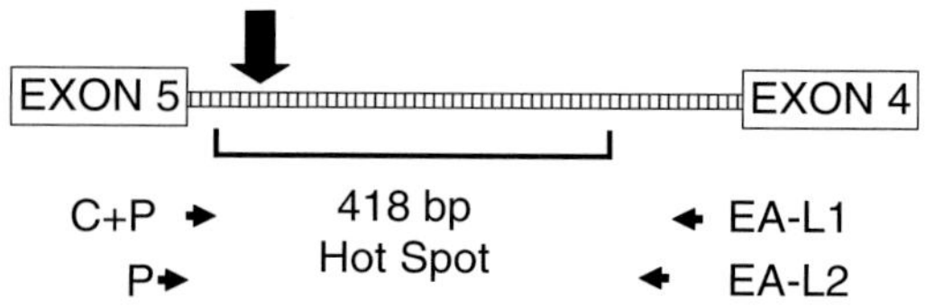

Fig. 11.2. Placement of primers in a case where a DNA break is located in one *Ea* molecule at the position marked by the large downward facing arrow. *Horizontal arrows* show the relative positions of the primers.

males) were also generated as controls. All the strains are available from The Jackson Laboratory (Bar Harbor, ME) but 9R and 13R are currently only stored as frozen embryos.

2. EKRB [Enriched Krebs Ringer Buffer, *(20)*]: 120.1 mM NaCl, 4.8 mM KCl, 25.2 mM $NaHCO_3$, 1.2 mM KH_2PO_4, 1.2 mM $MgSO_4$, 1.3 mM $CaCl_2$, 11.1 mM glucose, 1 mM glutamine, 20 mL of 50 × amino acids, 10 mL of 100 × nonessential amino acids and 10 mL of a 10,000 μg/mL streptomycin-10,000 units/mL penicillin solution in a final vol of 1 L. The amino acid supplements and antibiotics were from Invitrogen (Carlsbad, CA). Sterilize by passing through a 0.2 μm filter and store in a sterile container at 4°C. All reagent and buffer solutions should be made with high-quality deionized water.
3. EKRB-CO_2-BSA buffer: Place tissue culture flasks of EKRB with freshly added BSA (final concentration 0.5% (w/v)] in a tissue culture incubator (5% CO_2) overnight before use [see also *(20)*]. The final pH should be 7.2–7.3 after equilibration.
4. Crude collagenase. Store at –20°C.
5. Trypsin: dissolve 100 mg in 1 mL water and aliquot into microcentrifuge tubes. Store at –20°C.
6. DNase I: dissolve 5 mg in 5 mL water and aliquot into microcentrifuge tubes. Store at –20°C.
7. Nylon mesh, 70-μm (Small Parts Inc, Miami Lakes, FL).
8. BSA powder. Store at 4°C.

2.2. Agarose Plug Formation and Cell Lysis

1. InCert low-melting agarose (Cambrex Bio Science, Rockland Inc, Rockland, ME).
2. Proteinase K: dissolve 10 mg of powder in 1 mL of water and store in aliquots at –20°C.
3. Lysis buffer *(26)*: 0.5 M EDTA, pH 8.0, 100 μg/mL of Proteinase K in water and 1.0% Sarkosyl. Lysis buffer without Proteinase K can be stored at room temperature. Add Proteinase K just before use.
4. 75-mm^3 agarose plug molds (Bio-Rad, Hercules, CA).

2.3. Terminal Deoxynucleotidyl Transferase (TdT) Tailing

1. TE buffer: 10 mM Tris-HCl, 1 mM EDTA, pH 8.0.
2. Tailing buffer: 1 × New England Biolabs Restriction Enzyme Buffer # 4 (20 mM Tris-acetate pH 7.9, 10 mM magnesium acetate, 50 mM potassium acetate, 1 mM dithiothreitol), 2.5 mM $CoCl_2$, and 0.8 mM dGTP. To make 1 mL of tailing buffer, add 100 μL of 10 × New England Biolabs Buffer # 4, 100 μL of 25 mM $CoCl_2$, and 8 μL of 100 mM dGTP to 792 μL water. Prepare fresh for each experiment.
3. Terminal deoxynucleotidyl transferase: 20 units/μL (New England Biolabs, Ipswich, MA), store at –20°C.

2.4. DNA Extraction

1. Phenol saturated with TE buffer (10 mM Tris-HCl, 1 mM EDTA, pH 8.0). Stored at 4°C.
2. Chloroform.
3. 3 M Sodium acetate, pH 5.2.
4. Ethanol-100% and 70% (in water).

2.5. Primer Extension and PCR of dGTP-Tailed DNA Fragments

1. 10 × extension/PCR Premix: 100 mM Tris-HCl, pH 8.3, 500 mM KCl, 25 mM $MgCl_2$, 0.1% gelatin. Autoclave and store at 4°C.
2. dNTP Stocks: 100 mM each dATP, dCTP, dGTP, and dTTP.Store at –20°C.
3. Taq DNA polymerase. Store at –20°C.
4. 1 × extension/PCR Reaction Buffer: 1 × extension/PCR Premix solution, 50 μM each of dATP, dGTP, dCTP, and dTTP, 1 μL of sample DNA, 0.2 μM primer(s), and 1 unit of *Taq* DNA polymerase. PCR is carried out in 50-μL reactions.
5. Custom primers: Dissolve in water to a final concentration of 100 μM. Prepare 10 μM working dilutions for each primer. Store stocks at –20°C. For additional details on the primers used for analysis of the *Ea* recombination hot spot *see* **Note 1.**

2.6. Primer Labeling for PCR

1. Radioactively labeled primer option: *T4* polynucleotide kinase (10 units/μL) and [γ-^{32}P]ATP (4,500 Ci/mmol, 10 μCi/μL).
2. Fluorescently labeled primer option. Primers are obtained commercially, already labeled with an appropriate dye suitable for PCR and analysis based on the requirements of the available microcapillary DNA sequencing system.

2.7. Restriction Enzyme Digestion of PCR Product and Detection of DNA Breaks

1. 10 × TBE buffer.
2. Sample loading buffer: 98% formamide, 10 mM EDTA, pH 8.0, 0.25% xylene cyanol FF, and 0.25% bromophenol blue.

3. *Bsl* I (10,000 units/mL) The choice of enzyme will likely vary depending on the DNA region being examined.
4. pBR322 DNA-*Msp* I digest (New England Biolabs, Ipswich, MA).
5. 40% acrylamide-bisacrylamide solution (19:1).
6. Urea (ACS certified).
7. 0.45 μm NALGENE filter unit.
8. TEMED.
9. Ammonium persulfate: 10% solution in water.
10. Option for using fluorescently labeled primers. Refer to Manufacturer's instructions for reagents that are needed for use with microcapillary electrophoresis systems.

3. Methods

3.1. Testis Dissection and Germ Cell Preparation

1. An enriched germ cell preparation is obtained from adult mouse testes based on an already published procedure *(20)*. The protocol described below is to obtain germ cells from the two testes of a single 2–2.5-month old male mouse. Sacrifice the mouse by approved procedures. Lay the mouse with its ventral side facing up, expose the testis and remove the epididymis and associated fat, then remove the testis. Make a small cut at one end of each testis using a pair of fine dissection scissors. Hold each testis at one end with one forceps and with another forceps squeeze out the seminiferous tubules through the cut end into a 15-mL tube containing 12 mL EKRB-CO_2–BSA buffer. Just before dissection, add powered collagenase to the EKRB-CO_2-BSA solution to a final concentration of 1 mg/mL. Keep the tube closed to maintain the [CO_2] whenever possible.
2. Incubate the tube at 30°C while agitating it for 15 min to disperse the seminiferous tubules. This can be accomplished by inverting the tube up and down by hand continuously in a water bath. Alternatively, place the tube in a reciprocal shaking water bath at 110–120 cycles per min *(20)*. When finished, let the tube stand for one min until the seminiferous tubules settle to the bottom. Meanwhile, thaw the trypsin and DNase I stock solutions.
3. Use a plastic transfer pipette to gently remove the supernatant from the tube. Wash the seminiferous tubules two more times with fresh EKRB-CO_2-BSA, each time allowing the tubules to settle. Resuspend them in 12 mL EKRB-CO_2-BSA.

4. Add 30 μL of trypsin stock solution and 12 μL of DNase I stock solution and mix the contents by inverting the tube up and down. Incubate the tube at 30°C for 15 min while shaking the contents continuously to keep the tubules from settling; also invert the tube up and down a few times every minute.
5. Obtain a suspension of single testis cells by gently pipetting the tubule suspension up and down a few times (enough to separate the cells but not enough to damage them) with a soft plastic transfer pipette. Examine the preparation under a microscope to estimate the extent of cell dissociation.
6. Centrifuge the single cell suspension at 200*g* for 5 min. Discard the supernatant. Resuspend the cells in 12 mL cold (4°C) EKRB-CO_2-BSA and centrifuge again at 200*g* for 5 min. Resuspend in 1.5–2 mL cold EKRB-CO_2-BSA.
7. Pass the single cell suspension (using a soft plastic transfer pipette) through the 74-μm mesh filter. This step removes cell aggregates and incompletely digested tubules from the suspension so that the cell concentration can be accurately determined in the next step.
8. Dilute a small amount of the cell suspension 10-fold with cold EKRB-CO_2-BSA and count the cells using a hemocytometer. On the basis of the cell count, add extra EKRB (without the 5% CO_2 but with 0.5% BSA) to the bulk single cell suspension so that the cell concentration is 2×10^7 cells/mL. Immediately go on to the next step.

3.2. Agarose Plug Formation and Cell Lysis

1. Use EKRB with 0.5% BSA to make a 2.0% InCert low-melting-point agarose solution by heating in a microwave oven. Keep the agarose solution at ~40°C and the cell suspension at 37°C so that upon mixing the agarose will not solidify too quickly at room temperature. Warm an empty tube at 37°C as well.
2. Mix equal vol (500 μL) of cell suspension and agarose solution gently and thoroughly in the warmed tube to produce a final concentration of 10^7 cells/mL in 1.0% agarose. Immediately pipette the mixture into 75-mm^3 plug molds that are sealed at the bottom with paper tape. Fill the molds until they are slightly overflowing. Allow the agarose to solidify at 4°C for about 5 min.
3. Use an ethanol-wiped single-edge razor blade to cut the excess gel from the top of the mold. Remove the mold seal from the bottom and gently remove the agarose plugs by forcing them out the bottom using a plastic bar that was part of the mold but that could be cut from it. Add two plugs to a 1.7 mL microfuge tube containing about 1mL of lysis buffer. Incubate the tubes at 50°C for 40–48 h with occasional gentle agitation (*see* **Note 2**).

4. Pipette out the lysis buffer and carefully stand the tubes upside down on a piece of paper towel to remove as much lysis buffer as possible; the plugs usually adhere to the tube side. Be extremely careful not to let the plugs fall out. Wash the plugs 15 times (10 min each) with 1.5 mL of TE buffer each time at room temperature in the original lysis tubes using the same decanting procedure. During the 13th or 14th wash, components of the TdT tailing buffer can be thawed. Immediately go on to the next step.

3.3. Terminal Deoxynucleotidyl Transferase(TdT) Tailing

1. After the 15th wash with TE, separate each member of a pair of plugs into an individual 1.7 mL microfuge tube. Label one tube as "–TdT" and the other one as "+TdT". Add 300 μL tailing buffer to both tubes. Incubate at 4°C for 2–3 h.
2. Remove the tailing buffer by aspiration and add 300 μL of fresh tailing buffer to each tube. Add 18 μL of TdT (20 units/μL) to only the "+TdT" tube and gently mix. Place the tubes on ice in an ice bucket and transfer the bucket to a 4°C cold room for ~15 h. Remove the tubes and place them at 37°C for 2 h. Remove the tailing buffer and wash the plugs once with ~1.5 mL TE buffer. Remove the TE and add 350 μL fresh TE buffer to each tube. Incubate at 65°C for 5 min to melt the plug. (*see* also **Note 3**).

3.4. DNA Extraction

1. To extract the DNA from the plugs *(27)*, add 400 μL of phenol (warmed to ~37°C) to each tube (total volume now is 750 μL plus the melted 75-mm^3 plug) and shake the tube vigorously. Centrifuge at 12,000*g* for 10 min at room temperature and transfer aqueous layer to a clean tube.
2. Extract the ~350–400 μL of aqueous layer again with a mixture of phenol and chloroform (200 μL of each) followed by a final extraction using 400 μL chloroform.
3. Transfer the aqueous layer (about 400 μL) to a clean tube and add 40 μL of 3 M sodium acetate (pH 5.2) and then 900 μL of cold (–20°C) 100% ethanol. Shake well and place the tubes at –20°C for at least 30 min.
4. Centrifuge at 12,000*g* for 30 min at 4°C. Discard the supernatant.
5. Wash the DNA pellets once with 1 mL of 70% cold (–20°C) ethanol. Centrifuge at 12,000*g* for 10 min at 4°C. Discard the supernatant.
6. Air-dry the pellets at room temperature for 30–60 min. Dissolve the DNA overnight at room temperature in 75 μL of deionized water so that 1 μL should contain DNA from the equivalent of ~10^4 cells (~60 ng). Store at 4°C. We usually did not store DNA samples at either +4 or –20°C over long periods and so our knowledge of how breaks

between the tail and the 5′ end (rendering the fragments undetectable) might accumulate over time in samples stored in this way is limited.

3.5. Primer Extension and PCR

1. For each initial primer extension reaction, 10,000 cell equivalents are examined (~60 ng DNA; the ploidy of the germ cell population varies and on average there are 1.7 genomes per testis cell, *see (20)* for cell proportions). Add 1 μL of sample DNA and primer C+P (for a final concentration of 0.2 μM) to 50 μL of extension/PCR Reaction Buffer. Denature the DNA in each tube at 94°C for 4 min, incubate at 50°C for 3 min and then incubate at 72°C for 3 min (*see* **Note 4**). Immediately go on to the next step.
2. Initiate first-round PCR by adding 1 μL of a solution containing 10 μM primer P and 10 μM gene-specific primer EA-L1 and heating to 94°C. The final concentration of both primers is 0.2 μM. The cycling conditions for the *Ea* locus was denaturation at 94°C for 45 s, followed by annealing and extension at 66°C for 3 min for 30 cycles. The final extension step is for 5 min at 72°C (*see* **Note 5**).
3. Radioactively label primer EA-L2 (omit if using fluorescent detection; *see* **Step 5** below). Mix 2 μL of 20 μM gene-specific primer EA-L2, 1 μL (10 units) of T4 polynucleotide kinase, 5 μL of [γ-^{32}P]ATP, 1 μL of 10 × T4 kinase buffer, and 1 μL of H_2O (total vol 10 μL). Incubate the reaction at 37°C for 2 h, then inactivate the kinase by heating at 68°C for 10 min. We did not find it necessary to separate the free [γ-^{32}P]ATP from the labeled primer for the subsequent steps.
4. Prepare a second-round PCR master mix. For the typical number of reactions needed to fully load a single polyacrylamide gel, combine 457 μL of extension/PCR reaction buffer with the entire 10 μL EA-L2 primer radiolabeling mixture from **Step 3**, 3 μL of 20 μM unlabeled EA-L2, and 10 μL of 10 μM primer P (final volume 480 uL). For each individual second-round PCR reaction, combine 24 μL of this master mix with 1 μL of first-round PCR product. The final concentration in each reaction is 0.08 μM of ^{32}P-labeled EA-L2, 0.12 μM unlabeled EA-L2, and 0.2 μM of Primer P. Cycling conditions are as follows: initial denaturation step of 4 min at 94°C followed by 30 PCR cycles consisting of 45 s at 94°C and 1.5 min at 63°C. Perform a final extension step of 5 min at 72°C.
5. As an alternative to using the mixture of radiolabeled and unlabeled EA-L2 primer, primer EA-L2 can be fluorescently labeled and used directly at a final concentration of 0.2 μM in the second-round reaction (**Step 4**). This permits the

detection of breaks using automated DNA sequencing machines that allow DNA fragment analysis on many more samples to be run simultaneously (*see* **Note 4** and **Section 3.6.2**).

3.6. Restriction Enzyme Digestion of the Second-Round PCR Products and Detection of DNA Breaks

3.6.1. Radioactive Detection of DNA Breaks

1. For the analysis of the *Ea* recombination hot spot the second-round PCR products are analyzed with and without digestion by restriction enzyme *Bsl* I to distinguish between products derived from true breaks and those resulting from PCR artifact. Mix 2 μL of each second-round PCR product with 2 μL of a solution containing 2 × New England Biolabs Buffer 3 and four units of the restriction enzyme *Bsl* I (*see* **Note 6**). Mix another 2 μL from the same second-round PCR sample with 2 μL of the restriction enzyme digestion buffer lacking enzyme. Incubate both samples at 55°C for 2 h. Stop digestions by adding 16 μL of sample loading buffer (total volume of each sample = 20 μL).
2. Sizing the ^{32}P-labeled PCR products by polyacrylamide gel electrophoresis is performed as follows. Prepare the gel solution (6% acrylamide, 7.5 M urea) by mixing 15 mL of 40% acrylamide, 45 g of urea, 10 mL of 10 × TBE buffer, and deionized water to a final vol of 100 mL. Filter through a 0.45 μm NALGENE filter unit. Add 600 μL of 10% ammonium persulfate and 55 μL of TEMED and immediately pour the solution into a gel mold ~38 cm long by ~31 cm wide and 0.4 mm thick. Use two 20-sample shark-tooth combs to form 40 wells. Mix 5 μL of the stopped-enzyme digestion solution (**Step 1** above) with 11 μL of sample loading buffer and denature at 95°C for 10 min. Load 8 μL of this mixture into each well of the gel. Load a ^{32}P-labeled *Msp*I digest of pBR322 (prepared in the same way as for labeling primer EA-L2 and load 2–3 μL initially, more as the ^{32}P decays over a few weeks) as a molecular size marker in two lanes. For the *Ea* gene fragments run the gels at 1,000 V until the fastest running dye (bromophenol blue) is about 2 cm from the bottom of the gel (usually about 2 h). Autoradiography with freshly made probe usually takes about 2–3 d. Examples can be seen in **Fig. 11.3** as well as in *(9)*.

3.6.2. Fluorescence-Based Detection of DNA Breaks

Fluorescence-based detection can be performed using microcapillary electrophoresis. Numerous platforms for DNA sequencing and/or fragment analysis can be used. We provide no specific recommendations except to note that we used a CEQ2000 (or CEQ8000) machine with proprietary fragment analysis software (Beckman Coulter Inc, Fullerton, CA). Examples of results can be seen in **Fig. 11.4**. Because there are many manufacturers of such equipment which have different protocols we only provide general information on the details of this procedure (*see* **Note 7**).

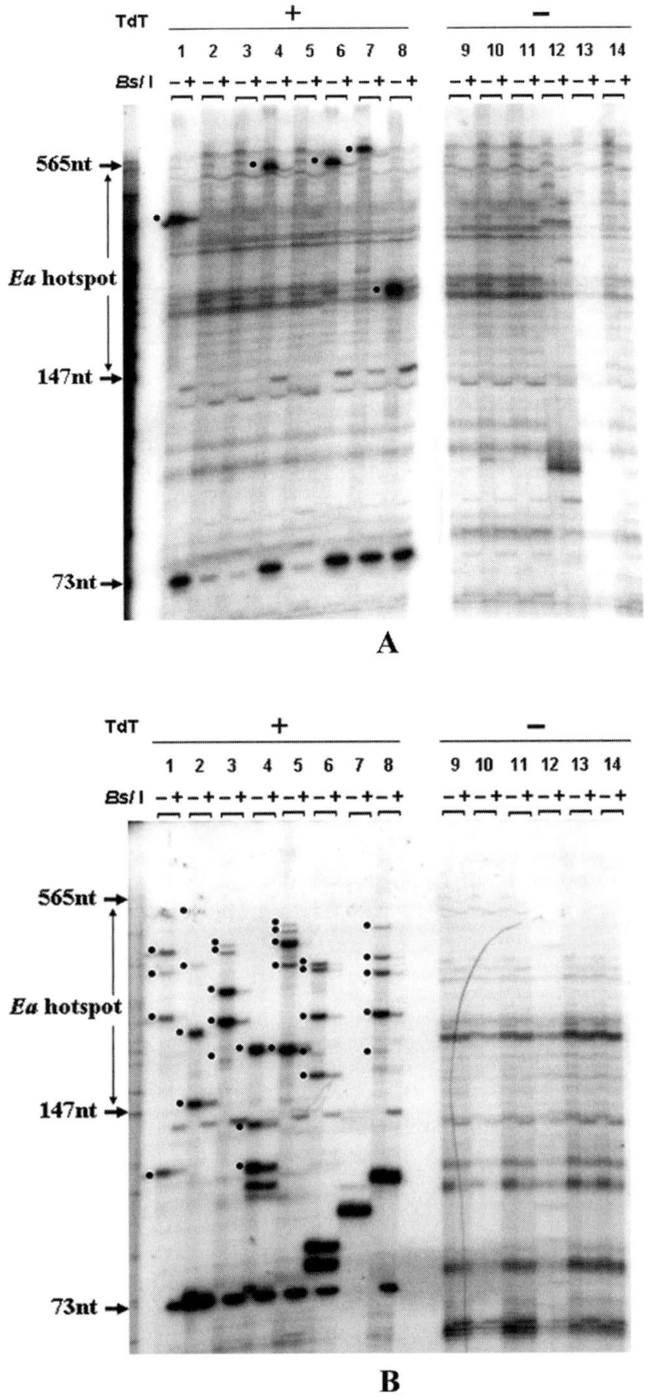

Fig. 11.3. Detection of DNA breaks in testis cells from strains 9R (panel A) and 13R (panel B). Representative data from a single experiment on a single mouse are shown in each case along with the no TdT control. In each autoradiograph, the boundaries of the recombination hot spot are indicated between two molecular weight markers. The lanes (each loaded with PCR product made using 10,000 cell equivalents of template) are arranged in pairs. The first member of the pair (–) contains second-round PCR products undigested by *Bsl* I. The second member of the pair (+) contains the same amplification product but after digestion with *Bsl* I. The 73 nt molecular weight marker shows the position of the labeled *Bsl* I digestion product derived from all DNA breaks at the *Ea* hot spot. The "thickness" of individual bands reflects not only the amount of PCR product produced, but the fact that DNA slippage can occur in the TdT-added G/C mononucleotide

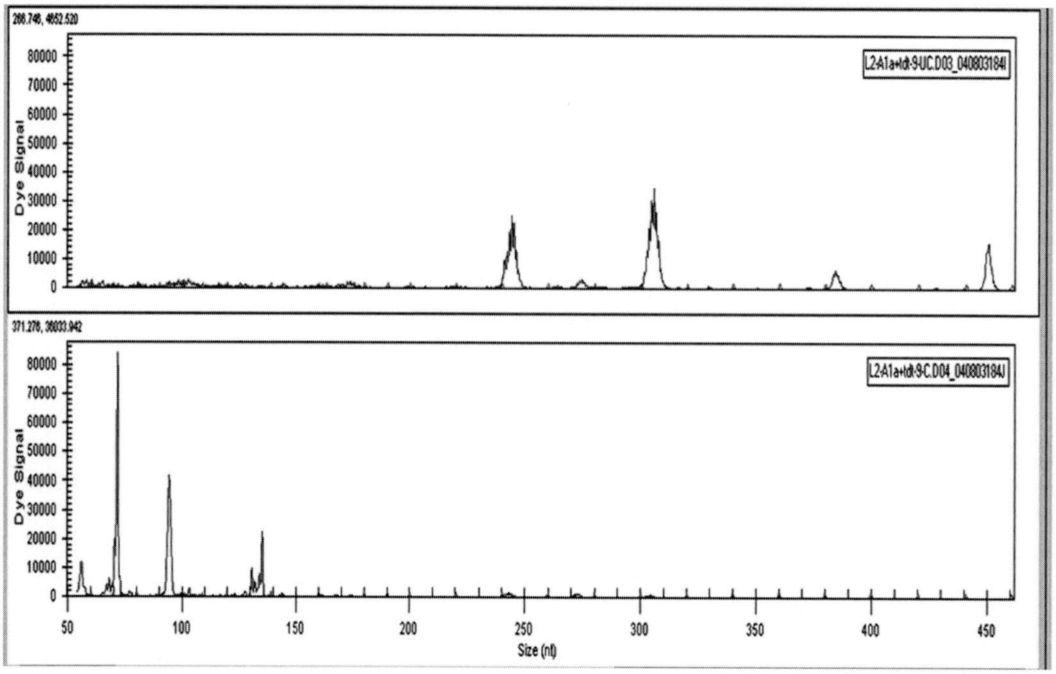

Fig. 11.4. Sample of *Ea* break data collected using a fluorescent dye-labeled EA-L2 primer and microcapillary electrophoresis on a CEQ 8000. In each panel the molecular size in nucleotides (x-axis) is plotted against the fluorescent dye signal (y-axis). The internal size markers (labeled with a dye different from the one used to detect the breaks) are not easily visible in this figure. The *upper panel* shows four peaks representing amplified DNA fragments taken from a second-round PCR product that was not digested with the restriction enzyme *BsI* I. Notice the stutter pattern (most easily seen in the two largest peaks). The *lower panel* shows the same sample but that was digested with *BsI* I. The peaks in the uncut panel were dramatically reduced by enzyme digestion. The expected digestion product (~73 nt) of *Ea* peaks was seen. The other bands in the cut panel were artifacts that were either smaller than 73 nt or were seen in all the other samples cut by the enzyme.

3.7. Analysis of Gel Electrophoresis Results

To consider a PCR product as coming from an authentic meiotic DNA break, three criteria need to be met. First, the size of the PCR product should be greater than or equal to a minimum size that represents the sum of the *Ea* second-round primer (EA-L2) size, the "C+P" primer size, the distance between the end of the EA-L2 binding site and the location of the chosen restriction enzyme digestion site (in our case *Bsl* I). Second, an authentic break molecule will contain the short mono-nucleotide repeat poly C:poly G (added by TdT).

Fig. 11.3. (continued) repeat tract. In each panel, a black dot is positioned next to each band that we consider to be an *Ea* specific break because the band in the "–" lane is cut down by *BsI* I (see "+" lane). Bands not labeled by a dot either are not cut by the enzyme or are also found in the no TdT control. Bands more than 565 nucleotides long were evaluated using the original autoradiographs (not shown) and not the figures as prepared for publication. Reprinted with permission of the ASM.

Therefore, the PCR product may contain minor "stutter" bands such that adjacent bands differ in size from one another by one nucleotide steps (*see* **Note 8**).

Third, all PCR products arising from mouse *Ea* templates with DNA breaks anywhere in the recombination hot spot area will be digested to a 73-bp fragment after *Bsl* I digestion since this site is located 73 bp from the 5′ end of EA-L2. Thus, any band greater than the expected minimum size in the uncut aliquot of a sample that is found to be digested to 73 bp in the cut aliquot qualifies as a site of a meiotic *Ea* DNA break [*see* **Note 9** and *(9)*].

The starting amount of DNA in each of the first-round PCRs is known based on the original cell counts (approximately 10,000 cell equivalents were added to each reaction). Thus the fraction of DNA breaks per cell in the *Ea* hot spot in each DNA sample can be estimated by counting the number of fragments in the "uncut" lanes that are digested by *Bsl* I digestion, and dividing this number by 10,000. This calculation assumes that all *Ea* breaks are detected with 100% efficiency and that each molecule with a break originally present in the 10,000 cell equivalent sample is also represented in the first- and second-round PCR products. Relative comparisons between the numbers of breaks detected in different mouse strains can be made without any efficiency assumptions. If desired, an estimate of the total number of genomes in a sample could be made using quantitative PCR with approximately two-fold precision. The positions of the breaks can be estimated with reasonably high resolution once adjustments are made for poly C:poly G length tract variation and the resolving power of the electrophoresis system. Second-round products can also be reamplified and sequenced *(9)*. Additional considerations are discussed in **Note 9**.

4. Notes

1. Four primers are needed for the analysis of breaks within any interval (*see* **Figs. 11.1 and 11.2** for primer placement and *(9)* for additional details). Two primers must be locus specific. The size of the interval can vary; we studied breaks in a region 400–500 bp long containing the *Ea* recombination hot spot. One primer is an extension primer, called C+P, with a polyC-containing 3′ end (used to target the sites of TdT addition of polyG tracts) and a 5′ end, called P, which contains an *E.coli* sequence identical to that of PCR primer P. The two additional primers restrict PCR product formation to the *Ea* gene. The EA-L1 PCR primer was used in the first amplification

round along with the P primer. EA-L2 and P were then used in a second round initiated by adding a small fraction of the first-round product as a template. EA-L2 was located between EA-L1 and the site of the DNA break tagged by the polyG tail.

C + P 5′ GTTAACCGCAACGTACCGTTGTTTGAG-CAGGCCCCCCCCCC 3′

P 5′ GTTAACCGCAACGTACCGTTGTTTGAGCAGG 3′

EA-L1 5′ GCGTTAAATGTGCTCAGAGACTGACA-GATGTGTG 3′

EA-L2 5′ CTGAGGGAGGAAAGCACGAGTG 3′

2. Plugs may be stored at this point at 4°C based on the experience of those who carry out pulse field gel electrophoresis. However, we have no first-hand knowledge of the effects of such storage in our assay and immediately went on to the next step.
3. TdT is a template-independent DNA polymerase capable of extending 3′ OH groups with an efficiency that depends on where the 3′ OH group is positioned (3′ overhang, blunt end, 3′ recessed end or internal nick), the base at the 3′ end, the length of the single strand, and the divalent cation used *(28–30)*. Here we insert a poly-dG tail to mark the sites of naturally occurring DNA strand breaks. Importantly, once the poly-dG tails are added to the DNA in the plug, any breaks incurred thereafter will not be detected by this assay. It is important to use a G tail since its final length seems to be quite limited compared to tails composed of C, T, or A *(29)*. For additional details see the New England Biolabs website .
4. In one analysis, we routinely studied 80,000 +TdT cell equivalents and 60,000 –TdT cell equivalents (a total of 14 primer extension/PCR samples) using a single polyacrylamide gel. The gel size was consistent with electrophoretic analysis of the 14 samples (14 uncut aliquots and 14 cut aliquots; see below) and the required molecular weight markers. Our choice of examining 10,000 total cell equivalents in each of the 14 individual samples was based on trying to limit the number of detectable DNA breaks in each lane of the gel. How many genomes can be used per extension/PCR reaction when other loci are examined will depend on the frequency of breaks in the region and needs to be established empirically.

   When using the fluorescent primer method to detect DNA breaks, more extension/PCR reactions can be set up per experiment. In the case of the CEQ machine we used at the time, 96 sequencing reactions were possible (each with its own internal molecular size marker). In addition to the six –TdT controls used in the polyacrylamide gel experiments, 42+TdT samples (420,000 cell equivalents) can be analyzed (all samples divided into an uncut and a cut aliquot = 96 loadings).

5. The melting temperature of the P primer should be matched with those of all locus specific primers; the P primer can easily be altered in length to affect its melting temperature. Because single molecule sensitivity must be achieved, the cycling conditions and primer concentrations must also be appropriate for the specific DNA sequence of interest and should be determined empirically.
6. If a different restriction enzyme is used, it is necessary to test whether the enzyme can cut the PCR product in the extension/PCR Reaction Buffer mix alone or when diluted with the appropriate restriction enzyme buffer. Otherwise some alternative approach (or enzyme) will be needed.
7. Restriction enzyme digestion is performed as in **Section 3.6.1, Step 1**. DNA size standards covering the appropriate length range and labeled with a different fluorophore (usually sold by the manufacturer of the sequencing system) are added to the sample before loading. In capillary electrophoresis, the total amount of sample plus size standards loaded in each capillary is critical to the quality of the separation. Two factors are most important. First, the amount of PCR product (and molecular size standards) loaded must not be less than needed to observe a signal or so great as to prevent an accurate distinction between adjacent peaks. Second, the ions present in the loaded sample (also containing PCR and restriction enzyme buffers) must not be so much as to reduce the efficiency of DNA loading. Since the PCR product is not radioactively labeled, desalting of PCR products can be easily accomplished by ethanol precipitation if the presence of excess salt interferes with the microcapillary electrophoresis step.

   In the beginning it might be wise to assess the amount of each second-round PCR product by running a small amount on a 2% agarose gel and staining with ethidium bromide. This will prevent microcapillary electrophoresis runs (which use expensive reagents) of a large number of samples that were not amplified successfully.

   The correct volume of sample to load can be worked out empirically. Dilute the PCR product with sample loading solution and load different amounts of PCR product (0.25 μL, 1 μL, and 2 μL) to find the amount needed to provide a fluorescent value within acceptable limits for the sequencing system.

   One of the advantages of microcapillary electrophoresis is that (depending on the particular machine available) 96 wells can be analyzed in an automated fashion. This means that 420,000 +TdT cell equivalents can be examined in one run instead of 80,000 in one gel. All sequencing machines need to have software available to estimate the size of the *Ea* DNA fragments based on the mobility of the molecular size standards.

   Loading samples into a 96-well plate in an eight microcapillary machine so that the cut and uncut aliquots of a sample are in adjacent columns would ensure that the same capillary is employed for running both the undigested and digested versions of the same PCR product, thereby controlling for possible inter-capillary variability.

8. A few PCR products with *Ea* breaks have been sequenced using reamplified, second-round PCR product bands extracted from a polyacrylamide gel *(9)*. Sequencing revealed a range of 8–13 guanines incorporated into the final PCR product.

9. There are additional factors that must be considered for analysis using a DNA sequencing machine. First, the size assigned by the software for the digested fluorescently labeled PCR product may differ from the expected size by a few nucleotides but will be consistent for all the samples. The assigned value may also change when a different batch of fluorescent primer is used.

   It is necessary to define certain rules for deciding whether a PCR product can be counted as a DNA break or not. Some of the rules that we used were: (A) The tallest peak is noted down as the size of the PCR product even though that peak may be part of a "stutter" band pattern. (B) Before comparing digested and undigested samples it is important to scale the y-axis (using the manufacturer's software). (C) The digested product peak size should be at least 50% less than the same peak in the run produced by undigested sample in order for it to be counted as an authentic break. Other DNA sequencing platforms might require other criteria.

Acknowledgments

The authors would like to thank Dr. Maria Jasin for providing the CHO cell lines that helped the development of this assay and Drs. Laura L. Richardson and Mary Ann Handel for providing the mouse strains carrying the *Ea* recombination hot spot. This work was supported in part by an NIH grant [NIGMS (GM36745)] to N.A.

References

1. Szostak, J. W., Orr-Weaver, T. L., Rothstein, R. J., and Stahl, F. W. (1983) The double-strand-break repair model for recombination. *Cell* **33**, 25–35.
2. Sun, H., Treco, D., Schultes, N. P., and Szostak, J. W. (1989) Double-strand breaks at an initiation site for meiotic gene conversion. *Nature* **338**, 87–90.
3. Sun, H., Treco, D., and Szostak, J. W. (1991) Extensive 3′-overhanging, single-stranded DNA associated with the meiosis-specific double-strand breaks at the *ARG4* recombination initiation site. *Cell* **64**, 1155–1161.
4. Keeney, S., Giroux, C. N., and Kleckner, N. (1997) Meiosis-specific DNA double-strand breaks are catalyzed by Spo11, a member of a widely conserved protein family. *Cell* **88**, 375–384.
5. Cao, L., Alani, E., and Kleckner, N. (1990) A pathway for generation and processing of double-strand breaks during meiotic

recombination in S. cerevisiae. *Cell* **61**, 1089–1101.

6. Cervantes, M. D., Farah, J. A., and Smith, G. R. (2000) Meiotic DNA breaks associated with recombination in *S. pombe*. *Mol. Cell* **5**, 883–888.
7. Romanienko, P. J., and Camerini-Otero, R. D. (2000) The mouse Spo11 gene is required for meiotic chromosome synapsis. *Mol. Cell* **6**, 975–987.
8. Keeney, S., Baudat, F., Angeles, M., Zhou, Z. H., Copeland, N. G., Jenkins, N. A., Manova, K., and Jasin, M. (1999) A mouse homolog of the *Saccharomyces cerevisiae* meiotic recombination DNA transesterase Spo11p. *Genomics* **61**, 170–182.
9. Qin, J., Richardson, L. L., Jasin, M., Handel, M. A., and Arnheim, N. (2004) Mouse strains with an active *H2-Ea* meiotic recombination hot spot exhibit increased levels of *H2-Ea*-specific DNA breaks in testicular germ cells. *Mol. Cell. Biol.* **24**, 1655–1666.
10. Guillon, H., Baudat, F., Grey, C., Liskay, R. M., and de Massy, B. (2005) Crossover and noncrossover pathways in mouse meiosis. *Mo.l Cell* **20**, 563–573.
11. Anderson, L. K., Reeves, A., Webb, L. M., and Ashley, T. (1999) Distribution of crossing over on mouse synaptonemal complexes using immunofluorescent localization of MLH1 protein. *Genetics* **151**, 1569–1579.
12. Froenicke, L., Anderson, L. K., Wienberg, J., and Ashley, T. (2002) Male mouse recombination maps for each autosome identified by chromosome painting. *Am. J. Hum. Genet.* **71**, 1353–1368.
13. Koehler, K. E., Cherry, J. P., Lynn, A., Hunt, P. A., and Hassold, T. J. (2002) Genetic control of mammalian meiotic recombination. I. Variation in exchange frequencies among males from inbred mouse strains. *Genetics* **162**, 297–306.
14. Kauppi, L., Jeffreys, A. J., and Keeney, S. (2004) Where the crossovers are: recombination distributions in mammals. *Nat. Rev. Genet.* **5**, 413–424.
15. Arnheim, N., Calabrese, P., and Tiemann-Boege, I. (2007) Mammalian meiotic recombination hot spots. *Annu. Rev. Genet.* **41**, 369–399.
16. Silver, L. M. (1995) Mouse genetics : concepts and applications, Oxford University Press, New York.
17. Petes, T. D. (2001) Meiotic recombination hot spots and cold spots. *Nat. Rev. Genet.* **2**, 360–369.
18. Lichten, M., and Goldman, A. S. (1995) Meiotic recombination hotspots. *Annu. Rev. Genet.* **29**, 423–444.
19. Paques, F., and Haber, J. E. (1999) Multiple pathways of recombination induced by double-strand breaks in *Saccharomyces cerevisiae*. *Microbiol. Mol. Biol. Rev.* **63**, 349–404.
20. Bellve, A. R. (1993) Purification, culture, and fractionation of spermatogenic cells. *Methods Enzymol.* **225**, 84–113.
21. Xu, F., and Petes, T. D. (1996) Fine-structure mapping of meiosis-specific double-strand DNA breaks at a recombination hotspot associated with an insertion of telomeric sequences upstream of the *HIS4* locus in yeast. *Genetics* **143**, 1115–1125.
22. Khambata, S., Mody, J., Modzelewski, A., Heine, D., and Passmore, H. C. (1996) Ea recombinational hot spot in the mouse major histocompatibility complex maps to the fourth intron of the *Ea* gene. *Genome Res.* **6**, 195–201.
23. Steinmetz, M., Uematsu, Y., and Lindahl, K. F. (1987) Hotspots of homologous recombination in mammalian genomes. *Trends Genet.* **3**, 7–10.
24. Shiroishi, T., Koide, T., Yoshino, M., Sagai, T., and Moriwaki, K. (1995) Hotspots of homologous recombination in mouse meiosis. *Adv. Biophys.* **31**, 119–132.
25. Lindahl, K. F. (1991) His and hers recombinational hotspots. *Trends Genet.* **7**, 273–276.
26. Birren, B. W., and Lai, E. H. C. (1993) Pulsed field gel electrophoresis : a practical guide, Academic Press, San Diego.
27. Sambrook, J., Fritsch, E. F., and Maniatis, T. (1989) Molecular cloning: a laboratory manual, Cold Spring Harbor Laboratory Press, Cold Spring Harbor, NY.
28. Deininger, P. L. (1987) Full-length cDNA clones: vector-primed cDNA synthesis. *Methods Enzymol.* **152**, 371–389.
29. Deng, G., and Wu, R. (1981) An improved procedure for utilizing terminal transferase to add homopolymers to the 3′ termini of DNA. *Nucleic Acids Res.* **9**, 4173–4188.
30. Kato, K. I., Goncalves, J. M., Houts, G. E., and Bollum, F. J. (1967) Deoxynucleotide-polymerizing enzymes of calf thymus gland. II. Properties of the terminal deoxynucleotidyltransferase. *J. Biol. Chem.* **242**, 2780–2789.

Chapter 12

End-Labeling and Analysis of Spo11-Oligonucleotide Complexes in *Saccharomyces cerevisiae*

Matthew J. Neale and Scott Keeney

Abstract

During meiosis Spo11 catalyzes the formation of DNA double-strand breaks, becoming covalently attached to the 5′ ends on both sides of the break during this process. Spo11 is removed from the DSB by single-stranded endonucleolytic cleavage flanking the DSB, liberating a short-lived species consisting of Spo11 protein covalently linked to a short oligonucleotide. The method presented here details how to detect these Spo11-oligo complexes in extracts made from meiotic yeast cells.

Key words: Spo11, meiosis, DNA double-strand break, end-labeling, terminal transferase.

1. Introduction

Meiotic recombination is initiated by DNA double-strand breaks (DSBs) created by the evolutionarily conserved protein, Spo11 *(1)*. Spo11 shares sequence similarity to subunit A of the type-II topoisomerase from Archaea, Topo VI, and is thought to perform DSB catalysis in a similar way. Specifically, Spo11 dimerization creates two intermolecular DNA cleavage sites each consisting of a Mg^{2+} coordination site (Toprim domain) and the 5Y-CAP domain where the catalytic tyrosine residue resides. Concerted DNA strand breakage creates a DSB with a two-nucleotide 5′ overhang with the 5′ ends covalently linked to the catalytic tyrosine present in the active site of each monomer (residue 135 in yeast). Mutation of this residue, or indeed others within the active site, abolishes or reduces DSB formation *(2, 3)*.

Scott Keeney (ed.), *Meiosis, Volume 1, Molecular and Genetic Methods, vol. 557*

DOI 10.1007/978-1-59745-527-5_12 Springerprotocols.com

Topo VI is an A_2B_2 heterotetramer, which functions in isolation. Spo11, on the other hand, requires at least nine other proteins of unknown stoichiometry to support DSB catalysis *(1)*. Most of these proteins have a poorly understood role in the catalytic process, and show poor evolutionary conservation [e.g., *(4)*]. Nevertheless, a pattern of essential interactions between them is slowly emerging *(4–6)*. Given that Spo11 shares homology with only the A subunit of Topo VI, it is possible that one or more of these proteins functions as an equivalent or substitute of the B subunit of Topo VI, which contains an essential ATPase domain.

In contrast to the transient lesions created by type II topoisomerases, Spo11-DSBs are not resealed by a reversal of the DNA cleavage reaction. Instead, Spo11-DSBs are channeled into a DNA repair pathway that involves $5' \rightarrow 3'$ single-stranded exonucleolytic resection, DNA strand invasion and extension of the exposed single-stranded ends, and finally Holliday junction resolution to yield intact, and recombinant, DNA molecules *(7)*. The first step in this processing pathway is the removal of the covalently bound Spo11 protein from the DSB end, a process that requires the Mre11 complex and the accessory protein, Sae2 *(1)*.

In cell lysates from both yeast and mouse, a fraction of Spo11 protein is detected covalently attached to short oligos that end with free 3′ OH termini *(8)*. Detection of these Spo11-oligo complexes is dependent on meiotic DSB formation and on the activity of the Mre11 complex and Sae2, mutation in either of which prevents release of Spo11 from DSB ends. In both organisms, two forms of the Spo11-oligo complexes are detected, differing in the length of the attached DNA. In yeast, the shorter form has oligos of less than 12 nt, whereas the longer form has oligos between 21 and 37 nt in length. In mouse, the length distribution of the shorter oligos is more heterogeneous (12–26 nt), whereas the larger form is more discrete (28–34 nt) *(8)*. In yeast, the two forms of Spo11-oligo complex are equally abundant, with kinetics of appearance and disappearance closely matching those of resected DSB molecules. Together these observations provide compelling evidence to support a model in which Spo11 is released from DSB ends via single-strand scissions flanking the DSB, a reaction that may be catalyzed by the Mre11 nuclease in combination with Sae2 *(8)*.

The relationship between the long and short forms of Spo11-oligo complex is not known. However, one idea, inspired by the observation that the two forms are equally abundant, is that the nicks are positioned asymmetrically flanking the DSB, thus yielding one longer and one shorter Spo11-oligo complex at every DSB (**Fig. 12.1**). Moreover, if the DSB end is duplex at the time when the releasing nicks are created, differential stability of the Spo11-oligo complexes on either side of the DSB (due to the extent of DNA base pairing) could create one free DSB end and one DSB

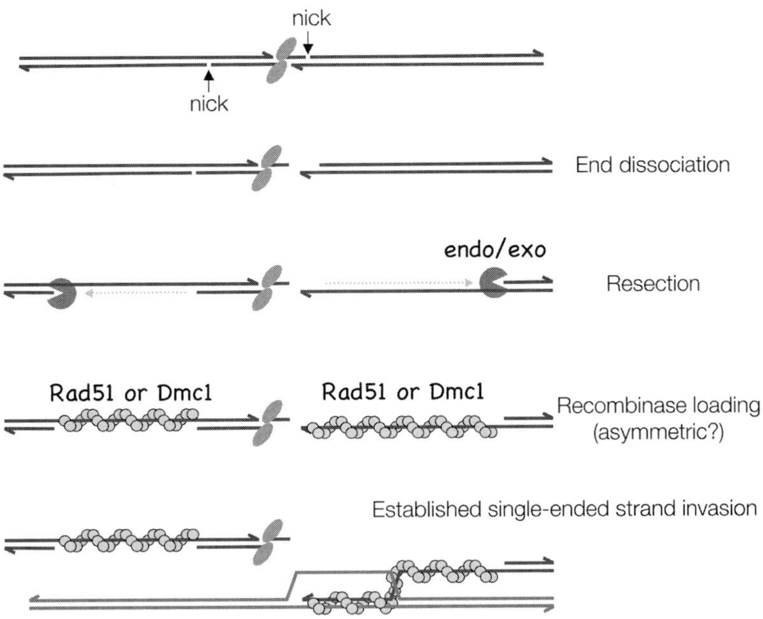

Fig. 12.1. Endonucleolytic release of Spo11 from DSB ends. Spo11 catalyzes DSB formation, becoming covalently attached to the 5′ ends at the break site. Single-stranded endonucleolytic cleavage flanking the Spo11-DSB permits release of Spo11 that is still covalently attached to a short oligonucleotide. Two equally abundant populations of Spo11-oligo complexes are detected, which differ in the length of the attached oligo, suggesting that the nicks could be positioned asymmetrically around each DSB. Adapted from **Ref.** *(8)*.

end blocked by the Spo11-oligo complexes. The lifespan of the Spo11-oligo complexes might then be expected to match that of unrepaired DSBs, as is observed. Direct physical evidence to support such a model is being sought.

Here, we present a method for detection of Spo11-oligo complexes from yeast (*see* **Fig. 12.2**). The methodology involves purifying Spo11 complexes from cell lysates by immunoprecipitation, followed by end-labeling any copurifying free 3′ OH DNA ends with radiolabeled nucleotide by using the template-independent polymerase, terminal deoxynucleotidyl transferase (TdT). End-labeled Spo11-oligo complexes are separated from any end-labeled DNA contaminants by fractionation on SDS-PAGE, and are detected by autoradiography. We prepare yeast lysates by breaking cells in the presence of the strong denaturant, trichloroacetic acid (TCA), and dissolving the resulting precipitate in Tris-buffered SDS. This method minimizes proteolysis, which is otherwise a significant problem when preparing meiotic protein extracts. However, care should be taken to keep temperatures low at all times, since the strongly acidic conditions can lead to reduction in Spo11-oligo labeling yield most likely due to acid-induced DNA depurination. The relative abundance of the Spo11-oligos appears to be a good measure of the relative frequency of DSB formation within the

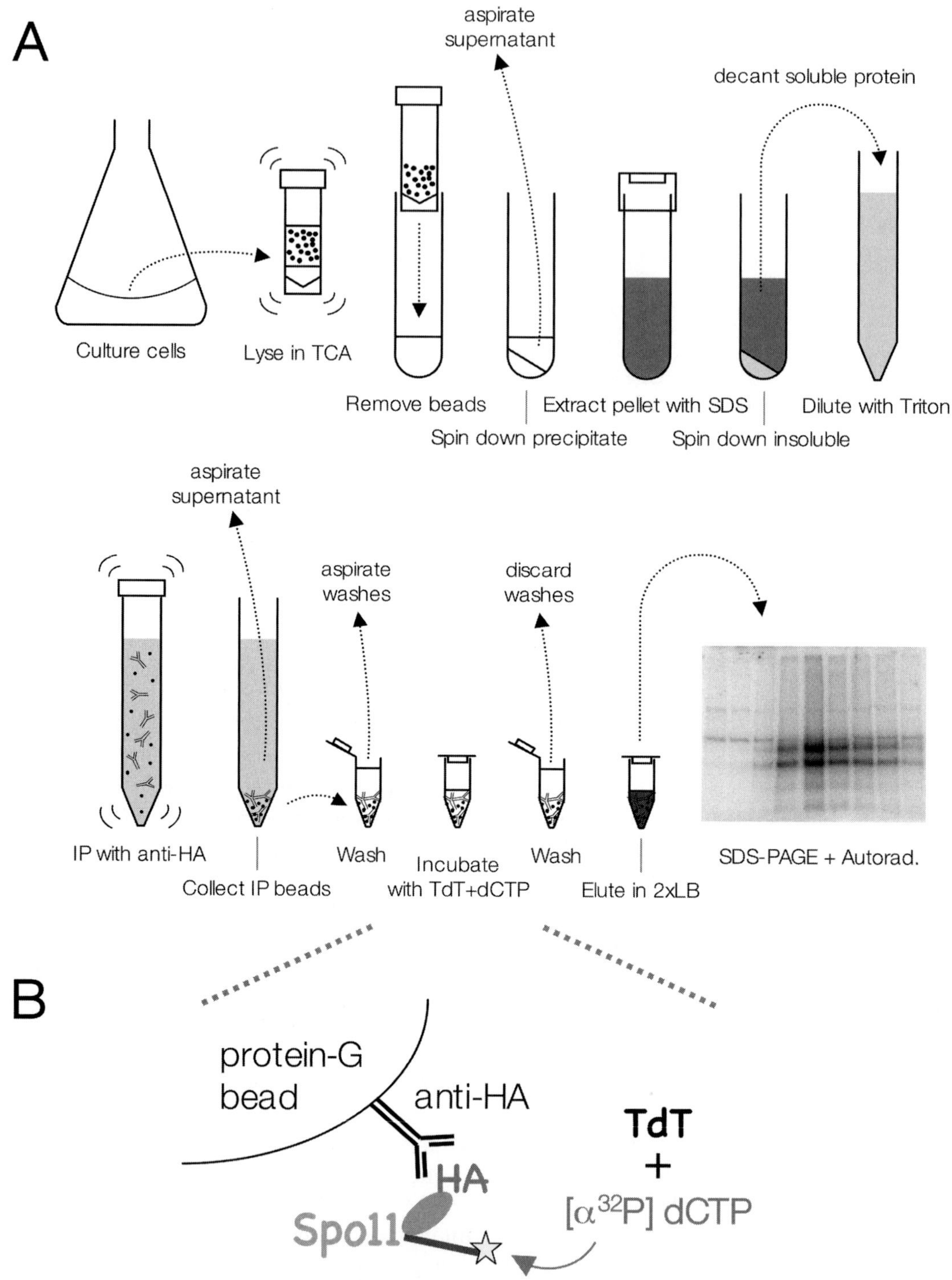

Fig. 12.2. Scheme for detecting Spo11-oligo complexes from meiotic yeast cells. (**A**) Meiotic cells are harvested and lysed in 10% TCA with zirconia beads. Precipitated protein in the extract is dissolved with 2% SDS and soluble protein diluted with Triton X100. Spo11-HA3His6 is immunoprecipitated with anti-HA antibody and protein-G-agarose beads. Immune complexes are washed, end-labeled using TdT and radioactive dCTP, then washed and eluted in gel-loading buffer. Spo11-oligo complexes are detected by SDS-PAGE and autoradiography [autoradiograph adapted from **Ref.** *(8)*]. (**B**) Schematic of the TdT end-labeling reaction.

cell. In principle, we believe that with suitable modification such methods could also be used to detect Spo11-oligo complexes in other organisms, and would thus provide a long-sought biochemical assay to measure DSB frequency.

2. Materials

2.1. Preparation of Denatured Yeast Cell Lysate

1. General yeast lab equipment, including: shaking incubator, temperature-controlled floor standing centrifuge and rotors, swinging bucket bench top centrifuge, vacuum aspirator, dry heat block, tube mixing platform, microcentrifuge, water bath.
2. Yeast growth media: YPD: 1% Yeast extract, 2% Peptone, 2% D-glucose. YPA: 1% Yeast extract, 2% Peptone, 2% potassium acetate, 0.001% antifoam (Sigma). SPM: 2% potassium acetate, 0.2 × supplements, 0.001% antifoam.
3. 100 × supplements: 1 mg/mL adenine, 1 mg/mL histidine, 3 mg/mL leucine, 1 mg/mL tryptophan, 1 mg/mL uracil. Dissolve in water and filter sterilize.
4. Bio-Spec mini bead beater 8 with 5-place, 7-mL tube holder, 7 mL polypropylene screw cap lysis tubes (Biospec Products Inc), and 0.5 mm zirconium/silica beads (BioSpec) (*see* **Note 1**).
5. 10% TCA. Dilute as needed from 100% stock.
6. Saturated bromophenol blue: approximately 0.5% bromophenol blue, dissolved and filtered.
7. β-mercaptoethanol (BME).
8. SDS extraction buffer: 2% SDS, 0.5 M Tris-HCl pH 8.1, 10 mM EDTA. Make up 1% of the volume with filtered saturated bromophenol blue (~0.005% final).

2.2. End-Labeling of Spo11-Oligo Complexes

1. 2 × IP buffer: 2% Triton X100, 30 mM Tris-HCl pH 8.1, 300 mM NaCl, 2 mM EDTA, 0.02% SDS. Prepare 1 × IP buffer by diluting with MilliQ water.
2. Protein-G-agarose beads (Roche).
3. Mouse monoclonal anti-HA antibody (clone F7, Santa Cruz) and mouse monoclonal anti-HA-HRP conjugated antibody (clone F7, SantaCruz).
4. 10 × TdT labeling buffer (500 mM potassium acetate, 200 mM Tris acetate, 100 mM magnesium acetate, 10 mM dithiothreitol, pH 7.9 at 25°C. Dilute to 1 × as required with MilliQ water. This 10 × buffer formulation is identical to 10 × NEB4 buffer (New England Biolabs).
5. dCTP (α-^{32}P, 6,000 Ci/mmol).

6. FPLC-pure terminal deoxynucleotidyl transferase (GE Healthcare).
7. 2.5 mM $CoCl_2$, dissolved in MilliQ water. Store at –20°C.
8. 2 × LB: 4% SDS, 100 mM Tris-HCl pH 6.8, 20% glycerol, 1 mM EDTA. Make up 10% of the volume with saturated bromophenol blue (~0.05% final).
9. Prestained protein molecular weight marker.
10. Standard SDS-PAGE running/transfer/gel drying apparatus and reagents (e.g. BioRad Mini-Protean).
11. Phosphorimaging system and screens (e.g. Fuji FLA7000), or film.
12. Whatman 3 MM paper.
13. DE81 paper (Schleicher and Schuell).
14. 1 × CAPS transfer buffer: 10 mM CAPS-NaOH pH 11, 10% methanol.
15. 1 × TBST: 25 mM Tris-HCl pH 7.5, 150 mM NaCl, 0.1% Tween-20.
16. Immobilon PVDF membrane (Millipore).
17. Powdered nonfat dry milk.
18. ECL+ reagent (GE Healthcare)

3. Methods

This protocol is optimized for sensitive detection of Spo11-oligo complexes from the synchronously sporulating *S. cerevisiae* strain, SK1. We use Spo11 that is fused at its C-terminus to the HA3-His6 epitope tag. In principle, we see no reason why it should not be possible to successfully detect Spo11-oligo complexes from other strain backgrounds or where Spo11 is fused to an alternative epitope tag, however, the efficiency of immunoprecipiation may require optimization. An outline of the various steps of the experiment is shown in **Fig. 12.2**.

3.1. Preparation of Denatured Yeast Cell Lysate

3.1.1. Preparing Cells for Synchronous Yeast Meiosis

1. From a –80°C glycerol stock, streak out yeast cells onto a YPD plate. Incubate 2 d at 30°C.
2. In a glass culture tube, inoculate 4 mL YPD liquid with a single colony taken from the YPD plate. Incubate 24 h at 30°C with shaking (250 rpm).
3. In a 2 L flask, inoculate 200 mL YPA to a cell density (OD_{600}) of 0.2 (this should require approx. 1:100 dilution of the YPD overnight culture). Incubate 14 h at 30°C with shaking (250 rpm; *see* **Note 2**).

4. Collect cells by centrifugation (3 min at 5,000*g*, 4°C) and drain supernatant. Wash with 200 mL water by sealing bottle and shaking vigorously. Pellet cells again, drain, and resuspend with 200 mL SPM prewarmed to 30°C. Return to a 2 L flask and incubate at 30°C with shaking (250 rpm).

3.1.2. Cell Lysis in TCA

1. Collect cells at 4 h in meiosis by centrifugation (3 min at 5,000*g*, 4°C) (*see* **Note 3**).
2. Resuspend pellet with 8 mL ice-cold water, and transfer equal amounts into two 7-mL lysis tubes.
3. Spin down cells (3 min at 5,000*g*) and aspirate supernatant. Cell pellets should be approximately 1 mL volume. Place tubes on ice.
4. Without disturbing the cell pellets, add an equal vol (1 mL) ice-cold 10% TCA. Pour chilled zirconia/silica beads into the tubes until the top of the beads is level with the *final* level of the TCA meniscus after it rises (approximately 2 mL of beads per tube). Tubes should be no more than half full.
5. Tightly secure tube caps, place into bead-beater and secure lid. Lyse for no longer than 30 s at full power. Quickly place tubes into an ice water bath and chill for 2 min, inverting a few times to speed the chilling process (*see* **Note 4**). Repeat this step twice more.
6. Invert lysis tubes. Flame a 25-gauge needle and pierce the bottom of one of the tubes 2–3 times. Push the pierced 7-mL lysis tube into the top of a 14-mL round bottom centrifuge tube and collect lysate by centrifuging for 1 min at 1,400*g* in a swinging bucket rotor. Return tube to ice.
7. Collect lysate from the second 7-mL tube in the same 14-mL round bottom tube by repeating **Step 6** (*see* **Note 5**).
8. Spin down precipitated protein 10 min, 16,000*g*, 4°C. Completely aspirate supernatant, and place tubes on ice. Proceed to the next section.

3.1.3. Protein Extraction from Denatured Yeast Lysate

1. Add 1 mL of SDS extraction buffer to the TCA precipitate. Break up the dense protein precipitate using a clean glass rod. Once a rough white paste is created, hold the glass rod loosely in your fingers and vortex the tube so that the rod spins inside the tube. A smooth paste should be created without significant foaming (*see* **Note 6**).
2. Add 1 mL of SDS extraction buffer and mix further with glass rod/vortex.
3. Add 4 mL of SDS extraction buffer, remove rod, and mix by gentle vortexing. Avoid foaming. The lysate should appear pale blue, homogeneous, and quite opaque.

4. Add 120 μL BME and mix.
5. Cap tube loosely, and heat for 5 min on a dry heat block at 95–100°C. Cool lysate on ice for 5 min.
6. Spin down insoluble cell debris 10 min, 16,000*g*, 4°C.
7. Pour supernatant into a 15-mL conical tube (approx. 6 mL total). Mix and place on ice (*see* **Note 7**).

3.2. End-Labeling of Spo11-Oligo Complexes

3.2.1. Immunoprecipitation of Spo11-Oligo Complexes

1. On ice, dilute 6 mL of cleared cell lysate with 6 mL of 2 × IP buffer and mix on a nutator for 10 min at 4°C (*see* **Note 8**).
2. Add 50 μL mouse monoclonal anti-HA antibody and 100 μL of protein-G-agarose bead suspension. Incubate overnight at 4°C on a nutator.
3. Collect immune complexes by gentle centrifugation (1 min at 500*g*). Aspirate supernatant.
4. Suspend beads with 1 mL of 1 × IP buffer and transfer to 1.5 mL Eppendorf tube. Gently pellet beads (1 min at 500*g*) and aspirate supernatant.
5. Wash beads with 1 mL of 1 × IP buffer, re-spin, and aspirate supernatant. Repeat.

3.2.2. End-Labeling of Spo11-Oligo Complexes Using TdT

1. Wash IP beads twice with 0.5 mL 1 × TdT labeling buffer, spin, and carefully remove supernatant (*see* **Note 9**).
2. Prepare TdT reaction buffer on ice as follows:

Component	Per reaction (μL)	Final concentration
10 × TdT labeling buffer	5	1 ×
2.5 mM $CoCl_2$	5	0.25 mM
MilliQ water	38	
TdT	1	
20 μCi dCTP (α-^{32}P, 6,000 Ci/mmol)	1	

3. Add 50 μL TdT reaction buffer to beads and mix. Incubate at 37°C for 60 min. Mix reaction periodically (*see* **Note 10**).
4. Pellet beads 10 s at 500*g*, and carefully remove reaction supernatant using a pipetman. This and subsequent wash supernatants will contain a lot of unincorporated radioactivity: discard as radioactive liquid waste (*see* **Note 11**).
5. Wash beads with 1 mL 1 × IP buffer. Pellet beads and remove

supernatant with pipetman. Discard with radioactive liquid waste. Alternatively, a vacuum aspirator dedicated to radioactive work may be used. Repeat this step twice.

6. Elute complexes by adding 50 μL 2 × LB and boiling for 5 min. Radiolabeled Spo11-oligo complexes can be stored at −20°C.

3.2.3. Detection of End-Labeled Spo11-Oligo Complexes

1. Pour a standard 1-mm-thick 7.5% SDS-PAGE gel, with 4% stacking gel. We use a premixed 30% acrylamide solution (29:1 acrylamide:bisacrylamide; BioRad).
2. Boil samples for 5 min, pulse spin, mix, re-spin, and load up to 20 μL of supernatant per lane. For best results, load into the bottoms of wells using gel-loading tips, and do not overload lanes. Load prestained marker to monitor band migration.
3. Run gel at 150 V for 60–70 min (approximately 5–10 min after dye front has migrated out of the gel, or until the 37 kDa marker is at the bottom of the gel).
4. Dismantle gel apparatus and discard the running buffer as radioactive liquid waste. Rinse gel with water and process for drying (**Step 5**) or transfer (**Step 6**).
5. Alternative 1: For drying, immerse gel in water, capture from beneath with a piece of Whatman 3 MM paper, and dry under vacuum over a sheet of DE81 paper at 80°C for 30 min.
6. Alternative 2: For transfer, soak gel 10 min in 1 × CAPS transfer buffer, and transfer onto PVDF membrane in 1 × CAPS transfer buffer. Transfer at 100 V for 60 min (approximately 300–500 mA) with buffer circulation and cooling. Following transfer, allow PVDF membrane to completely dry (about 15 min at room temperature).
7. Mark the position of protein standards with a radioactive pen. Wrap the dried gel or PVDF blot with plastic wrap and expose to a phosphor screen or to film (*see* **Note 12**). A sample labeling experiment is shown in **Fig. 12.3**.

3.2.4. Western Detection of Spo11 (Optional)

Spo11 can be detected on the PVDF membrane (**Section 3.2.3, Step 6**) using mouse monoclonal anti-HA-HRP conjugated antibody.

1. Rewet the membrane in methanol, rinse with 1 × TBST, and block for 15 min in 1 × TBST containing 5% nonfat dry milk.
2. Incubate the membrane for 30 min with mouse monoclonal anti-HA-HRP conjugated antibody diluted 1:10,000 in 1 × TBST containing 1% nonfat dry milk (*see* **Note 13**).
3. Wash the membrane three times with 1 × TBST for 10–15 min each.
4. Incubate the membrane with ECL+ reagent and expose to film (5–120 s).

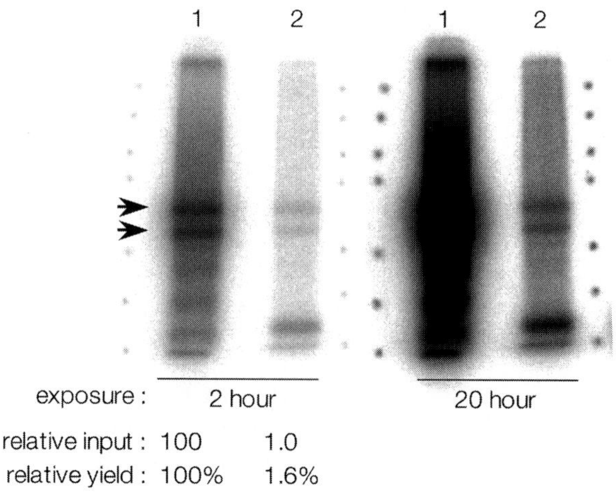

Fig. 12.3. Sample experimental result demonstrating sensitivity of the labeling assay. Lane 1 shows the signal obtained from a cell extract prepared from 200 mL of wild-type culture. Lane 2 shows the signal obtained from a mixture of extract from 2 mL of wild-type culture mixed with extract prepared from 198 mL of culture of a *spo11(Y135F)* mutant, which produces no DSBs and contains no Spo11-oligo species. The assay is sensitive enough to quantitatively detect 1% or less of the wild-type level of Spo11-oligo species. Two exposures are shown. In the longer (20 h) exposure, the signal in lane 1 is heavily saturated. Arrowheads mark Spo11-oligo species.

3.2.5. Sampling Multiple Time Points

Although the methods presented above can be used to sample multiple time points by starting with a large (e.g., 1 L) culture, it is more convenient to sample smaller aliquots of cells from a 200 mL culture. The procedure works well (although with reduced sensitivity) starting with 20 mL aliquots of meiotic cell culture, and as such, permits performing cell lysis and the entire immunoprecipitation/labeling protocol in 1.5-mL tubes. All lysis and immunoprecipitation reagent volumes should be scaled accordingly, and extra care should be taken to keep things cold during cell lysis. Reagent volumes used for the radiolabeling steps should remain unchanged. Expect sensitivity to be reduced ~10-fold.

3.3. Potential Problems

Failure to keep the cell lysate cool during TCA lysis can dramatically affect the detection of Spo11-oligo species. **Figure 12.4** illustrates this point (compare lanes 1 and 2 with lanes 3 and 4). On the basis of Western blot analysis of the recovery of Spo11 protein (data not shown), the reduced signal in lanes 3 and 4 does not appear to be a consequence of increased proteolysis. We surmise that acid-induced depurination, which would be greater at elevated temperature, renders the Spo11-associated oligo refractory to extension by TdT. Thus, it is important to use short bursts in the bead beater (<30 s) and to chill on ice in between.

Different TdT preparations contain different protein contaminants that can be labeled with TdT and α-^{32}P-labeled nucleotide. This point is also illustrated in **Fig.12.4**, where parallel labeling

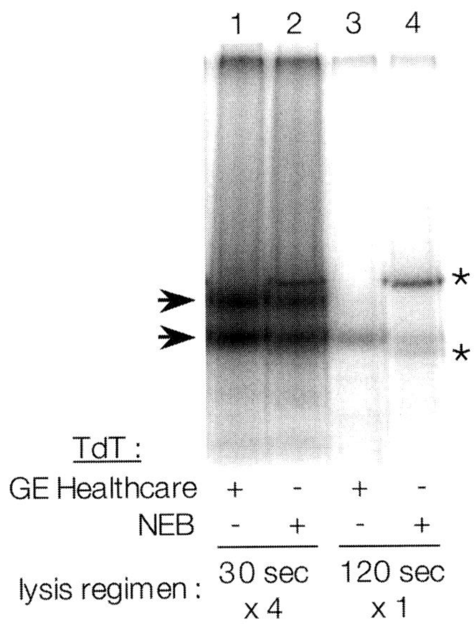

Fig. 12.4. Potential problems. Identical cell pellets were lysed with four bursts of 30 s each with chilling on ice ($\geq$ 2 min) between bursts (lanes 1, 2), or were lysed continuously for 2 min in the bead beater at 4°C, but with no other cooling (lanes 3, 4). Subsequent immunoprecipitation and labeling steps were performed in parallel as described in this protocol. Spo11-oligo species are marked with arrowheads. Overall signal was reduced >10-fold in lanes 3 and 4, with complete loss of the upper Spo11-oligo band. In lanes 1 and 3, end-labeling with [α-^{32}P]dCTP was carried out using FPLC-pure TdT from GE Healthcare, whereas labeling in lanes 2 and 4 was carried out with recombinant TdT from New England Biolabs. The asterisks mark end-labeled contaminants from the latter preparation. See text (**Section 3.3**) for further discussion.

was carried out with TdT from different commercial sources (compare lanes 1 and 3 with lanes 2 and 4). The source of these labeling artifacts is not known, although we speculate that the contaminant in the New England Biolabs preparation is a covalent complex of *E. coli* topoisomerase I and a short oligo (the TdT enzyme was expressed in recombinant form and purified from *E. coli*). These results show that care should be exercised in choosing a source of TdT.

4. Notes

1. If you do not have access to a dedicated bead-beating apparatus, cell lysis can also be effected by vortexing vigorously in glass-walled tubes with zirconium beads. Perform multiple rounds of vortexing, 30 s at a time, with chilling on ice for >1 min in between. To assess cell lysis efficiency, *see* **Note 5**. Continue until >95% of cells appear lysed.

2. After 14 h of growth, cell density should be ~5 × 10^7 cells/mL (OD_{600} of a 10-fold dilution should be 0.3–0.6), with short chains of 5–10 enlarged, round cells with few (<10%) small buds. If cell density is significantly different, scale all subsequent volumes accordingly.
3. Four hours normally represents the peak of both DSB formation and Spo11-oligo abundance. The protocol can be appropriately scaled to sample a culture at multiple time points.
4. It is essential not to allow the tubes to heat up significantly during the cell lysis step as high temperature reduces the observed signal, a process we believe to be due to acid-catalyzed depurination of the Spo11-oligo chain (*see* **Fig. 12.4**).
5. Cell lysis efficiency can be checked by mixing 1 μL of TCA lysate with 10 μL of SDS extraction buffer on a microscope slide. Lysed cells appear as translucent ghosts and cell wall fragments. Intact cells will be phase dark. Lysis efficiency is usually >95%.
6. The addition of too much buffer too soon can lead to partially solubilized protein aggregates that are subsequently difficult to completely resuspend, reducing yield.
7. Cleared lysate can be stored at –20°C but should be reheated briefly and, if necessary, re-clarified by centrifugation prior to continuing with the immunoprecipitation.
8. It is possible to detect Spo11-oligo complexes from as little as 60 μL of cell lysate (equivalent to just 2 mL of meiotic culture; *see* **Fig. 12.3**). However, when performing this assay for the first time, or when assaying strains expected to contain fewer Spo11-oligo complexes, we suggest using 3–6 mL of lysate per assay as described here. When performing the assay for the first time, it may be useful to use half of the lysate for a mock (no antibody) control. If using less than 6 mL of lysate, scale reagent volumes proportionally in **Section 3.2.1**, **Steps 1** and **2**. In the original protocol *(8)*, the cell lysate was diluted 10-fold to final concentrations of 0.2% SDS, 1% Triton. However, for immunoprecipitation of Spo11-HA3His6 with anti-HA, we find that just 2-fold dilution (to 1% SDS, 1% Triton) is sufficient to achieve the same or higher yield, most likely due to the increased concentration of antibody and protein-G beads. Optimization of the immunoprecipitation (particularly the extent of dilution) will be necessary if attempting to use alternative antibody/affinity tag combinations.
9. When in 1 × TdT labeling buffer (1 × NEB4), the beads are easy to disturb during aspiration. We recommend using an aspirator tipped with an ultrafine sequencing gel-loading tip. A second wash may not be necessary if: (a) the volume of

protein-G-beads used is low (due to scaling down the immunoprecipitation), and (b) care is taken to remove all supernatant after the first wash.

10. We generally mix the reactions about every 15 min by holding the tube in a pair of long tweezers and agitating gently for a few seconds. The signal can be increased by incubation for longer than 60 min (e.g., 2–4 h). For accurate sizing of oligo complexes, use a chain-terminating nucleotide such as cordycepin triphosphate.
11. Placing the reaction tube in a small hand-held acrylic block during this and the subsequent wash steps minimizes exposure to the radioactive buffer. A six-place micro-centrifuge dedicated to radioactive work is also useful.
12. Specific labeling should be apparent when monitoring the gel or blot with a hand-held detector, and give readily detectable bands with 1–2 h exposure to a phosphor screen. For accurate quantification, vary the length of exposure to achieve an image with the best dynamic range without saturation.
13. Optimal antibody concentration will depend on the scale of the immunoprecipitation. Use of an HRP-conjugated antibody prevents cross-reaction with the heavy chain of the antibody used for immunoprecipitation, which migrates at the same position as Spo11-HA3His6.

References

1. Keeney, S. (2001) Mechanism and control of meiotic recombination initiation. *Curr. Top. Dev. Biol.* **52**, 1–53.
2. Bergerat, A., de Massy, B., Gadelle, D., Varoutas, P. C., Nicolas, A., and Forterre, P. (1997) An atypical topoisomerase II from Archaea with implications for meiotic recombination. *Nature* **386**, 414–417.
3. Diaz, R. L., Alcid, A. D., Berger, J. M., and Keeney, S. (2002) Identification of residues in yeast Spo11p critical for meiotic DNA double-strand break formation. *Mol. Cell. Biol.* **22**, 1106–1115.
4. Maleki, S., Neale, M. J., Arora, C., Henderson, K. A., and Keeney, S. (2007) Interactions between Mei4, Rec114, and other proteins required for meiotic DNA double-strand break formation in *Saccharomyces cerevisiae*. *Chromosoma* **116**, 471–486.
5. Arora, C., Kee, K., Maleki, S., and Keeney, S. (2004) Antiviral protein Ski8 is a direct partner of Spo11 in meiotic DNA break formation, independent of its cytoplasmic role in RNA metabolism. *Mol. Cell* **13**, 549–559.
6. Li, J., Hooker, G. W., and Roeder, G. S. (2006) *Saccharomyces cerevisiae* Mer2, Mei4 and Rec114 form a complex required for meiotic double-strand break formation. *Genetics* **173**, 1969–1981.
7. Neale, M. J. and Keeney, S. (2006) Clarifying the mechanics of DNA strand exchange in meiotic recombination. *Nature* **442**, 153–158.
8. Neale, M. J., Pan, J., and Keeney, S. (2005) Endonucleolytic processing of covalent protein-linked DNA double-strand breaks. *Nature* **436**, 1053–1057.

Chapter 13

Detection of SPO11-Oligonucleotide Complexes from Mouse Testes

Jing Pan and Scott Keeney

Abstract

The SPO11 protein generates programmed DNA double-strand breaks (DSBs) that initiate meiotic recombination. Endonucleolytic cleavage 3′ to the DSB sites releases SPO11 from DNA, leaving SPO11 covalently associated with an oligonucleotide. This chapter describes detection of the release product, SPO11-oligonucleotide complexes, from mouse testis lysates. The method for determining the size of SPO11-associated oligonucleotides is also provided.

Key words: SPO11, mouse, testes, terminal transferase, DNA double-strand break.

1. Introduction

To detect mouse SPO11-oligonucleotide complexes, SPO11 is immunoprecipitated from testis lysates and immunoprecipitates are labeled with terminal transferase (TdT) in the presence of radioactive nucleotides. This protocol is largely derived from the corresponding protocol from budding yeast (**Chapter 12** in this volume). Unlike synchronous meiotic cultures of yeast, only a small fraction of cells in adult testes are spermatocytes in early meiotic prophase *(1)*, the stage in which SPO11-oligonucleotide complexes are likely to be present based on results from budding yeast *(2)*. In addition, the amount of SPO11-oligonucleotide complexes is likely to be comparable per meiotic cell in mouse and budding yeast due to similar numbers of SPO11-induced double-strand breaks *(3)*, but the mouse genome ($\sim 2.5 \times 10^9$ bp) is about 200 times the size of the budding yeast genome ($\sim 1.2 \times 10^7$ bp). The small amount of SPO11-oligonucleotide complexes versus the large amount of genomic DNA in mouse poses a challenge for

Scott Keeney (ed.), *Meiosis, Volume 1, Molecular and Genetic Methods, vol. 557*
© Humana Press, a part of Springer Science+Business Media, LLC 2009
DOI 10.1007/978-1-59745-527-5_13 Springerprotocols.com

detection. To increase the signal-to-noise ratio, a simple step of ultracentrifugation is used to pellet most genomic DNA from crude testis lysates, and SPO11-associated oligonucleotide complexes are then immunoprecipitated from the supernatant. To determine the size of SPO11-associated oligonucleotides, SPO11-oligonucleotide complexes are purified using SDS-PAGE and then deproteinized. DNA is then precipitated and resolved on a sequencing gel.

2. Materials

2.1. Preparation of Testis Lysates

1. 4 1/2" and 3 1/2" dissecting scissors, two fine-pointed dissection forceps.
2. PBS: 137 mM NaCl, 2.7 mM KCl, 10 mM NaH_2PO_4 / Na_2HPO_4, pH 7.4.
3. Lysis buffer: 1% Triton X-100, 400 mM NaCl, 25 mM HEPES-NaOH, pH 7.4, 5 mM EDTA. Supplement with Complete, Mini, EDTA-free protease inhibitor cocktail (Roche; use one tablet per 10 mL lysis buffer) and 2 mM dithiothreitol immediately before use.
4. Plastic pestles to fit 1.5 mL centrifuge tubes.
5. Benchtop ultracentrifuge (or floor ultracentrifuge for large scale).
6. Beckman TLA100.2 rotor or equivalent for 1-mL capacity ultracentrifuge tubes (or for large scale, Sorvall AH-650 rotor or equivalent for 3.5 mL or 5 mL ultracentrifuge tubes).
7. 1 mL-capacity polycarbonate ultracentrifuge tubes (or for large scale, 3.5-mL or 5.5-mL capacity ultracentrifuge tubes).

2.2. Immunoprecipitation

1. Antibody against mouse SPO11 (Kamiya Biomedical Company, clone 129/180).
2. Protein A agarose beads (Roche).
3. Wash buffer: 1% Triton X-100, 150 mM NaCl, 15 mM Tris-HCl, pH 7.4.

2.3. TdT Labeling

1. Terminal transferase (FPLC*pure* from GE Healthcare, or recombinant from New England Biolabs).
2. 10 × Labeling buffer (equivalent to 10 × NEB4 from New England Biolabs): 20 mM Tris-acetate, pH 7.9, 50 mM potassium acetate, 10 mM magnesium acetate, 1 mM dithiothreitol.
3. [α-^{32}P] dCTP (6,000 Ci/mmol, 20 mCi/mL).

2.4. Detection of Labeled Mouse SPO11-Oligonucleotide Complexes

1. 2 × SDS-PAGE sample buffer: 100 mM Tris-HCl, pH 6.8, 4% SDS, 20% glycerol, 10% β-mercaptoethanol, 10% saturated bromophenol blue.
2. SDS-PAGE reagents and apparatus.
3. PVDF or nitrocellulose membrane.

4. CAPS buffer: 10 mM cyclohexylaminopropane sulfonic acid, pH 11, 10% methanol. Adjust 100 mM CAPS stock to pH 11 with NaOH. To make CAPS buffer, mix 1 vol of 100 mM CAPS stock, 1 vol of methanol, and 8 vol of water.
5. Phosphor imaging plate and analyzer.

2.5. Detection of SPO11 Protein by Western Blotting

1. PBST: PBS containing 0.1% Tween 20.
2. Blocking solution: 2% BSA and 0.05% sodium azide in PBST.
3. Horse radish peroxidase (HRP) -conjugated anti-mouse secondary antibody (GE Healthcare or equivalent).
4. ECL plus Western blotting detection system (GE Healthcare).

2.6. Determination of the Length of SPO11-Associated Oligonucleotides

1. Low-retention centrifuge tubes, 2 mL and 1.5 mL.
2. Protease K elution mix: 50 mM Tris-HCl, pH 8.0, 0.5% SDS, 1 mM EDTA, 1 mM $CaCl_2$, 100 μg/mL protease K.
3. 9 M ammonium acetate.
4. Glycogen.
5. Absolute ethanol and 70% ethanol (kept at –20°C).
6. Sequencing gel loading buffer: 80% deionized formamide, 10 mM EDTA, pH 8.0, 0.5 mg/mL xylene cyanol FF, 10% saturated bromophenol blue.
7. Sequencing gel reagents and apparatus.
8. 10 bp DNA ladder (Invitrogen).
9. T4 polynucleotide kinase.
10. [γ-^{32}P] ATP (3,000 Ci/mmol, 10 mCi/mL).
11. 0.4 M dithiothreitol.
12. Sequencing gel fixing solution: 10% methanol, 7% acetic acid in water.
13. Cellulose paper, DE81 paper.

3. Methods

3.1. Preparation of Testis Lysates

1. Prepare four wild-type mice (for positive and negative controls) and two mice from other strains of interest (*see* **Note 1**). Sacrifice animals using carbon dioxide and collect testes in ice-cold PBS.
2. Rinse testes with PBS a few times. Tear apart and remove the outer membrane (*tunica albuginea*) using two fine-pointed dissection forceps. Drop de-capsulated testes into 1.5 mL tubes placed on ice, one testis per tube.
3. Add 450 μL cold lysis buffer to each tube. Grind testes with plastic pestles. About half a minute of grinding is needed to obtain a viscous crude lysate without visible cell clumps (*see* **Note 2**).

4. Immediately transfer crude lysates into cold, 1-mL capacity ultracentrifuge tubes. Each tube holds lysates from two testes. Use P1000 pipette tips with about 3 mm removed from ends to pipette the viscous lysates.
5. Balance the centrifuge tubes using lysis buffer. Spin the crude lysates at 100,000 rpm (355,040*g*) in a TLA 100.2 rotor (or equivalent) on a benchtop ultracentrifuge for 15 min at 4°C. Pellets contain insoluble material and most of the genomic DNA. Carefully decant supernatants into 15 mL tubes, one strain per tube (*see* **Note 3**).

3.2. Immunoprecipitation

1. Divide cleared lysate from wild-type animals equally into two 15 mL tubes. To one tube of wild-type lysate, add 40 μL PBS (mock immunoprecipitation as the negative control). To the other tube of wild-type lysate (positive control), add 40 μL SPO11 antibody (200 μg/mL, use at 2 μg/testis). Also add 40 μL SPO11 antibody to lysates from other strains. Incubate all reactions at 4°C with end-over-end rotation for 1 h (*see* **Note 4**).
2. Add 60 μL slurry of protein A agarose beads (15 μL slurry/testis) to all reactions (*see* **Note 5**). Incubate at 4°C for 3 h to overnight with end-over-end rotation.
3. Spin beads down at 3,000 rpm (~1,700*g*) for 3 min in a table-top clinical centrifuge. Remove supernatants.
4. Add 0.5 mL cold wash buffer to each tube. Transfer resuspension into 1.5 mL tubes. Spin beads down in a table-top microcentrifuge at 5,000 rpm (~ 2,300*g*) for 1 min. Remove supernatants. Wash beads three times more with cold wash buffer.

3.3. TdT Labeling

1. Wash beads twice with 300 μL cold 1 × labeling buffer. After the last wash, remove as much liquid as possible without disturbing the beads using a P20 pipette.
2. Set up TdT reaction mix. Use fresh [α-^{32}P] dCTP. (If the size of SPO11-associated oligonucleotides is to be determined (**Section 3.6.**), use radioactive cordycepin triphosphate for labeling. *See* **Note 6**.)

	1 × reaction mix (μL)
10 × labeling buffer	5
[α-^{32}P] dCTP	2
TdT (~20,000 U/mL)	1
Water	42
Total volume	50

3. Add 50 μL TdT reaction mix to each tube. Incubate at 37°C for 45 min. Mix reactions every 5 min.
4. Spin beads down. Remove supernatants. Wash beads three times with 0.5 mL cold wash buffer. Remove as much liquid as possible with a P20 pipette after the last wash.

3.4. Detection of Labeled Mouse SPO11-Oligonucleotide Complexes

1. Add 30 μL of 2 × SDS sample buffer to each tube. Mix well, incubate at 95°C for 2 min, and chill immediately on ice.
2. Resolve samples on an 8% SDS-PAGE gel. For each sample, load 20 μL of supernatant excluding beads (about 2.6 testes equivalent). Also load a lane of prestained protein size standard. Stop the run when the bromophenol blue dye front has just migrated out of the gel (about 1 h 8 min at 150 V on a Bio-Rad Mini-protean electrophoresis apparatus).
3. Transfer the gel to a PVDF or nitrocellulose membrane in CAPS transfer buffer (1 h at 100 V at 4°C on a Bio-Rad Mini-protean electrophoresis apparatus for a 0.75 mm gel) (*see* **Note** 7).
4. Cover the membrane with plastic wrap. Put a piece of Scotch tape over the protein size standard. Make a hot pen by dipping a Sharpie marker into radioactive liquid waste, and use the pen to mark the size standard on the tape. Cover the radioactive marks with another piece of Scotch tape. (If the region containing SPO11-oligonucleotide complexes is to be excised for sizing the oligonucleotides (**Section 3.6.**), also mark the four corners of the membrane with the hot pen for orientation purposes).
5. Expose the membrane to a phosphor imaging plate from 4 h to overnight. SPO11-oligonucleotide complexes from wild-type mice migrate between 47–59 kD (**Fig. 13.1**) (*see* **Notes 8** and **9**).

3.5. Detection of SPO11 Protein by Western Blotting (skip this section and go directly to Section 3.6. if the size of SPO11-oligonucleotide is to be determined.)

1. Unwrap the membrane and incubate it with blocking solution on a rocking platform for 30 min at room temperature.
2. Pour off blocking solution and replace with SPO11 antibody diluted 1:500 in blocking solution. Incubate on a rocking platform at room temperature for 1 h.
3. Wash the membrane three times with PBST, 5 min per wash.
4. Incubate the membrane with HRP-conjugated anti-mouse secondary antibody (for the antibody from GE Healthcare, dilute 1:10,000 in PBST containing 5% nonfat dry milk) for 30 min at room temperature.

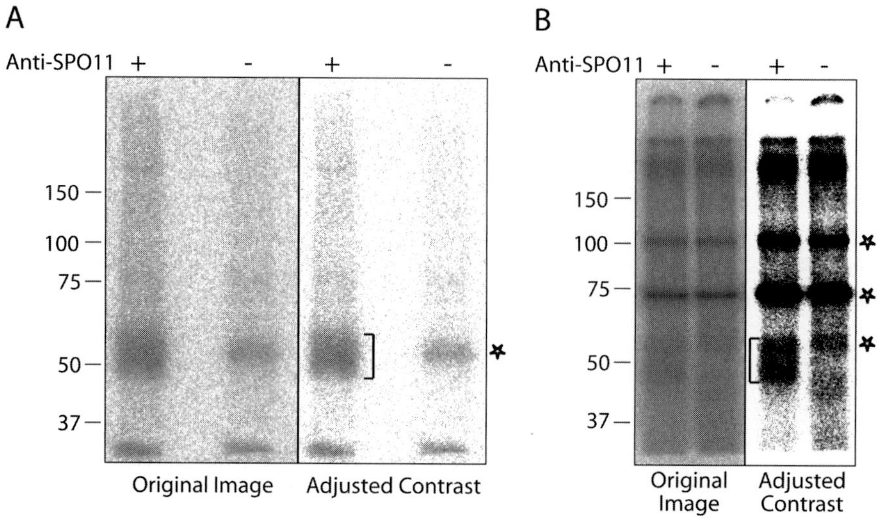

Fig. 13.1. Examples of labeling of mouse SPO11-oligonucleotide complexes. SPO11-oligonucleotide complexes were immunoprecipitated from wild-type testis lysate and labeled with TdT and [α-^{32}P] dCTP. Labeling specific to SPO11-oligonucleotide complexes is marked with brackets, and nonspecific labeling is marked with asterisks. In both A and B, the *left panel* is the original image and in the *right panel* contrast has been adjusted with Photoshop. A. Labeling reactions were resolved on a 7.5% Ready gel (Bio-Rad). Nonspecific labeling at around 55 kD in the lane without SPO11 antibody (mock immunoprecipitation control) is associated with contaminant(s) from the TdT (GE Healthcare). Input to lanes: about 2.6 testes equivalent. B. Labeling reactions were resolved on an 8% SDS gel. Nonspecific bands at 75 and 100 kD, as well as the smear at about 60 kD in the lane without SPO11 antibody is associated with contaminants from the TdT (New England Biolabs). Input to lanes: about 3 testes equivalent.

5. Wash the membrane three times with PBST, 5 min per wash.
6. Detect with ECL+ Western blotting detection system according to manufacturer's instructions. Mouse SPO11 is expressed as multiple isoforms *(4, 5)*. Western blotting detects two bands of free SPO11 migrating at about 40 and 44 kD, respectively (**Fig. 13.2**).

3.6. Determining the Length of SPO11-Associated Oligonucleotides

1. Print out the image obtained from exposure to the phosphor imaging plate at 100% size. On a light box, align the Sharpie marks on the membrane with the marks on the print-out. Draw a box on the membrane around the regions that contain the SPO11-oligonucleotide complexes, and the corresponding region in the control lane.
2. Cut out boxed regions, and put these small pieces of membrane into 2 mL low-retention centrifuge tubes, one strain per tube. Add 0.5 mL protease K elution mix to each tube. Seal the tubes with Parafilm. Elute at 50°C with mixing overnight (*see* **Note 10**).

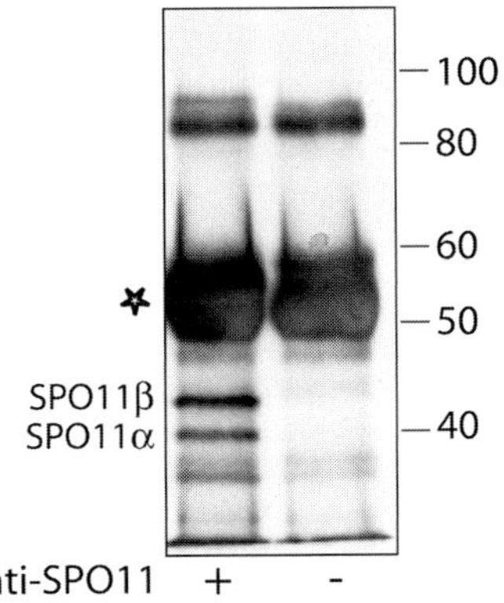

Fig. 13.2. SPO11 protein detected by Western blotting. Anti-SPO11 or mock immunoprecipitates from wild-type testis lysates were probed with SPO11 antibody. Input to lanes: ~2.6 testes equivalent. The antibody heavy chain is marked with an asterisk.

3. Take the membrane out of the tubes using clean forceps. Add 150 μL 9 M ammonium acetate, 1 μL 10 mg/mL glycogen, and 1.25 mL absolute ethanol to each tube. Precipitate on dry ice for 1 h. Spin at 13,200*g* in a microcentrifuge at 4°C for 15 min. Rotate tubes 180°C and spin for another 15 min. Rotate tubes 180°C again and spin for the last 15 min to collect most precipitates at the bottom of the tubes. Remove supernatants very carefully with a P1000 pipette. Add 250 μL 70% ethanol (−20°C) to each tube, spin at 13,200*g* in a microcentrifuge at room temperature for 1 min. Remove supernatants very carefully. Air dry pellets completely.
4. Add 20 μL of sequencing gel loading buffer to each pellet. Incubate at 37°C for 20 min with constant mixing (*see* **Note 11**). Samples can be stored at −20°C.
5. Label some 10 bp DNA ladder as molecular weight standards for the sequencing gel (*see* **Note 12**).
6. Incubate sequencing gel samples and the labeled 10 bp DNA ladder at 65°C for 2 min and then rapid chill on ice. Load all 20 μL onto a 20% sequencing gel (*see* **Note 13**). Run until the bromophenol blue dye front has reached the bottom of the gel.
7. Incubate the gel in fixing solution for 10 min with mild agitation. Transfer the gel to water and incubate for another 10 min. Dry the gel onto a piece of cellulose paper backed by a piece of DE81 paper (*see* **Note 13**). Expose to a phosphor imaging plate overnight. SPO11-associated oligonucleotides from wild-type mouse testes span the region ~14–36nt **(Fig. 13.3)**.

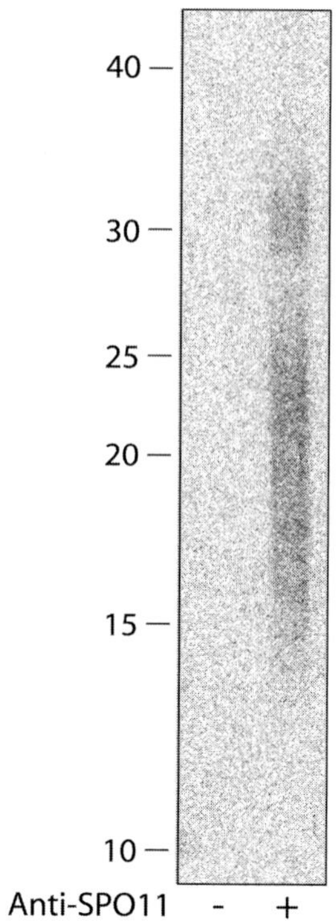

Fig. 13.3. Size of SPO11-associated oligonucleotides in wild-type mice. SPO11-associated oligonucleotides were labeled with [α-^{32}P] cordycepin triphosphate and resolved on a 20% sequencing gel. Input to lanes: ~14 testes equivalent.

4. Notes

1. We use the same number of testes when comparing amounts of SPO11-oligonucleotide complexes among strains. Although SPO11-oligonucleotide complexes can be detected from two wild-type testes of an adult animal, we recommend using more material for robust signals. Four testes of each strain to be tested are used in this protocol for detection of SPO11-oligonucleotide complexes on an SDS gel, and experiments can be scaled up accordingly. If the size of SPO11-associated oligonucleotides is to be determined (**Section 3.6.**), at least eight testes of each strain should be used.
2. Fast, up-and-down movement is most effective to disrupt tissue and lyse cells using a plastic pestle. Take care to avoid spilling over during the grinding. As an alternative to plastic pestles, glass tissue grinders of a proper size can be used.

3. Do not take up supernatants by pipetting, because pipetting disturbs the pellets and causes a high background in the TdT labeling reaction. When the volume of the crude lysates is more than 2 mL, centrifugation is more conveniently carried out using either 3.5 mL or 5.5 mL-capacity ultracentrifuge tubes in a Sorvall AH-650 rotor (or equivalent) at 48,000 rpm (235,687*g*) for 30 min.
4. If precipitates appear after the incubation, spin briefly in a table-top centrifuge. Use supernatants for the following steps.
5. A slurry of protein A beads is made up with beads and buffer in approximately equal volume. To ensure that an equal amount of beads is added to each reaction, dispense aliquots of slurry into 1.5 mL tubes first. Spin beads down and make adjustment if there is an obvious difference in the volume of beads among tubes. Transfer the beads to immunoprecipitation reactions by resuspending the beads with 200 μL lysate.
6. We have also tested [α-^{32}P]-labeled dGTP, dATP, and the chain terminator cordycepin triphosphate in the TdT labeling reaction. There is no obvious difference in the migration of mouse SPO11-oligonucleotide complexes labeled by dCTP, dGTP, or cordycepin triphosphate on the 8% SDS gel, and any of these nucleotides can be used in the labeling reaction. Do not use [α-^{32}P] dATP because labeled SPO11-oligonucleotide complexes migrate slower and become less compact due to long and heterogeneous dA tailing. To determine the exact size of SPO11-associated oligonucleotides as described in **Section 3.6.**, use cordycepin triphosphate in the labeling reaction.
7. Alternatively, the gel can be dried onto a piece of cellulose paper backed by a piece of DE81 paper. The dried gel is then wrapped and exposed to the imaging plate as described in **Section 3.4**, **Steps 4** and **5**.
8. We observed different background labeling associated with TdT from different commercial sources, likely due to impurities in the TdT preparation. We recommend using FPLC *pure* TdT from GE Healthcare or TdT from New England Biolabs for this highly sensitive assay. A background smear at around 55 kD is present in the mock reaction when TdT from GE Healthcare is used in the labeling reaction (**Fig. 13.1A**). Two nonspecific bands (about 75 and 100 kD) are present in the mock reaction when TdT from New England Biolabs is used (**Fig. 13.1B**).
9. We occasionally observe high lane background due to TdT labeling of contaminating DNA that is present in the immunoprecipitates (**Fig. 13.1**). Nevertheless, minor adjustment

of brightness and contrast of the original image should reveal clear, specific signals of SPO11-oligonucleotide complexes. If SPO11-oligonucleotide complexes do not stand above background even after image adjustment, try repeating the experiment with more starting material. Alternatively, a second round of immunoprecipitation can reduce contaminating DNA significantly. Briefly, after the first round of immunoprecipitation (**Section 3.2**), elute SPO11 by boiling the protein A beads in an equal volume of 2 × SDS buffer. Save the SDS supernatant. Rinse the beads with 40 vol of wash buffer. Combine the washes with the SDS supernatant: this is eluate from the first immunoprecipitation. Perform the anti-SPO11 immunoprecipitation with the eluate, followed by TdT labeling, as described in **Sections 3.2** and **3.3.**

10. Tubes are fixed with tapes on the carousel in a hybridization oven and rotated slowly. Adjust the angle of the tubes relative to the rotating axis to make sure that the membrane has full contact with solution during the elution.
11. Tap the tubes to let the drop of liquid cover the rim above the cone bottom of the 2 mL tube, because much of the precipitate accumulates around the rim.
12. Label the 10-bp DNA ladder using T4 polynucleotide kinase (T4 PNK).

10 bp ladder (1 μg/ μL)	1 μL
10 × T4 PNK buffer	1 μL
[γ-^{32}P] ATP	2 μL
T4 PNK	1 μL
0.1 M DTT	0.4 μL
Water	4.6 μL
Total volume	10 μL

Incubate at 37°C for 2 h. Ethanol precipitate the DNA and dissolve precipitates in 100 μL water. Transfer 1 μL to a 1.5 mL tube and use a Geiger counter to estimate radioactivity. Dilute the labeled 10 bp ladder with sequencing gel loading buffer to ~100 cpm/20 μL.

13. We run 27 cm long, 0.4 mm thick sequencing gels with 6 mm wide wells using a custom-made apparatus. For detailed procedure on how to make, run, fix and dry a sequencing gel, please refer to Molecular Cloning *(6)*.

References

1. Bellve, A.R. (1993) Purification, culture, and fractionation of spermatogenic cells. *Methods Enzymol.* **225**, 84–113.
2. Neale, M.J., Pan, J., Keeney, S. (2005) Endonucleolytic processing of covalent protein-linked DNA double-strand breaks. *Nature* **436**, 1053–1057.
3. Baudat, F., de Massy, B. (2005) Regulating double-stranded DNA break repair towards crossover or non-crossover during mammalian meiosis. *Chromosome Res.* **15**, 565–77.
4. Keeney, S., Baudat, F., Angeles, M., Zhou, Z.H., Copeland, N.G., Jenkins, N.A., Manova, K., Jasin, M. (1999) A mouse homolog of the *Saccharomyces cerevisiae* meiotic recombination DNA transesterase Spo11p. *Genomics* **61**, 170–182.
5. Romanienko, P.J., Camerini-Otero, R.D. (1999) Cloning, characterization, and localization of mouse and human *SPO11. Genomics* **61**, 156–169.
6. Sambrook, J., Russel, D.W. (2001) Molecular Cloning. Chapter 12. Protocols 8, 11, and 12. Cold Spring Harbor Laboratory Press, Cold Spring Harbor, New York.

Chapter 14

Stabilization and Electrophoretic Analysis of Meiotic Recombination Intermediates in *Saccharomyces cerevisiae*

Steve D. Oh, Lea Jessop, Jessica P. Lao, Thorsten Allers, Michael Lichten, and Neil Hunter

Abstract

Joint Molecule (JM) recombination intermediates result from DNA strand-exchange between homologous chromosomes. Physical monitoring of JM formation in budding yeast has provided a wealth of information about the timing and mechanism of meiotic recombination. These assays are especially informative when applied to the analysis of mutants for which genetic analysis of recombination is impossible, i.e. mutants that die during meiosis. This chapter describes three distinct methods to stabilize JMs against thermally driven dissolution as well as electrophoretic approaches to resolve and detect JMs at two well-characterized recombination hotspots.

Key words: Meiosis, homologous recombination, double-strand break, joint molecule, Holliday junction, D-loop, strand-exchange, electrophoresis.

1. Introduction

Current models of homologous recombination include the formation of joint molecule (JM) intermediates where parental DNA molecules are linked by one or more Holliday junctions *(1–3)* (**Figs. 14.1C and 14.2A**). A JM consisting of full-length duplex molecules linked by two Holliday junctions (double-Holliday junction, dHJ), has been identified as a central intermediate in processes that lead to the formation of crossover recombinants during budding yeast meiosis *(4–6)*. In addition, a single-end invasion intermediate (SEI), resulting from strand-exchange between one of the two ends of a double-strand break and a full-length duplex molecule, has also been identified as a dHJ precursor *(7)*. Stabilization methods are required to isolate all of these

Scott Keeney (ed.), *Meiosis, Volume 1, Molecular and Genetic Methods, vol. 557*

DOI 10.1007/978-1-59745-527-5_14 Springerprotocols.com

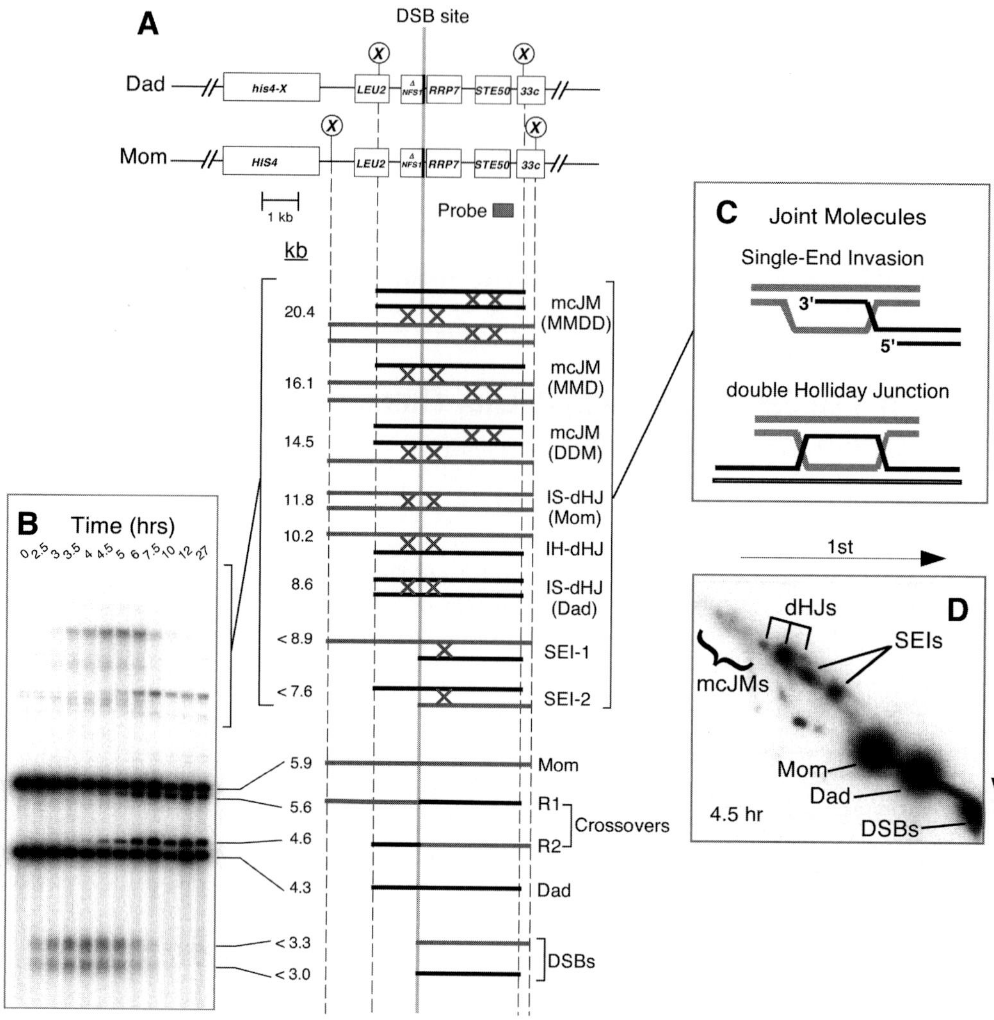

Fig. 14.1. *HIS4LEU2* Physical Assay System for Monitoring Recombination. (**A**) Map of the *HIS4LEU2* locus showing diagnostic restriction sites and position of the probe, Probe 4 *(10)*. Lollipops indicate *Xho*I restriction sites. DNA species detected following Southern hybridization are shown below. SEI-1 and SEI-2 are the two major SEI species detected with Probe 4 *(7)*. IH-dHJs, interhomolog double-Holliday junctions; IS-dHJs, intersister double-Holliday junctions; mcJMs, multichromatid joint molecules containing three or four chromatids; M, Mom chromatid; D, Dad chromatid. (**B**) Image of one-dimensional (1d) gel hybridized with Probe 4 showing the DNA species detailed in (**A**). (**C**) Predicted structures of SEI and dHJ joint molecule intermediates. (**D**) Image of native/native 2d gel hybridized with Probe 4. Species detailed in (**A**) are highlighted. For details *see (7, 18, 21)*.

intermediates, as they readily dissolve via spontaneous, thermally driven Holliday junction movement called branch migration *(8)*. This chapter describes methods for JM stabilization, DNA purification, and JM detection by one-dimensional and two-dimensional agarose gel electrophoresis. It combines methods used in the Hunter and in the Lichten laboratories. So that technical inquiries can be properly directed, the laboratory of origin for each method will be identified.

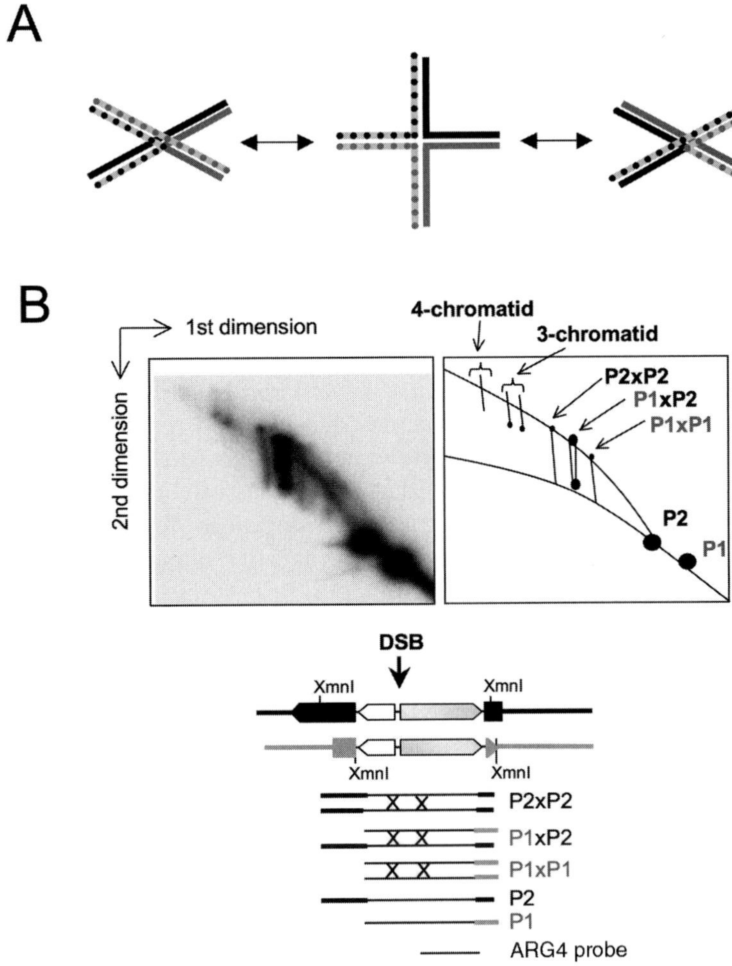

Fig. 14.2. Holliday junction stabilization and detection on 2-d gels. (**A**) Holliday junctions (HJs) populate two types of structures. In the presence of multivalent cations or high salt, HJs populate alternative stacked-X structures (left and right), which preserve base stacking through the junction on both duplexes and which do not branch migrate. In the absence of cations and in low salt, repulsion between negative charges on the phosphate backbone causes HJs to adopt an open cruciform structure, where base stacking is interrupted at the junction. This structure branch migrates readily. (**B**) Example of 2-d neutral/neutral gels to detect joint molecules. The principal of separation is very similar to that of Bell and Byers *(14)*, except that gels contain Mg^{+2} to stabilize HJs. Interhomolog (P1 × P2), intersister (P1 × P1; P2 × P2) and multi-chromatid JMs (3 and 4-chromatid) can be detected. Details of the recombination reporter system used (black—*HIS4*; gray—*LEU2*; white—*URA3*; hatched—*ARG4*) can be found in Jessop et al. *(17)*.

1.1. DNA Purification

DNA purification generally involves the use of protein-denaturing reagents, as well as nuclease-inhibiting magnesium-chelators, both of which tend to accelerate branch migration causing loss of JMs. Therefore, agents that constrain branch migration are essential to stabilize JMs during DNA purification and electrophoresis. Two types of strategies have been utilized in our laboratories, and these are described below.

1.1.1. Psoralen Crosslinking (Hunter)

Inter-strand crosslinking by psoralen, added to cells before lysis, can be used to constrain branch migration during DNA purification, restriction enzyme digestion, and electrophoresis *(9, 10)* (**Fig. 14.1**). This method has distinct advantages. First, because it creates a covalent crosslink between DNA strands, psoralen-linked DNA can be purified, restriction enzyme-digested and separated on agarose gels using standard methods. Second, because inter-strand crosslinking will covalently stabilize any molecules that are held together by Watson–Crick base pairing, a wide variety of JMs, including those that do not contain full Holliday junctions, can in theory be stabilized and analyzed. Practical considerations, including the limited solubility of psoralen, mean that the maximum achievable crosslink density is about one per 300–400 nucleotides (Wolf Heyer, personal communication). In practice it is generally ~1–2 crosslinks per kb, and branch migration can occur in the region between crosslinks. Also, if JM strand composition is to be analyzed, psoralen crosslinks must be removed by alkali-treatment at 65° C before strands in JMs can be separated.

1.1.2. Preventing Holliday Junction Isomerization and Branch Migration (Lichten)

Holliday junctions can adopt either an open cruciform or two alternate stacked-X conformations *(11)* (**Fig. 14.2A**). Branch migration is much slower under conditions (high salt or multivalent cations) that favor the stacked-X conformation *(8)*. In addition, conditions that prevent free arm movement, and thus the transition between the two stacked-X conformations, also retard branch migration *(12, 13)*. These principles form the basis of two methods for DNA isolation that preserve JMs. The first, DNA isolation in agarose plugs, has the advantage of technical simplicity, but has not been shown to result in quantitative JM recovery. In addition, because it involves restriction enzyme digestion of DNA in agarose, it can be used with only a subset of restriction enzymes. The second method *(12)* uses a multivalent cationic detergent (cetyltrimethylammonium bromide, CTAB) to denature proteins and constrain JM movement during initial purification steps, and magnesium and spermidine to stabilize JMs during later purification steps, restriction enzyme digestion, and gel electrophoresis. This method has the advantage of preserving JMs in a form that is readily denatured for component strand analysis, but has the disadvantage of being more technically demanding. In addition, it may not preserve as many kinds of JMs as psoralen crosslinking.

1.2. JM detection and Analysis Using Agarose Gel Electrophoresis

Inter-chromosomal JMs were first detected using two-dimensional (2–d) agarose gel electrophoresis *(9, 14, 15)*, where the first electrophoretic dimension separates mainly on the basis of molecular weight, and a second, orthogonal dimension separates on the basis of combined size and shape (**Figs. 14.1D and 14.2B**). This remains the method of choice, as it provides the greatest separation of JMs from parental and recombinant duplex DNA molecules, and also allows insight into JM structure. However, because a limited number of samples can be analyzed on 2-d gels, JM detection by single-dimension agarose gel electrophoresis is also a useful approach, although it offers more limited resolution (**Fig. 14.1B**). Both methods are generally used in conjunction with a recombination test locus where parental homologs are marked with restriction site polymorphisms; these allow recombinant products to be distinguished from parental homologs, and distinction of joint molecules formed between homologs from those formed between sister chromatids *(7, 10, 16, 17)* (**Figs. 14.1 and 14.2B**).

If parental homologs are marked with the appropriate restriction site polymorphisms, a second type of 2-d gel (native-denaturing) can be used to probe JM structure, and to detect heteroduplex DNA present in both JMs and recombinant products *(4, 5, 7, 18, 19)* (**Fig. 14.3**). The first electrophoretic dimension is performed under standard conditions, separating on the basis of size, and the second, orthogonal electrophoretic dimension is performed under DNA denaturing conditions, thus separating duplex and joint molecules into their single-strand components. This approach has been used to determine the strand composition of SEIs *(7)*, to show that the major JM species contain an even number of Holliday junctions *(4)*, and to probe the structures of JMs that contain more than two chromatids *(18)*. If the two parental chromosomes differ by a restriction site polymorphism that creates a correction-resistant mismatch, 2-d native-denaturing gels can also be used to detect heteroduplex DNA *(5, 19)*.

2. Materials

Supplier of materials is given only when deemed important; in the absence of an indication, any high-quality source will do.

2.1. Pregrowth and Sporulation

2.1.1. Sporulation in 1% Potassium Acetate (Lichten)

1. YEPD broth: 1% (w/v) Bacto^{TM} yeast extract (BD, Franklin Lakes, NJ), 2% (w/v) Bacto^{TM} peptone (BD), 2% D-glucose, 0.004% adenine. Adjust pH to 5.5 with 1 N HCl. Prepare broth without glucose as a 1.1 × solution and autoclave; mix

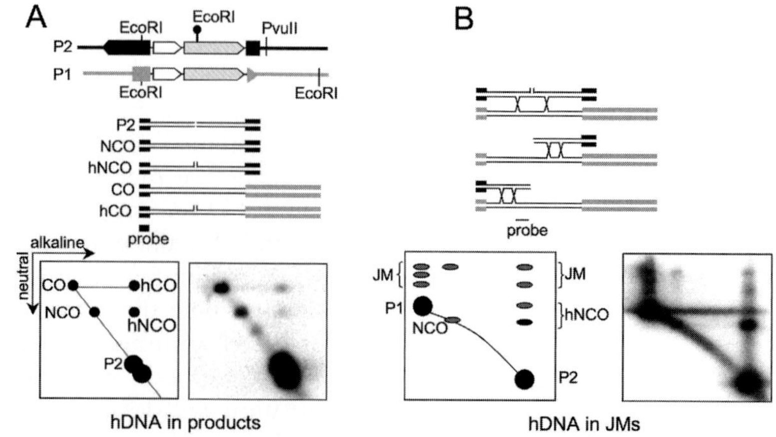

Fig. 14.3. Heteroduplex detection and strand analysis on 2-d neutral/alkaline gels. (**A**) Detection of heteroduplex on neutral/alkaline gels. A *URA3-ARG4* recombination reporter is inserted at the *HIS4* and *LEU2* loci on chromosome *III* (black—*HIS4*; gray—*LEU2*; white—*URA3*; hatched—*ARG4*). For details, *see (19)*. The copy of *ARG4* at *HIS4* is marked with a palindrome (lollipop) that contains an *Eco*RI site. Recombinant products detected upon *Eco*RI/*Pvu*II digestion and probing with *HIS4* sequences are shown (P2—parental; NCO—noncrossover gene conversion; CO—crossover). When this palindrome is present in heteroduplex DNA, *Eco*RI cutting produces a nick in the palindrome-containing strand (hNCO—noncrossover with palindrome in heteroduplex; hCO—crossover with palindrome in heteroduplex). Separation on 2-d neutral/alkaline gels and probing with the indicated probe produces on-diagonal spots for parental, NCO and CO products, and off-diagonal spots for heteroduplex-containing hNCO and hCO products. For further details, *see (19)*. (**B**) Strand analysis and heteroduplex detection in JMs. Recombination intermediates from the same recombination reporter configuration are detected using neutral, Mg^{+2}-containing conditions in the first dimension, and alkaline conditions in the second. Structures produced by *Eco*RI/*Pvu*II digestion of three types of JMs, which differ in terms of HJ location, are shown, along with a representative gel showing detection of the three structures. For further details, *see (5)*.

this with sterile 20% (w/v) glucose in a 9:1 ratio before use. YEPD agar is identical but contains 2% BactoTM agar (BD), and is prepared by autoclaving all components together.

2. Pre-sporulation broth (SPS): 1% (w/v) Bacto^{TM} peptone, 0.5% (w/v) Bacto^{TM} yeast extract, 1% (w/v) potassium acetate, 0.17% (w/v) Difco^{TM} yeast nitrogen base without amino acids, 1% (w/v) ammonium sulfate, 0.5% (w/v) potassium hydrogen pthalate. Adjust pH to 5.5 with 10 N KOH as sodium inhibits sporulation. Make fresh and autoclave.
3. 2.8 L triple-baffled Fernbach flask.
4. Sporulation medium (KAc): 1% (w/v) potassium acetate supplemented with nutrients according to the auxotrophic requirements at 1/5 levels used in vegetative growth medium (*See* **Table 10.1** of this volume) and with 0.001% polypropylene glycol 2000 as an anti-clumping agent.

2.1.2. Pregrowth, Sporulation and Psoralen Crosslinking (Hunter)

Materials for sporulation are the same as in **Section 2.1.1** except:

1. YPG agar: 1% (w/v) Bacto™ yeast extract, 2% (w/v) Bacto™ peptone, 1.5% Bacto™ agar, 2% glycerol.
2. 500 mL centrifuge bottles.
3. DAPI stock: 1 μg/mL 4′,6-diamidino-2-phenylindole.
4. Sorbitol/ethanol solution: 0.1 M sorbitol in 40% ethanol.
5. 60 mm Petri dishes.
6. 10% (w/v) sodium azide.
7. 5X psoralen stock solution: 0.5 mg/mL Trioxalen dissolved in ethanol (200 proof). Store at 4°C wrapped in aluminum foil. To completely dissolve prior to use, shake at room temperature overnight.
8. 1X psoralen working stock: 50 mM Tris-HCl pH 8.0, 50 mM EDTA pH 8.0. Keep on ice, wrapped in aluminum foil. Make fresh on the day of the experiment.
9. 50 mM EDTA, 50 mM Tris-HCl pH 8.0.
10. Long-wave UV transilluminator box.

2.2. DNA Extraction and Purification

2.2.1. Guanidine/sarkosyl Preparation of Psoralen Cross-Linked DNA (Hunter)

1. Spheroplasting buffer: 1 M sorbitol, 50 mM potassium phosphate pH 7.0, 10 mM EDTA pH 7.5.
2. β-mercaptoethanol.
3. Zymolyase 100T.
4. Guanidine/sarkosyl solution: 4.5 M guanidine-HCl, 0.1 M EDTA, 0.15 M NaCl, 0.05% sodium lauryl sarcosinate (Sarkosyl).
5. RNase stock solution: 10X TE pH 8.0 (100 mM Tris-HCl, 10 mM EDTA pH 8.0), 50 μg/mL RNase (DNase-free).
6. Proteinase-K solution: 20 mg/mL proteinase-K, 20 mM $CaCl_2$, 10 mM Tris-HCl pH 7.5, 50% glycerol. Store at –20°C.
7. 2.0 mL microfuge tubes.
8. Phenol pH 8.0/chloroform/isoamyl alcohol mixed in a 25:24:1 ratio.
9. 200 proof ethanol.
10. 3 M Sodium acetate pH 7.2.
11. 37°C water bath; 65°C water bath.

2.2.2. Yeast DNA Preparation with CTAB (Lichten)

1. 50% (v/v) glycerol, 0.5% (w/v) sodium azide. Store at room temperature.
2. Spheroplasting solution with glycerol: 1 M sorbitol, 50 mM potassium phosphate, 10 mM EDTA, and 5 mM hexamine cobalt trichloride (CoHex), 20% glycerol, pH 7.5. Filter sterilize and store at 4°C.

3. β-mercaptoethanol.
4. 5 mL round bottom polypropylene tubes.
5. Spheroplasting solution with CoHex: 1 M sorbitol, 50 mM potassium phosphate, 10 mM EDTA, 5 mM CoHex, pH 7.5. Filter sterilize and store at 4°C.
6. Zymolyase 100T.
7. CTAB Extraction Solution: Preparation method is critical, as 2 M NaCl, 3% CTAB is too viscous to be filtered. Weigh out 11.69 g NaCl and 0.54 g CoHex, add 10 mL 1 M Tris-HCl pH 7.5 and 5 mL 0.5 M EDTA pH 8. Make up to 50 mL with H_2O. Add 10 mL 10% (w/v) polyvinylpyrrolidone, average MW ~40,000 (PVP-40) to dissolve NaCl. Filter sterilize the following solutions into the same receiving vessel (the order is important): 10 mL 10% (w/v) PVP-40; 60 mL NaCl/CoHex/Tris/EDTA/PVP-40 from above; 30 mL 10% (w/v) CTAB. Store at 37°C to prevent CTAB precipitation. If CTAB precipitates, discard.
8. RNase A: RNase A, DNase free, High Concentration (Roche).
9. Proteinase K: 20 mg/mL in 10 mM Tris-HCl pH 7.5, 20 mM $CaCl_2$, and 50% (v/v) glycerol. Store aliquots at −20°C.
10. Chloroform:isoamyl alcohol (24:1).
11. CTAB Dilution Solution: 1% (w/v) CTAB, 50 mM Tris-HCl, 10 mM EDTA, 4 mM CoHex, pH 7.5. Filter sterilize and store at room temperature.
12. Fine-tip transfer pipettes (Samco, San Fernando, CA).
13. 0.4 M NaCl in TECoHex: 0.4 M NaCl, 10 mM Tris-HCl, 1 mM EDTA, 1 mM CoHex, pH 7.5. Filter sterilize and store at room temperature.
14. 1.42 M NaCl in TECoHex: 1.42 M NaCl, 10 mM Tris-HCl, 1 mM EDTA, 1 mM CoHex, pH 7.5. Filter sterilize and store at room temperature.
15. 95% (v/v) ethanol.
16. 70% (v/v) ethanol, 0.3 mM CoHex.
17. TMN: 10 mM Tris-HCl, 10 mM $MgCl_2$, 200 mM NaCl, pH 7.5. Filter sterilize and store at 4°C.
18. 70% (v/v) ethanol, 3 mM $MgCl_2$.
19. TMSpe: 10 mM Tris-HCl, 2 mM $MgCl_2$, 50 μM spermidine, pH 7.5. Filter sterilize and store at 4°C.
20. NanoDrop *ND*-1000 Spectrophotometer (Wilmington, DE).

2.2.3. Yeast DNA Preparation in Agarose Plugs (Lichten)

1. 50 mM EDTA, pH 7.5.
2. Plug casting blocks (BioRad, Hercules, CA).

3. SCE: 1 M sorbitol, 0.1 M sodium citrate, 60 mM EDTA, pH 7.0. Filter sterilize, store at 4°C.
4. LMP agarose mix: 1% (w/v) low melting-point agarose (SeaPlaque GTG, Lonza, Basel, Switzerland), 125 mM EDTA, pH 7.5. Measure volume, heat in microwave until agarose is dissolved, re-adjust volume. Store at 55°C.
5. Solution 1: SCE + 5% (v/v) β-mercaptoethanol + 1 mg/mL Zymolyase 100T. Mix briefly at 40°C just prior to use.
6. Solution 2: 0.45 M EDTA, 10 mM Tris-HCl, 7.5% (v/v) β-mercaptoethanol, 10 μg/mL DNase-free RNase. Make just prior to use.
7. Solution 3: 0.25 M EDTA, 10 mM Tris-HCl, 1% (w/v) n-lauryl sarcosine, 1 mg/mL proteinase K. Warm to 50°C just prior to use to dissolve components.
8. Plug storage solution: 50 mM EDTA, 50% (v/v) glycerol, pH 7.5.
9. TE: 10 mM Tris-HCl, 1 mM EDTA, pH 7.5

2.3. Electrophoresis

2.3.1. 1-d Agarose Gels to Detect JMs (Lichten)

1. SeaKem GTG agarose (Lonza, Basel, Switzerland).
2. TBE: 90 mM Tris base, 90 mM boric acid, 2 mM EDTA. Prepare as 10x stock.
3. 1 M $MgCl_2$.
4. 5X loading buffer + Mg: 15% Ficoll 400, 25 mM Tris-HCl, pH 7.4, 0.25 mM EDTA, 25 mM $MgCl_2$, 0.1 mg/mL Orange G dye (Sigma).
5. 0.5 M EDTA, pH 7.5–8.0.
6. 0.25 N HCl.
7. Alkaline transfer buffer: 1.5 M NaCl, 0.5 N NaOH.

2.3.2. 2-d Native/Native to Detect JMs (Lichten)

1. Gel tray with Plexiglas lane dividers, made by modifying standard submarine electrophoresis apparatus. *See* **Fig. 14.4** and legend for more details.
2. SeaKem Gold agarose (Lonza).
3. 10 mg/mL ethidium bromide.
4. Piece of developed X-ray film, gel lifter, or other stiff plastic to aid manipulation of gel strips.
5. Horizontal electrophoresis apparatus with recirculation and cooling capabilities.
6. Cooling circulating pump.

2.3.3. 2d Native/Denaturing Electrophoresis (Lichten)

1. Alkaline running buffer: 50 mN NaOH, 1 mM EDTA. Make fresh for each use.

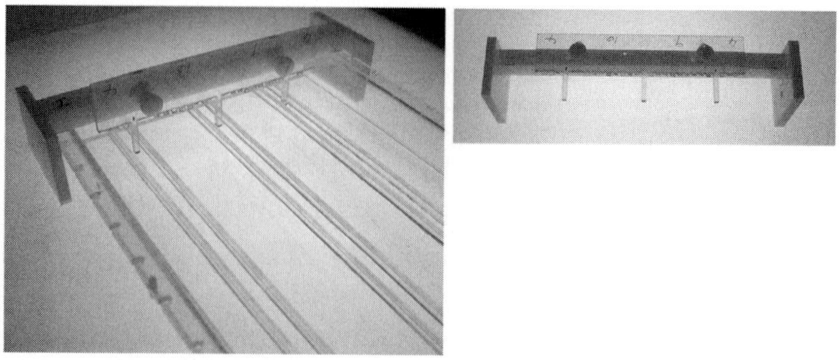

Fig. 14.4. Modified gel tray and comb for 1st-dimension electrophoresis. Gel strips are defined by 4 × 4 mm Plexiglas strips glued onto the gel tray; wells are formed with a comb made by breaking all but the desired teeth from a standard well-forming comb. The distance between Plexiglas strips is determined by the width of the comb teeth. In this figure, a modified BioRad (Hercules, CA) Sub-Cell 15 × 25 cm gel tray and modified 30-tooth comb and adjustable holder are shown, but any similar apparatus can be used.

2.3.4. 1d Native Gel Electrophoresis (Hunter)

1. Buffer Puffer™ gel box, Model A5 (Owl Separation Systems) with a 36-well comb.
2. SeaKem LE agarose (Lonza).
3. TBE: 90 mM Tris base, 90 mM boric acid, 2 mM EDTA. Prepare as 10x stock.
4. Loading buffer: 100 μl 6X loading dye + 60 μl 10 × NEB3 restriction enzyme buffer (New England Biolabs; extra salt prevents the samples from drifting out of the wells).
5. 6X loading dye: 0.25% (w/v) bromophenol blue, 0.25% (w/v) xylene cyanol FF, 15% (w/v) Ficoll in water.
6. λ*Bst*EII and λ*Hind*III DNA molecular weight markers (New England Biolabs).
7. 10 mg/mL Ethidium bromide in dH_2O. Store at room temperature wrapped in aluminum foil.
8. Alkaline Southern transfer buffers: 0.25 M HCl and 0.4 M NaOH.

2.3.5. 2-d Native/Native Gel Electrophoresis (Hunter)

1. Buffer Puffer™ gel box, Model A5 (Owl Separation Systems) with a 36-well comb.
2. SeaKem LE Agarose (Lonza); SeaKem Gold Agarose (Lonza).
3. TBE: 90 mM Tris base, 90 mM boric acid, 2 mM EDTA. Prepare as 10x stock.
4. λ*Bst*EII and λ*Hind*III DNA molecular weight markers (New England Biolabs).
5. Long-wave UV transilluminator.
6. Clean razor blades.
7. Fluorescent ruler.

2.3.6. Removal of Psoralen Cross-Links and 2d Native/ Denaturing Electrophoresis (Hunter)

1. Buffer Puffer™ gel box, Model A5 (Owl Separation Systems) with a 36-well comb.
2. SeaKem LE Agarose (Lonza); SeaKem Gold Agarose (Lonza)
3. TBE: 90 mM Tris base, 90 mM boric acid, 2 mM EDTA. Prepare as 10x stock.
4. λ*Bst*EII and λ*Hind*III DNA molecular weight markers (New England Biolabs).
5. Shaking water bath at 65°C.
6. 10 mg/mL Ethidium bromide in dH_2O. Store at room temperature wrapped in aluminum foil.
7. Long-wave UV transilluminator box.
8. Clean razor blades.
9. Equilibration Buffer: 5 mM Na_3PO_4. Warm solution to 65°C and adjust to pH 12 using 50% NaOH.
10. Crosslink Reversal Buffer: 1 mM Na_3PO_4, 3 M urea. Warm solution to 65°C and adjust to pH 12 using 50% NaOH.
11. 5X Alkaline Running buffer: 0.25 M NaOH, 5 mM EDTA.
12. 1X Alkaline Running buffer: 50 mM NaOH, 1 mM EDTA.

3. Methods

3.1. Pregrowth and Sporulation

3.1.1. Pregrowth, Sporulation and Psoralen Crosslinking (Hunter)

Day 1:

1. Patch cells from −80°C glycerol stock onto YPG plates.

Day 2:

2. Streak for single colonies on YEPD plates.

Day 4:

3. Inoculate single diploid colonies into 5 mL YEPD broth. Grow overnight at 30°C.
4. Prepare SPS (700 mL for each strain to be sporulated, plus a little extra to use as a blank in the spectrophotometer), 1% SPM (1 L per strain), and one 1 L and two 500 mL graduated cylinders. Autoclave to sterilize.

Day 5:

5. Measure 350 mL SPS into each of two 2.8 L baffled flasks. Add two drops of antifoam from a P200 micropipette.
6. At 3:00 pm, dilute cells from YEPD overnight cultures 1/500 and 1/1,000 into 350 mL SPS.
7. Incubate with vigorous shaking (350 rpm) at 30°C for 18 h. Make sure that the flasks are tightly clamped.

8. Measure 600 mL SPM into a 2.8 L baffled flask (one for each strain to be sporulated). Add two drops of antifoam. Prewarm at 30°C and keep remaining 1% SPM at 30°C.
   Day 6:
9. At 9:00 am, measure the OD_{600} of each culture. Select cultures with an OD_{600} ~1.0–1.1 (*see* **Note 1**). Transfer 0.5 mL of selected culture into a 1.5 mL microfuge tube, harvest the cells by centrifugation and resuspend in 0.5 mL ice-cold Sorbitol/Ethanol solution. Stain cells with DAPI by mixing 5 μL cells with 5 μL DAPI stock. Place 4–5 μL stained cells on a slide, cover with a coverslip, and view cells by fluorescence microscopy. Select cultures with ≥80% of G1/G0 cells (defined as cells with no bud and with a single DAPI-staining body).
10. Transfer selected cultures into 500 mL centrifuge bottles and harvest cells at 3,200*g*, 28°C, for 3 min.
11. Wash cells by resuspending in an equal volume of 1% SPM. Harvest as in **Step 10**. Resuspend thoroughly in SPM from the sporulation flask. Return cells to sporulation flask, take the zero time-point sample (*see* **Step 12**) and incubate the culture with vigorous shaking at 30°C.
12. We typically take samples after 0, 2.0, 2.5, 3.0, 3.5, 4.0, 4.5, 5.0, 6.0, 7.0, 8.0, 9.0, 10.0, and 24 h, but this varies depending on the specific strain being analyzed. For each time point sample, collect 45 mL of culture in a 50 mL conical tube. Add 0.45 mL 10% sodium azide. Also take 0.5 mL of culture into a 1.5 mL microfuge tube, harvest the cells by centrifugation and resuspend in 0.5 mL ice-cold Sorbitol/Ethanol solution. Freeze at −20°C. These small samples will be used for DAPI staining to monitor the progression of meiotic divisions.
13. Spin the large sample, in the 50 mL tube, at 3,200*g*, 3 min. Pour off the supernatant and drain the tubes on a paper towel.
14. Resuspend the cell pellet in 2.5 mL ice-cold 1X psoralen solution.
15. Transfer to a 60 mm Petri dish, place on the long-wave UV box and expose for 10 min on the high setting. Swirl the cells twice during this period.
16. Pipette the cells back into the original 50 mL tube. Recover any cells remaining in the dish by rinsing with 2 mL ice-cold 50 mM EDTA/50 mM Tris-HCl solution and transferring to the 50 mL tube.
17. Harvest the cells at 3,200*g*, 3 min. Drain well and freeze at −20°C.

3.1.2. Pregrowth and Sporulation (Lichten)

Day 1:

1. Set up a 5 mL overnight culture of a diploid strain in YEPD broth, using a fresh colony (no more than 2 days growth on YEPD agar plates).
2. Prepare 1 L SPS, 1 L 1% KAc, one 500 mL and two 250 mL graduated cylinders, and one 250 mL centrifuge bottle. Autoclave to sterilize.

Day 2:

1. Using one of the sterile 250 mL graduated cylinders from Day 1, aliquot 230 mL SPS each to four 2 L Erlenmeyer flasks.
2. About 18 h before the desired time of sporulation initiation, dilute the overnight culture in 2-fold increments between 1/500 and 1/2,000 into 230 mL SPS and incubate overnight, shaking vigorously (~300 rpm) at 30°C (do not discard remainder of SPS). The goal is to have one of the cultures reach 2×10^7 cells/mL on the morning of Day 3.
3. Using the sterile 500 mL graduated cylinder from Day 1, aliquot 400 mL of 1% KAc into a 2.8 L triple-baffled Fernbach flask. Add 400 μL 1% polypropylene glycol and required supplements. Place flask in shaking water bath at 30°C. Keep remainder of 1% KAc at 30°C.
4. For CTAB DNA preparation, add 8.4 mL of 50% glycerol + 0.5% sodium azide to a 50 mL Falcon tube for each time point to be harvested.

Day 3:

1. In the morning, measure the OD_{600} of the four cultures (*see* **Note 1**). Pick the culture with an OD_{600} of 1.35–1.4 (~2×10^7 cells/mL).
2. Pellet the selected culture in a 250 mL centrifuge bottle at 3,200*g*, 28°C, 3 min.
3. Resuspend in an equal volume (230 mL) of pre-warmed 1% KAc without supplements.
4. Take a 30 mL sample (0 h) and process for DNA preparation in agarose plugs (**Section 3.2.3**) or for DNA preparation in CTAB (**Section 3.2.2**).
5. Pellet remaining 200 mL, at 3,200*g*, 28°C, 3 min. Resuspend in 400 mL 1% KAc + supplements (from the sporulation flask).
6. Return to sporulation flask and incubate shaking vigorously at 30°C in a shaking water bath.
7. Take 30 mL samples at desired time points as in **Step 4**. Up to 13 samples may be taken.

3.2. DNA Extraction and Purification

3.2.1. Guanidine/sarkosyl Preparation of Psoralen Cross-Linked DNA [Hunter; Modified from (20)]

1. Thoroughly resuspend the cell pellets from **Section 3.1.1** in 0.8 mL spheroplasting buffer. Add 0.1 mL of freshly made and well-mixed 4 mg/mL 100T zymolyase (made up in spheroplasting buffer). Then add 9 μL of β-mercaptoethanol. Incubate at 37°C for 15 min. Gently mix by swirling; repeat this mixing twice during the incubation.
2. Harvest spheroplasted cells by centrifugation for 4 min, 1,500*g*. Carefully remove as much of the supernatant as possible.
3. Add 2.5 mL guanidine/sarkosyl solution and resuspend the stringy spheroplast pellet by gentle "finger-vortexing." Place at 65°C for 20 min. Finger vortex several times during this incubation.
4. Cool tubes on ice. Add 2.5 mL ethanol. Mix tubes well by inversion and store at −20°C overnight (alternatively, incubate at −20°C for >20 min before proceeding to **Step 5**).
5. Pellet precipitated material by centrifugation for 15 min at 3,200*g* in a benchtop centrifuge. Drain well. Briefly centrifuge and remove the last traces of supernatant with a pipette.
6. Add 0.7 mL RNase solution. Disperse the pellet with the pipette tip. Incubate on a roller drum at 37°C for 1 h. Finger vortex several times during this incubation to help disperse the pellet.
7. Add 25 μL proteinase-K solution and incubate at 65°C for 1 h. Finger vortex several times during this incubation. It is okay to stop and freeze samples overnight at this point.
8. Transfer samples to 2 mL microfuge tubes. Extract *twice* with 0.7 mL phenol/chloroform/isoamyl alcohol (25:24:1). Mix well but do not vortex. Centrifuge at full speed in a microcentrifuge for 10 min. Carefully remove the upper, aqueous layer to a fresh 2 mL microfuge tube.
9. Precipitate the DNA: add 35 μL of 3 M sodium acetate and 1.4 mL of ethanol and mix well by inversion. Let the samples stand for 20 min. DNA should form a visible precipitate. Centrifuge briefly and decant the supernatant. If no spool of precipitated DNA is visible or if the spool is very small, centrifuge for 5 min at full speed in a microcentrifuge.
10. Wash the DNA pellets by adding 1.5 mL 70% ethanol, centrifuge briefly, decant the supernatant, and drain the tubes on a rack that is wadded with paper towels. Finally, pulse-spin in the centrifuge and remove the last traces of ethanol with a pipette.

11. Air dry the DNA pellets for ~10 minutes (or until pellet looks dry) and then resuspend in 100 μL 1X TE. Allow DNA to hydrate in the refrigerator overnight. Mix well by flicking the tube. Do not vortex.
12. Determine DNA concentrations by measuring OD_{260} with a NanoDrop spectrophotometer.
13. Store samples at −20°C.

3.2.2. Yeast DNA Preparation with CTAB (Lichten)

This method achieves separation of nucleic acids from polysaccharides and proteins by exploiting the insolubility of cetyltrimethylammonium-nucleic acid complexes at low salt concentrations. It is optimized for 30–50 mL of a meiotic culture, and must be adjusted for other volumes or cell densities.

1. Add 30 mL culture samples to 50 mL conical tubes containing 8.4 mL of 50% glycerol + 0.5% sodium azide. Pellet cells, 3,200*g*, 4°C, 2 min.
2. Resuspend pellet in 2 mL spheroplasting solution plus 20% glycerol.
3. Transfer to 5 mL round bottom tubes.
4. Pellet cells, remove supernatant thoroughly, and freeze cells on dry ice. Store at −80°C.
5. Thaw the frozen cell pellet from **Step 4** on ice. Prepare zymolyase solution. For each sample, use 500 μL spheroplasting solution with CoHex (without glycerol), with 5 μL β-mercaptoethanol and 0.25 mg zymolyase 100T. Invert to dissolve and keep on ice prior to use.
6. Resuspend cell pellet in 500 μL zymolyase solution.
7. Incubate at 37°C for ~5 min. Check degree of spheroplasting by visual inspection in a phase-contrast microscope. When ~80% of the cells have spheroplasted (*see* **Note 2**), continue to **Step 8**.
8. Pellet spheroplasts at 3,200*g* for 2 min at room temperature and thoroughly remove the supernatant. Resuspend pellet in 500 μL CTAB extraction solution by vortexing gently.
9. Add 12.5 μL proteinase K (20 mg/mL) to 0.5 mg/mL final. Mix by vortexing gently.
10. Add 1 μL RNase and mix by vortexing gently.
11. Incubate at 37°C for 15 min, vortexing gently every ~5 min.
12. Transfer to a 1.5 mL conical microcentrifuge tube containing 300 μL chloroform:isoamylalcohol. Vortex at full speed for 10 s, shake thoroughly, and vortex for 5 s (*see* **Note 3**).

13. Promptly remove upper (aqueous) layer to a fresh 5 mL round bottom tube, carefully avoiding the white interface, which contains protein, polysaccharides, and excess CTAB.
14. Layer 1.5 mL of CTAB dilution solution on top; if cell extract volume differs significantly from 500 μL, use 3 volumes of CTAB dilution solution.
15. Mix by inverting gently about five times and leave undisturbed at room temperature for 10 min. The solution should be fairly cloudy, with discrete localized precipitates.
16. Invert about 20 times, until a precipitate is observed. If no precipitation is observed, add an additional 0.5 mL of CTAB dilution solution and repeat the mixing and 10 min rest at room temperature. Leave at room temperature for 2 more minutes. A discrete white precipitate should sink to the bottom of the tube.
17. Carefully remove the supernatant using a thin-bore capillary transfer pipette. Pulse spin in centrifuge if necessary. However, excessive centrifugation will result in recovery of excess CTAB, which will interfere with subsequent steps. Do not let the pellet dry, as it will then be impossible to resuspend.
18. Immediately add 2 mL ice-cold 0.4 M NaCl in TECoHex. Vortex briefly until the pellet floats.
19. Carefully remove the supernatant with a transfer pipette, taking care to prevent drying. This wash may be repeated if desired.
20. Promptly add 0.5 mL ice-cold 1.42 M NaCl in TECoHex. Gently shake the tube until the pellet becomes translucent.
21. Transfer the suspension to a 1.5 mL conical microcentrifuge tube. Add 1 mL 95% ethanol (at room temperature). Invert until the DNA is completely precipitated, then allow the precipitate to settle to the bottom of the tube. Pulse spin in microcentrifuge if necessary.
22. Remove supernatant with a transfer pipette and add 1 mL 70% ethanol-0.3 mM CoHex (at room temperature).
23. Vortex briefly until pellet floats. Pulse spin in microfuge if necessary.
24. Remove ethanol completely with transfer pipette. If solid white particles of CTAB are visible, repeat **Steps 20–24**.
25. Completely remove ethanol and allow pellet to dry briefly. The pellet should still look damp, but ethanol should have evaporated. Over-drying will make resuspending difficult.
26. Add 100 μL ice-cold TMN to the pellet. Allow pellet to soak in TMN on ice for at least 30 min before proceeding.

27. Add 200 μL 95% ethanol (at room temperature). Invert until DNA is completely precipitated.
28. Remove supernatant and add 200 μL 70% ethanol-3 mM $MgCl_2$ (at room temperature).
29. Vortex briefly. Repeat **Step 28** if desired.
30. Remove ethanol completely. Allow pellet to dry briefly. Pellet should still look damp, but ethanol should have evaporated.
31. Resuspend promptly in 50–100 μL ice-cold TMSpe (*see* **Note 4**). It is important to completely resuspend the pellet, which may require prolonged incubation on ice, with gentle mixing.
32. Determine nucleic acid concentrations by measuring OD_{260} with a NanoDrop spectrophotometer. In successful preparations, the majority of nucleic acid species will be high molecular weight DNA.

3.2.3. Yeast DNA Preparation in Agarose Plugs (Lichten)

The following protocol makes about nine blocks.

1. Take 30 mL samples from sporulating cultures into 50 mL conical tubes. Pellet cells, 3,200*g*, 2 min, room temperature. Resuspend in 30 mL 50 mM EDTA; pellet again; resuspend in 15 mL 50 mM EDTA, pellet again, and drain.
2. While washing cells, mix 0.83 mL 1% LMP agarose with 0.17 mL Solution 1. Equilibrate at 40°C.
3. Resuspend cells from **Step 1** in 0.3 mL 50 mM EDTA, incubate at 40°C for 2 min.
4. Mix cells with 0.6 mL LMP agarose/Solution 1 mix. Pipette into plug mold, following manufacturer's instructions; each well holds 90 μL.
5. Chill blocks at 4°C at least 30 min. Blocks can remain at 4°C while other samples are being taken and processed.
6. Place 10 mL Solution 2 into a 15 mL screw-cap tube for each time point. Express agarose blocks into Solution 2 and incubate at 37°C, 1 h with gentle mixing.
7. Pour off Solution 2 and replace with 10 mL Solution 3. Incubate overnight at 50°C.
8. Pour off solution 3, and drain well. Add 10 mL storage buffer and incubate at room temperature, 1 h. Pour off storage buffer, drain well, and add 10 mL fresh storage buffer. Store at –20°C.

3.3. Restriction Enzyme Digestion

3.3.1. Psoralen Cross-Linked DNA (Hunter)

1. Digest 2 μg of DNA in 80 μL total volume with a 4-fold excess of restriction enzyme.
2. Precipitate digested DNA by adding 190 μL ethanol and 5 μL 3 M sodium acetate pH 7.2. Let sit for 20 min at room temperature.

3. Centrifuge for 10 min at full speed in a microcentrifuge.
4. Decant the supernatant and wash the pellet by adding 100 μL 70% ethanol.
5. Centrifuge for 5 min at full speed in a microcentrifuge.
6. Decant, pulse spin tubes and pipette off residual ethanol.
7. Air dry pellets for ~10 min at room temperature.
8. Resuspend pellets in 15 μL 1X TE.

3.3.2. Yeast DNA Prepared with CTAB (Lichten)

In general, 0.5–1 μg of DNA is digested in 20 μL, using buffer conditions suggested by the manufacturer. Because spermidine and residual CTAB inhibit some restriction enzymes, it may be necessary to try different vendors and increase the amount of enzyme added to the reaction. If digests are to be analyzed for JMs, reactions are supplemented with spermidine to a final concentration of 0.1 mM (*see* **Note 5**).

3.3.3. Yeast DNA Preparation in Agarose Plugs (Lichten)

1. Place about 1/3 (~30 μL, or 30 μg by weight) of a plug in a 15 mL screw-cap microfuge tube. Soak in the following buffers at room temperature, with gentle mixing by inversion or rotation, removing buffers completely with a capillary transfer pipette.
2. 5 mL TE, 30 min, two times.
3. 5 mL 1 × restriction enzyme buffer (as will be used in digest), 30 min, two times.
4. Transfer plug to a 1.5 mL microfuge tube, and remove all buffer with a capillary transfer pipette.
5. Heat at 65°C, 5 min to melt agarose.
6. Incubate at 37°C, 5 min.
7. Add 20 units desired restriction enzyme (*see* **Note 6**). Incubate at 37°C, 2 h.
8. Add another ten units restriction enzyme. Incubate at 37°C, 2 h.
9. Heat to 50°C briefly, to make the solution easy to pipette.
10. Load into the well of a "dry" agarose gel. Let solidify 2–3 min.
11. Immerse gel in buffer and commence electrophoresis.

3.4. Electrophoresis

3.4.1. 1-d Agarose Gels to Detect JMs (Lichten)

Use only plastic- and glassware that has never been exposed to ethidium bromide. These electrophoresis conditions are for digests where parental fragments are ~4.5 kb:

1. Gel: 0.5% SeaKem GTG agarose, prepared in TBE + 4 mM $MgCl_2$.
2. Running buffer: TBE + 3 mM $MgCl_2$.

3. After digests are complete, add 1/5 volume 5x loading buffer + Mg. Load onto gel.
4. Run gel at 1.9 V/cm for 24 h with buffer recirculation (*see* **Note 7**).
5. Wash gel twice with gentle agitation, 15 min, with >2 gel vol 10 mM EDTA.
6. Wash gel twice with gentle agitation, 15 min, with >2 vol 5 mM EDTA (*see* **Note 8**).
7. Depurinate DNA by washing gel with gentle agitation, 20 min, with >2 vol 0.25 N HCl.
8. Equilibrate gel with gentle agitation, 20 min, in alkaline transfer buffer.
9. Transfer gel contents to nylon membranes by capillary or vacuum transfer, using alkaline transfer buffer.

3.4.2. 2-d Native/Native Electrophoresis to Detect JMs (Lichten)

Use only plastic- and glassware that has never been exposed to ethidium bromide in preparing reagents for and performing the first electrophoretic dimension. These electrophoresis conditions are for digests where parental fragments are ~4.5 kb:

Day1

1. Melt 0.4% SeaKem Gold agarose in TBE + 4 mM $MgCl_2$ and pour into a 15 × 25 cm gel tray with dividers (*see* **Note 9**). While gel is solidifying, prepare TBE + 3 mM $MgCl_2$ running buffer and cool to 4°C.
2. After digests are complete, add 1/5 volume 5x loading buffer + Mg. Load onto gel.
3. Run the gel ~43 h at 0.7 V/cm at 4°C with buffer recirculation.
4. While gel is running, prepare 4 L TBE + 3 mM $MgCl_2$ + 0.4 μg/mL ethidium bromide and cool to 4°C.

Day 3

1. Melt 0.9% SeaKem GTG agarose in 250 mL TBE + 4 mM $MgCl_2$. Add ethidium bromide to 0.4 μg/mL, final concentration. Cool to 55°C before pouring.
2. Remove the gel from *Day 1* from the electrophoresis tank and remove excess buffer from the wells with a capillary transfer pipette. Using a small amount of molten agarose, fill the wells of the gel. Once agarose has solidified, carefully remove lane strips to a glass tray, being careful to preserve orientation.
3. Soak lane strips in 1.35 L of 1 × TBE + 3 mM $MgCl_2$ + 0.4 μg/mL ethidium bromide, 30 min, at 4°C. Do not agitate.
4. Arrange strips in an empty gel tray, 90° relative to the direction of electrophoresis.

5. Carefully pour the molten 0.9% agarose (from **Step 1**) around the strips and allow to solidify. Once the gel has set, transfer to cold room and allow to cool for ~15 min before placing in electrophoresis apparatus and adding TBE + 3 mM $MgCl_2$ + 0.4 μg/mL ethidium bromide running buffer.
6. Run the gel for 5 h at 6 V/cm with recirculation and with cooling to 4°C.
7. Rinse gel briefly with water and photograph, if desired.
8. Wash in EDTA, depurinate and equilibrate with transfer buffer as described in **Section 3.4.1**, **Steps 5–9**.

3.4.3. 2d Native/Denaturing Electrophoresis (Lichten)

Use only plastic- and glassware that has never been exposed to ethidium bromide in preparing reagents for the first electrophoretic dimension. These electrophoresis conditions are for digests where parental fragments are ~ 4.5 kb:

Day 1

1. Melt 0.4% SeaKem Gold agarose in TBE + 4 mM $MgCl_2$ and pour into a 15 × 25 cm gel tray with lane dividers (*see* **Note 9**). When gel has solidified, place in electrophoresis apparatus with TBE + 3 mM $MgCl_2$, room temperature.
2. After digests are complete, add 1/5 volume of 5 × loading buffer + Mg. Load onto gel.
3. Run the gel ~24 h at 1.7 V/cm at room temperature.
4. Prepare enough 1x alkaline running buffer for running gel and chill to 4°C.

Day 2

1. Wash gel twice with gentle agitation, 20 min, in 1 L of 10 mM EDTA.
2. Carefully remove lane strips and soak in 1 L of 5 mM EDTA for 20 min without shaking (*see* **Note 8**).
3. Arrange strips in an empty gel tray 90° relative to the direction of electrophoresis.
4. Carefully pour molten 0.5% SeaKem GTG agarose, prepared in H_2O and cooled to 50°C before pouring, and allow to solidify.
5. Soak gel in >2 volumes 250 mM NaOH, 5 mM EDTA for 30 min.
6. Soak gel in >2 volumes 50 mM NaOH, 1 mM EDTA for 30 min.
7. Move gel to cold room, place in electrophoresis apparatus, cover with 1x alkaline running buffer and allow gel to cool for 15 min.
8. Run for ~48 h, 1 V/cm in the cold room.

Day 4

1. Rinse gel in water, depurinate and equilibrate with transfer buffer as described in **Section 3.4.1**, **Steps 5–9**.

3.4.4. 1-d Native Gel Electrophoresis (Hunter)

Day 1:

1. Prepare a 0.6% agarose gel: add 2.1 g SeaKem LE agarose to 350 mL 1 × TBE (*without* ethidium bromide). Dissolve completely by heating in the microwave and boiling steadily for 30 s. Cool to 50°C and pour into a clean, level gel tray in the cold room. Allow to harden for at least 30 min.
2. Add 5 μL loading buffer to each digested DNA sample, mix gently.
3. Carefully load gel and run in ~2 L 1X TBE (*without* ethidium bromide) at 70 V (~2 V/cm) for 26 h at room temperature.

Day 2:

4. Stain gel in 1 L dH_2O with 0.5 μg/mL ethidium bromide in a clean Pyrex dish with gentle shaking for 30 min.
5. View gel on UV box to check that DNA digestion is complete.
6. Blot the gel by alkaline transfer overnight.

3.4.5. 2-d Native/Native Gel Electrophoresis (Hunter)

Day 1:

1. Prepare a 0.4% agarose gel: add 1.4 g SeaKem Gold agarose to 350 mL 1X TBE (*without* ethidium bromide). Dissolve completely by heating in the microwave and boiling steadily for 30 s. Cool to 50°C and pour into a clean, leveled gel tray in the cold room. Allow to harden for at least 30 min.
2. Add 5 μL loading buffer to each digested DNA sample, mix gently.
3. Carefully load the gel, leaving at least one lane space between samples. Also load λ*Bst*EII and λ*Hind*III DNA ladder on either side of the gel. Run the first dimension gel for 21 h at 35 V (1 V/cm) at room temperature.
4. Prepare two 5-L batches of 1X TBE plus 0.5 μg/mL ethidium bromide and put in the cold room to chill overnight.

Day 2:

5. Stain gel in 1 L pre-chilled 1X TBE + 0.5 μg/mL ethidium bromide for 30 min at room temperature with gentle shaking. To analyze 12 samples (a standard time course) prepare two 400 mL 0.8% agarose gels: add 3.2 g SeaKem LE agarose to 400 mL 1X TBE. Dissolve completely by heating in the microwave and boiling steadily for 30 s. Add ethidium bromide to 0.5 μg/mL, mix and place to cool in 50°C water bath.
6. Lay a piece of Saran Wrap on top of the *long-wave* UV box and carefully slide the gel on top. Visualize the DNA and carefully cut ~9.5 cm slices from the lanes to cover the size range of interest; typically ~1 cm from the wells down to the 2.2 kb

band of the λ*Bst*EII ladder. Start by cutting off the top and bottom of the gel and then excise each lane as cleanly and as quickly as possible.

7. Using a flexible plastic ruler, place the excised lanes in three rows of two in a 20 × 25 cm gel tray (rows spaced ~8 cm apart). In the cold room, carefully pour the 0.8% agarose around the slices to *just* cover them. Allow to harden for 30 min.
8. Run the gels in the 4°C cold room using the pre-chilled 1X TBE with ethidium bromide at 170 V (6 V/cm) for 6 h.
9. Blot gel by alkaline transfer overnight.

3.4.6. Removal of Psoralen Cross-Links and 2d Native/ Denaturing Electrophoresis (Hunter)

Day 1:

1. Digest 4–10 μg DNA (depending on the abundance of the joint molecule of interest) with a 4-fold excess of restriction enzyme as in **Section 3.3.1**. Ethanol precipitate, dry and resuspend DNA in 15 μl 1 × TE. Add 5 μL loading buffer and load a 0.4% SeaKem Gold agarose gel prepared as in **Section 3.4.5**. Leave two lane spaces between samples and include λ*Bst*EII and λ*Hind*III DNA ladders on either side of the gel. Note that each sample requires its own second dimension gel. We typically analyze two samples at a time, i.e. 2 s dimension gels.
2. Run gel at 35 V (1 V/cm) for 30 h.
3. Chill 6 L dH_2O in the cold room (for alkaline running buffer). Also chill 1 L dH_2O + 0.5 μg/mL ethidium bromide.

Day 2:

1. Prewarm a shaking water bath to 65°C. Prepare 500 mL Equilibration Buffer and 500 mL Crosslink Reversal Buffer for **Steps 5–16**.
2. Stain gel in 1 L pre-chilled dH_2O + 0.5 μg/mL ethidium bromide for 30 min at room temperature with gentle shaking.
3. Lay a piece of Saran Wrap on top of the *long-wave* UV box and carefully slide the gel on top. Visualize the DNA and carefully cut slices from the lanes to cover the size range of interest. For the *HIS4LEU2* locus, we typically excise a ~15 cm slice encompassing DNA in the 2.3 to ~30 kb range. Cut off one corner of the gel slice to keep track of the orientation. Weigh the gel slices to estimate their volume.
4. In a small plastic container (one per gel slice), rinse the gel slices in ≥10 vol of cold dH_2O with gentle shaking for 15 min (≥10 vol are used in all subsequent rinses and incubations in **Steps 7–14**). Very carefully decant the wash and repeat.

5. Incubate as above with pre-warmed Equilibration Buffer in the 65°C water bath. Repeat this wash. During **Steps 5** and **6**, the gel slices become very fragile and slippery and great care must be taken when decanting buffers.
6. Incubate in pre-warmed Crosslink Reversal Buffer for 2–4 h at 65°C. Repeat this incubation.
7. Prepare a 1% alkaline agarose gel: dissolve 4 g SeaKem LE agarose in 380 mL 1 mM EDTA and cool in a 50°C bath (keep a stir bar in the flask). Separately, prepare 20 mL of 1 M NaOH; this will be added later to the agarose for a final volume of 400 mL (Note: the agarose percentage may need to be optimized to resolve the size range of interest. 1% is optimal for analysis of component DNA strands at the *HIS4LEU2* locus).
8. Briefly rinse gel slice in 500 mL of dH_2O at room temperature.
9. Wash twice with $\geq$10 vol of ice-cold dH_2O for 15 min to remove remaining urea.
10. Wash twice, for 15 min, with ice-cold 5X alkaline running buffer to denature the DNA.
11. Wash twice, for 15 min, with ice-cold 1X alkaline running buffer to equilibrate gel slices.
12. Place gel slices in gel tray perpendicular to first dimension. Add the 20 mL of 1 M NaOH to the cooled 1% gel (for a final concentration of 50 mM NaOH) while stirring.
13. *Promptly* pour the agarose to *just* cover the gel slice. Allow the gel to harden in the cold room for 30 min.
14. Add 1X alkaline running buffer to just cover gel. Lay Saran Wrap on top of buffer to exclude air. Run at 52 V (~1.7 V/cm) for 50 h with a change of buffer after ~24 h. The run time may need to be optimized to resolve molecules of interest.
15. Rinse the gel twice in 1 L of dH_2O with gentle shaking.
16. Blot gel by alkaline transfer overnight.

4. Notes

1. Because different spectrophotometers have different source-cuvette-detector geometries, this light-scattering measurement may not be an accurate measure of cell density. In general, spectrophotometers with at least 1 cm distance between the cuvette and the detector will give an OD_{600} of ~1.4 at the desired cell concentration. Other spectrophotometers will require calibration. SK1 cultures grown in SPS to

this concentration will be at the transition between exponential growth and stationary phase; the corresponding transition point will need to be empirically determined if other strains and/or other pregrowth regimes are used.

2. Monitor the extent of spheroplasting every few min by mixing 2 μL of spheroplast solution with 2 μL of water on a microscope slide. Optimal spheroplasting is achieved when about 80% of cells are lysed (ghosts) in water. Over or under spheroplasting will drastically reduce DNA yields.
3. Failure to vortex thoroughly leads to poor DNA recovery.
4. DNA yields should be ≥75% of expected or about 0.8 μg of DNA per mL of meiotic culture. Yields will be lower for time points early in meiosis, before DNA replication, and in late meiosis as an increasing fraction of the DNA becomes sequestered in spores. Approximate volumes (in μL) for resuspending that will give roughly equal DNA concentrations, assuming normal DNA replication and sporulation timing, are: 0 h (100), 1–2 h (50), 2.5 h (70), 3 h (80), 3.5–6.5 h (100), 7 h (75), 8 h (50).
5. Some restriction enzymes are sensitive to the spermidine and other cations present in these DNA preps, and often an enzyme from one manufacturer can perform better than the same enzyme from others. We have found the following to reliably give complete digests: *Eco*RI (50 U/μL; Fermentas, Glen Burnie, MD; use 50 U/digest); *Xho*I (50 U/μL; Fermentas; use 50 U/digest); *Xmn*I (New England Biolabs, Ipswich, MA; use 30 U/digest). Care should be taken that the concentration of Mg^{2+} and spermidine in digests does not exceed 10 and 0.1 mM, respectively, as higher concentrations may lead to smearing and incomplete transfer.
6. Some restriction enzymes are not compatible with digestion in agarose. While manufacturers' recommendations are useful, it is important to test different lots of restriction enzymes for complete digestion before committing precious samples. To minimize glycerol concentration and sample cooling, it is helpful to use the greatest enzyme concentration available.
7. Exceeding 2 V/cM will cause heating, which promotes branch migration of HJs. A 24 h run time is optimized for parental fragments of ~4.5 kb, with a 25 cm-long gel. Run times for other digests and gel lengths will need to be determined empirically.
8. It is important to chelate all Mg^{2+} with EDTA prior to treatment with alkali, as $Mg(OH)_2$ is insoluble and will trap DNA in the gel.
9. SeaKem Gold agarose (Lonza) is the only agarose with sufficient gel strength to permit manipulation of gel slices.

Acknowledgments

Work in the Hunter lab is supported by NIH NIGMS grant GM074223; work in the Lichten lab is supported by the Intramural Research Program of the National Cancer Institute. The Hunter lab protocols are modified from protocols developed by Tony Schwacha and Nancy Kleckner *(4, 10)*.

References

1. Hunter, N. (2006) Meiotic recombination, in *Molecular Genetics of Recombination* (Aguilera, A., and Rothstein, R., eds.), Springer-Verlag, Heidelberg, pp. 381–442.
2. Bishop, D. K. and Zickler, D. (2004) Early decision; meiotic crossover interference prior to stable strand exchange and synapsis. *Cell* **117**, 9–15.
3. Szostak, J. W., Orr-Weaver, T. L., Rothstein, R. J., and Stahl, F. W. (1983) The double-strand-break repair model for recombination. *Cell* **33**, 25–35.
4. Schwacha, A. and Kleckner, N. (1995) Identification of double Holliday junctions as intermediates in meiotic recombination. *Cell* **83**, 783–791.
5. Allers, T. and Lichten, M. (2001) Differential timing and control of noncrossover and crossover recombination during meiosis. *Cell* **106**, 47–57.
6. Börner, G. V., Kleckner, N., and Hunter, N. (2004) Crossover/noncrossover differentiation, synaptonemal complex formation, and regulatory surveillance at the leptotene/zygotene transition of meiosis. *Cell* **117**, 29–45.
7. Hunter, N. and Kleckner, N. (2001) The single-end invasion. An asymmetric intermediate at the double- strand break to double-Holliday junction transition of meiotic recombination. *Cell* **106**, 59–70.
8. Panyutin, I. G. and Hsieh, P. (1994) The kinetics of spontaneous DNA branch migration. *Proc. Natl. Acad. Sci. U.S.A.* **91**, 2021–2025.
9. Bell, L. R. and Byers, B. (1983) Homologous association of chromosomal DNA during yeast meiosis. *Cold Spring Harb. Symp. Quant. Biol.* **47**, 829–840.
10. Schwacha, A. and Kleckner, N. (1994) Identification of joint molecules that form frequently between homologs but rarely between sister chromatids during yeast meiosis. *Cell* **76**, 51–63.
11. Duckett, D. R., Murchie, A. I., and Lilley, D. M. (1990) The role of metal ions in the conformation of the four-way DNA junction. *EMBO J.* **9**, 583–590.
12. Allers, T. and Lichten, M. (2000) A method for preparing genomic DNA that restrains branch migration of Holliday junctions. *Nucleic Acids Res.* **28**, e6.
13. Collins, I. and Newlon, C. S. (1994) Meiosis-specific formation of joint DNA-molecules containing sequences from homologous chromosomes. *Cell* **76**, 65–75.
14. Bell, L. and Byers, B. (1983) Separation of branched from linear DNA by two-dimensional gel electrophoresis. *Anal. Biochem.* **130**, 527–535.
15. Bell, L. and Byers, B. (1979) Occurrence of crossed strand-exchange forms in yeast DNA during meiosis. *Proc. Natl. Acad. Sci. U.S.A.* **76**, 3445–3449.
16. Schwacha, A. and Kleckner, N. (1997) Interhomolog bias during meiotic recombination: meiotic functions promote a highly differentiated interhomolog-only pathway. *Cell* **90**, 1123–1135.
17. Jessop, L., Allers, T., and Lichten, M. (2005) Infrequent co-conversion of markers flanking a meiotic recombination initiation site in *Saccharomyces cerevisiae*. *Genetics* **169**, 1353–67.
18. Oh, S. D., Lao, J. P., Hwang, P. Y., Taylor, A. F., Smith, G. R., and Hunter, N. (2007) BLM ortholog, Sgs1, prevents aberrant crossing-over by suppressing formation of multichromatid joint molecules. *Cell* **130**, 259–272.
19. Allers, T. and Lichten, M. (2001) Intermediates of yeast meiotic recombination contain heteroduplex DNA. *Mol. Cell* **8**, 225–231.

20. Holm, C., Meeks-Wagner, D. W., Fangman, W. L., and Botstein, D. (1986) A rapid, efficient method for isolating DNA from yeast. *Gene* **42**, 169–173.

21. Lao, J. P., Oh, S. D., Shinohara, M., Shinohara, A., and Hunter, N. (2008) Rad52 promotes postinvasion steps of meiotic double-strand-break repair. *Mol. Cell* **29**, 517–524.

Chapter 15

Using *Schizosaccharomyces pombe* Meiosis to Analyze DNA Recombination Intermediates

Randy W. Hyppa and Gerald R. Smith

Abstract

The fission yeast *Schizosaccharomyces pombe* has many biological characteristics that make it an ideal model organism for the study of meiosis. A nearly synchronous meiosis is one of the most important. Under certain environmental and genetic conditions, large cultures of *S. pombe* can be induced to undergo meiosis in a timely and predictable manner that allows for changes in the DNA to be observed and analyzed by gel electrophoresis. Initiation of meiotic recombination via programmed DNA double-strand breaks, the formation of joint molecule recombination intermediates, and the resolution of these intermediates into crossover DNA products can all be seen with consistent timing during the progression of a synchronous meiotic induction. The timing of recombination events, the genetic requirements for the formation and disappearance of recombination intermediates, and the analysis of the DNA structures of those intermediates allow a comparison of meiotic recombination in fission yeast with that in the only other species similarly studied, the budding yeast *Saccharomyces cerevisiae*.

Key words: Fission yeast, *Schizosaccharomyces pombe*, meiotic induction, DNA double-strand breaks, joint molecules, Holliday junctions, intersister, interhomolog, crossover, pulsed-field gel electrophoresis, two-dimensional gel electrophoresis.

1. Introduction

Synchronous meiosis in *S. pombe* can be studied in two genetic backgrounds, with slightly different methods. (**i**) A diploid heterozygous both at the mating-type locus (h^+/h^-) and for complementing auxotrophic markers is cultured in minimal medium to mid-log phase and then starved of nitrogen. This starvation initiates meiosis in the cells *(1)*, though not in a very synchronous manner. (**ii**) Greater synchrony can be achieved with a temperature-sensitive mutation (*pat1-114*) in the *pat1* gene, which encodes a protein kinase that represses meiosis *(2, 3)*. Even haploid *pat1-114* cells

Scott Keeney (ed.), *Meiosis, Volume 1, Molecular and Genetic Methods, vol. 557*

DOI 10.1007/978-1-59745-527-5_15 Springerprotocols.com

initiate meiosis, form double-strand breaks (DSBs) with the same timing and position, and proceed at least as far as the formation of joint molecules (JMs). *S. pombe pat1-114* haploid or diploid cells (homozygous for *h*, to prevent starvation-induced meiosis) are grown to mid-log phase in minimal medium and then starved for nitrogen to arrest the cells in G1 of the cell cycle. The reintroduction of nitrogen concurrent with a shift to high temperature allows the arrested cells to progress through a highly synchronous meiosis. This enables the study of meiotic events in an entire population. Although the sites of DSBs are the same in *pat1*$^+$ and *pat1-114 (4)*, many of the meiotic events happen at such a low frequency that they cannot be visualized in a *pat1*$^+$ (starvation-induced meiosis) background. Most studies have, therefore, been done with *pat1-114* mutants.

Shortly after meiosis begins (at 2–3 h after induction) a round of DNA replication takes place (5). The branched DNA replication intermediates can be visualized at this time using two-dimensional (2D) gel electrophoresis, which separates DNA primarily by mass in the first dimension and shape in the second *(6)*. After replication is complete (at 3–4 h), meiotic DNA double-strand breaks are introduced by the Rec12 protein, a Spo11 homolog that cuts the DNA and becomes covalently bound to the 5' ends *(7,8)*. The DSBs are repaired and are no longer visible 5–6 h after meiosis begins *(9)*. A mutation in the *rad50* gene, called *rad50S,* prevents Rec12 from being removed from the DNA; hence, the DSBs cannot be repaired, and they accumulate *(10)*. During the repair of the DSBs (at 4–5 h in *rad50*$^+$ strains) a different set of branched DNA intermediates is formed – meiotic JMs called Holliday junctions (HJs). HJs are formed following the invasion of one homolog (or sister) by the other during DSB repair mediated by homologous recombination and are visualized using 2D gel electrophoresis and electron microscopy *(11)*. Particularly useful for visualizing these intermediates are strains with a deletion of the HJ resolvase gene *mus81*, as accumulation of HJs in these mutants allows for easier analysis. Resolution of the HJs (at 5–6 h in *mus81*$^+$ strains) leads to the formation of crossover DNA products.

We describe here the analysis, by gel electrophoresis and Southern blot hybridization, of each of these intermediates – DSBs, intersister (IS) and interhomolog (IH) JMs (HJs), and final crossover products.

2. Materials

2.1. *S. pombe* Cell Culture

1. Yeast extract liquid (YEL) medium: 5 g of yeast extract (Difco), 30 g of glucose. Make to 1 L with water and autoclave. For yeast extract agar (YEA) add 2% agar (Difco) before autoclaving.

2. EMM2* (modified Edinburgh minimal medium 2): 50 mL of 20 × EMM2 salts, 25 mL of 20% (w/v) NH_4Cl, 25 mL of 0.40 M Na_2HPO_4, 12.5 mL of 40% glucose (w/v), 1 mL of 1,000 × vitamins, 0.1 mL of 10,000 × trace elements. Make to 1 L with water and filter sterilize.
3. 20 × EMM2 salts: 30.6 g of potassium phthalate (monobasic), 10 g of KCl, 5 g of $MgCl_2$, 0.1 g of Na_2SO_4, 0.1 g of $CaCl_2$. Make to 500 mL with water and autoclave.
4. 1,000 × vitamins: 1 mg of biotin, 10 mg of calcium pantothenate, 1 g of nicotinic acid, 1 g of myoinositol. Make to 100 mL with water and autoclave.
5. 10,000 × trace elements: 0.5 g of H_3BO_3, 0.4 g of $MnSO_4$, 0.4 g of $ZnSO_4{\bullet}7H_2O$, 0.2 g of $FeCl_3{\bullet}6H_2O$, 0.15 g of Na_2MoO_4, 0.1 g of KI, 0.04 g of $CuSO_4{\bullet}5H_2O$, 1 g of citric acid. Make to 100 mL with water and filter sterilize.
6. Pombe minimal (PM) medium: as EMM2 but with 20 mL of 0.40 M Na_2HPO_4 and 50 mL of 40% glucose per liter.

2.2. Meiotic Induction and DNA Extraction

1. 50 mM ethylenediamine tetraacetic acid (EDTA) pH 8.0, kept at 4°C during the induction.
2. Spheroplasting Buffer (CPES): 0.40 M EDTA pH 8.0, 120 mM Na_2HPO_4, 40 mM citric acid (free acid), 1.2 M sorbitol, 10 mM sodium azide. Store at room temperature.
3. Lysing Enzymes, from *Trichoderma harzianum* (Sigma).
4. Lyticase, from *Arthrobacter luteus*, crude (Sigma).
5. 1.0 M DTT.
6. Lysis Buffer: 0.5 M EDTA pH 8.0, 10 mM Tris-HCl pH 7.5, 10 mM sodium azide, 1% N-Lauroylsarcorine sodium salt.
7. Proteinase K (Invitrogen): 20 mg/mL in 10 mM Tris-HCl pH 7.5, 20 mM $CaCl_2$, 50% glycerol. Store at –20°C.
8. 2% Low Melting Point agarose (BioRad) in 0.25 M EDTA pH 8.0, 1.2 M sorbitol.
9. Plastic plug molds and ejectors (BioRad).
10. TE: 10 mM Tris-HCl pH 7.5, 1 mM EDTA pH 8.0.
11. 100 mM PMSF (phenylmethylsulfonyl fluoride). Store at −20°C.
12. 5 × psoralen stock: 0.5 mg/mL Trioxalen (4,5',8 Trimethylpsoralen) in 100% ethanol, stored at 4°C and wrapped in foil.
13. Working psoralen solution: 1 × psoralen stock, 50 mM Tris-HCl pH 8.0, 50 mM EDTA pH 8.0.
14. 50 mM trisodium citrate.
15. Propidium iodide: 4 mg/mL in 50 mM trisodium citrate.

16. 70% ethanol.
17. 1.5 mL microcentrifuge tubes.

2.3. Gel Electrophoresis of Recombination Intermediates

1. CHEF Mapper (BioRad) or other pulsed-field gel electrophoresis (PFGE) unit.
2. Chromosomal Grade Agarose (BioRad).
3. PFGE gel casting platform, 21 × 14 cm (BioRad).
4. PFGE markers (New England Biolabs).
5. Agarose.
6. β-agarase (New England Biolabs).
7. 5 × TBE concentrated stock: 54 g of Tris base, 27.5 g of boric acid, 20 mL of 0.50 M EDTA pH 8.0. Make to 1L with water.
8. 50 × TAE concentrated stock: 242 g of Tris base, 57.1 mL of glacial acetic acid, 100 mL of 0.50 M EDTA pH 8.0 in 1L of water.
9. Ethidium bromide (EtBr): 10 mg/mL. Store in metal foil-wrapped tube or bottle at 4°C.
10. 6 × gel loading buffer: 0.25% bromophenol blue, 0.25% xylene cyanol, 30% glycerol in water
11. 3.0 M sodium acetate pH 5.2.
12. 100% ethanol.
13. 70% ethanol.
14. UV box with camera.

2.4. Southern Blotting, Hybridization, and Analysis

1. Alkaline Transfer Solution: 1.5 M NaCl, 0.50 M NaOH.
2. 0.25 N HCl.
3. UV Stratalinker 1800 (Stratagene) or equivalent.
4. Zeta-probe GT Genomic Tested blotting membrane (BioRad #162-0197) or equivalent.
5. 20 × SSC: 3.0 M sodium chloride, 300 mM trisodium citrate pH 7.0.
6. 1.0 M Tris-HCl pH 7.5.
7. Church Buffer: 7% SDS, 0.5 M NaH_2PO_4 pH 7–7.2, 1% BSA, 2 mM EDTA pH 8.0.
8. 5 × labeling Buffer (Promega #U115A).
9. dATP, dGTP, dTTP mix (0.5 mM each).
10. [α-^{32}P] dCTP, 3,000 Ci/mmol.
11. 10 × BSA: 1 mg/mL in water. Store at –20°C.
12. Klenow, DNA PolI Large Fragment (New England Biolabs #M0210S).
13. Microspin S-200 HR columns (GE Healthcare #27-5120-01).

14. Hybridization oven and bottles.
15. 10% sodium dodecyl sulfate (SDS).
16. Plastic wrap (Saran or equivalent).
17. X-ray film cassettes.
18. BioMax MS-Film (Kodak #829-4985).
19. Film processor.
20. Typhoon Trio variable mode imager (GE Healthcare) or equivalent.
21. Storage phosphor screens (GE Healthcare).
22. Image Quant software (GE Healthcare) or equivalent.
23. Microsoft Excel.

3. Methods

When inducing an *S. pombe* strain for meiosis, it is important to keep in mind the timing of events and the focus of the induction. This will dictate how much culture to induce, what time points to take, and what strains to use. Most meiotic events can be observed in a 0–6 h induction, time points being taken in hourly intervals. Half-hour time points can be taken for finer temporal analysis. This is useful when looking for replication and recombination intermediates (adding the 2.5 and 4.5 h points, respectively) or for the earliest sign of DSBs (adding the 3.5 h point). When mutants have delayed DSB initiation or repair, induction time points past 6 h should be taken. The induction is time-consuming, and it is important to have all tubes, plug molds, and reagents prepared and labeled ahead of time.

DNA from a meiotic induction can be analyzed in several ways for physical recombination intermediates. Meiotic DSBs can be visualized from DNA molecules in a large size-range. Pulsed-field gel electrophoresis reveals DSBs genome-wide on whole chromosomes **(Fig. 15.1A)**, on DNA hundreds of kb long after cutting with rare restriction enzymes **(Fig. 15.1B)**, and, with more frequently cutting enzymes, on DNA tens of kb long to reveal DSBs with higher resolution **(Fig. 15.1C)**. Standard electrophoresis can be used to achieve separation of DNA up to about 20 kb long for fine-scale mapping. The most intense meiotic DSB hotspots, those with more than a few percent DNA breakage, can be used to observe the formation of both recombination intermediates and products by the creation or destruction of restriction sites in heterozygous condition to form DNA with different strand lengths, as has been described for the DSB hotspot *mbs1 (11)* **(Fig. 15.2A)**. Gel electrophoresis of meiotic DNA digested with

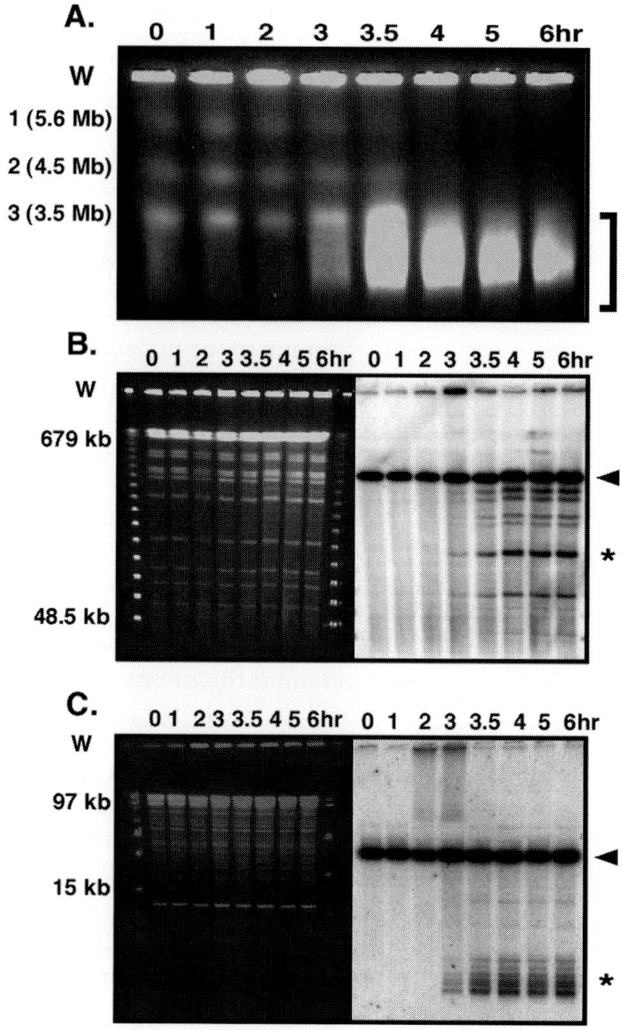

Fig. 15.1. Pulsed-field gel electrophoresis (PFGE) of *S. pombe* meiotic DNA to assay DSBs. A *pat1-114 rad50S* haploid strain was induced for meiosis (**Sections 3.1.1** and **3.2**). (**A**) PFGE of whole chromosomes (labeled 1, 2, and 3) (**Section 3.3.1.1**), stained with EtBr (0.5 μg/mL). The bracket on the right highlights the broken chromosomal DNA. (**B**) PFGE of DNA from the same induction digested with *NotI*, stained with EtBr (*left panel*) (**Section 3.3.1.2**). The size markers are Lambda Ladder PFG markers (NEB). A Southern blot of the gel (*right panel*) was hybridized with a 1 kb probe specific to the left end of the 501 kb *NotI* fragment J (*arrowhead*). DNA fragments broken at the meiotic DSB hotspot *mbs1* are indicated with an asterisk. (**C**) PFGE of DNA from the same induction digested with *EagI*, stained with EtBr (*left* panel) (**Section 3.3.1.2**). The size markers are Mid-Range 1 (*left*) and Low-Range PFG markers (*right*) (NEB). A Southern blot of the gel (*right panel*) was hybridized with a 1 kb probe specific to the left end of the 60.3 kb *EagI* fragment (*arrowhead*) of the *NotI* fragment J. DNA fragments broken at the meiotic DSB hotspot *mbs1* are indicated with an *asterisk*. In each panel W indicates the wells into which the plugs with DNA were placed.

the chosen restriction enzymes will separate the crossover DNA formed via meiotic recombination (**Fig. 15.2B**); isolation of the JM intermediates – interhomolog and intersister – requires electrophoresis in a second dimension (**Fig. 15.3A and B**).

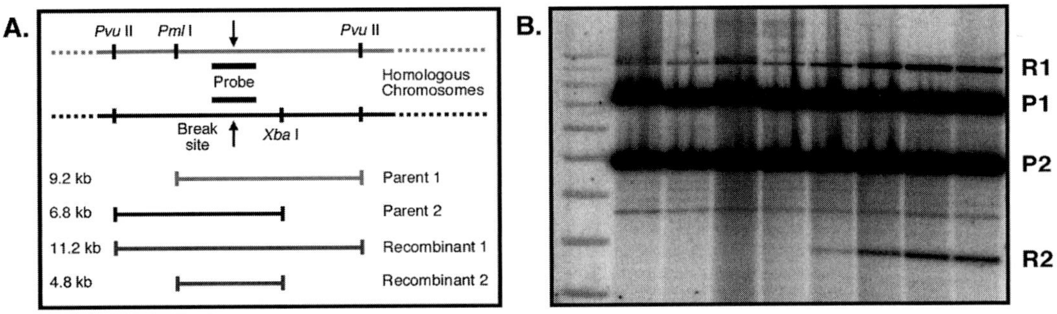

Fig. 15.2. Standard gel electrophoresis of *S. pombe* meiotic DNA to assay crossover DNA. (**A**) The DNA construct to assay JMs and crossover DNA at *mbs1*. The *PmlI* and *XbaI* sites are heterozygous in the diploid strain and flank the DSB hotspot *mbs1*. Digestion at the flanking *PvuII* sites in combination with *PmlI* and *XbaI* digestion produces parental DNA 9.2 kb long (parent 1; P1) and 6.8 kb long (parent 2; P2). Recombination (meiotic crossing-over or conversion of one marker) produces DNA 11.2 kb long (recombinant 1; R1) and 4.8 kb long (recombinant 2; R2). (**B**) Southern blot of the gel described in **Section 3.3.2**, probed as depicted in **Fig. 15.2A**. Note that incomplete digestion produces a fragment identical to R1. Therefore, quantitation of recombination is 2(R2)/[P1+P2+2(R2)]. Left most lane contain a 1 kb ladder; the bottom marker is 4 kb. (Modified from **Ref.** *(11)*).

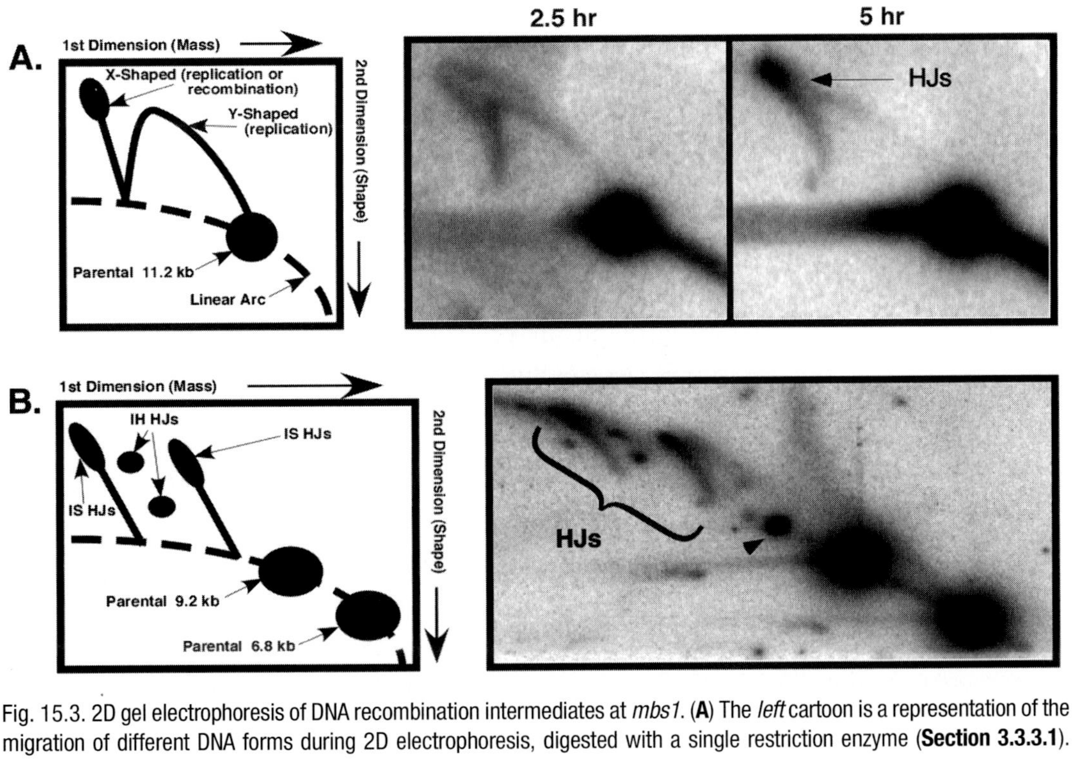

Fig. 15.3. 2D gel electrophoresis of DNA recombination intermediates at *mbs1*. (**A**) The *left* cartoon is a representation of the migration of different DNA forms during 2D electrophoresis, digested with a single restriction enzyme (**Section 3.3.3.1**).

3.1. *S. pombe* Cell Culture for Meiotic Induction

3.1.1. Conditions for *pat1-114* Strains

1. Streak an *S. pombe pat1-114* strain from a −80°C glycerol stock onto a YEA plate with appropriate supplements (100 μg/mL). Incubate the plate at 25°C.
2. When grown (about 4–5 d), pick an isolated colony to 5 mL of YEL with appropriate supplements (100 μg/mL; *see* **Note 1**) and incubate on a roller drum at 25°C until saturated (2–3 d).
3. Use the starter culture to inoculate (1:100) a 50 mL EMM2* culture with supplements (75 μg/mL). Grow at 25°C with moderate shaking until the OD_{600} is between 1.5 and 2.0 (2–3 d). Do not let the OD_{600} become much higher than this, as subsequent growth may be affected.
4. Dilute the EMM2* culture into 500 mL of EMM2* to an OD_{600} of 0.10 in a 2-L flask. This needs to be done approximately 14–16 h before starvation is to occur (*see* **Note 2**). Incubate at 25°C with moderate shaking.
5. When the culture reaches an OD_{600} of 0.3–0.4, centrifuge the cells for 5 min at 4,000 rpm (2,800*g*) and wash once in sterile water. Resuspend the cells in 500 mL of EMM2* without NH_4Cl and with supplements limited (10 μg/mL). Grow at 25°C with moderate shaking for 18–22 h.
6. The OD_{600} will have approximately doubled after starvation, to about 1×10^7 cells/mL. Add 12.5 mL of 20% (w/v) NH_4Cl and supplements to 75 μg/mL, and place the culture in a 34°C water bath with moderate shaking. This time marks the initiation of meiotic induction.

3.1.2. Conditions for *pat1⁺* Strains

1. Streak an *S. pombe* $pat1^+/pat1^+$ h^+/h^- diploid strain from a glycerol stock to an unsupplemented YEA plate (*see* **Note 3**) and incubate at 30°C.
2. When grown (3–4 d), pick an isolated colony to 10 mL of YEL (add guanine to 80 μg/mL if the diploid is maintained by *ade6-M210/ade6-M216* complementation; only Ade^+ cells can grow in the presence of guanine). Incubate on a roller drum at 30°C until nearly saturated (1–2 d).

Fig 15.3 (continued) The *left* image is a Southern blot of DNA from a *pat1-114 mus81Δ* diploid strain, which was meiotically induced for 2.5 h. The branched intermediates are replication structures. The *right* image is from the same induction at 5 h. The X-shaped spikes are HJs arising from meiotic recombination (indicated by *arrow*). Both time points were digested with *Pvu*II and probed as in **Fig. 15.2A**. (**B**) The *left* cartoon depicts the migration of different DNA forms from a diploid strain with heterozygous flanking restriction markers (**Fig. 15.2A**) triply digested and run in a 2D gel. The *right* image is of DNA prepared at 4.5 h after induction of the strain in (**A**), digested with *Pvu*II, *Pml*I, and *Xba*I, and assayed by 2D gel electrophoresis (**Section 3.3.3.2**). The IH HJs run as two species (due to their asymmetric structure) at a position between the two IS HJ spikes (highlighted by bracket). *See* **Ref.** *(11)* for further explanation. The black arrowhead points to a partial digestion fragment. [Modified from **Ref.** *(11)*].

3. Use the starter culture to inoculate (1:100) a 500 mL PM culture in a 2-L flask (*see* **Note 4**) and incubate at 30°C with moderate shaking for 12–14 h.
4. When the OD_{600} has reached 0.5–0.7, centrifuge the cells for 5 min at 4,000 rpm (2,800*g*) and wash once in sterile water. Resuspend the cells in 500 mL of PM without NH_4Cl and with glucose reduced to 1.0%. Place at 34°C and shake *vigorously* (*see* **Note 5**). This time marks the initiation of meiotic induction. The timing of a *pat1*$^+$ induction is 3–4 h slower than that of *pat1-114* (**Fig. 15.1B**), so allow for the extra time.

3.2. Meiotic Induction and DNA Extraction

1. Starting at 0 h – and at appropriate time points thereafter – centrifuge 30 mL of culture at 5,000 rpm (3,000*g*) at 4°C for 5 min. While this is spinning, put 1 mL of culture into a 1.5 mL microcentrifuge tube, spin the cells down at 13,000 rpm (16,000*g*) for 15 s, wash once in water, and resuspend in cold 70% ethanol. Store at –20°C until use for flow cytometric analysis (*see* **Step 10** below). Wash the cell pellet from the 30 mL sample with 30 mL of cold 50 mM EDTA pH 8.0 and spin at 5,000 rpm (3,000*g*) at 4°C for 5 min. Decant and put on ice (*see* **Note 6**).
2. In the same tube, resuspend the cells in 300 μL of spheroplasting buffer (CPES) with DTT (5 mM), Lysing Enzymes (5 mg/mL), and Lyticase (1 mg/mL). Mix these components just prior to induction and store on ice.
3. Place the tube containing the cells in CPES in a 50°C water bath for 1 min. Quickly add 400 μL of 2% low melting point agarose (previously melted and kept at 50°C) and rapidly pipette up and down to mix thoroughly and quickly. Keep the tubes in the water bath while mixing to prevent the agarose from solidifying. Pipette the cell-agarose mixture into prepared plug molds and cool at 4°C for 3–5 min to solidify the agarose. This should make 8–10 plugs of 100 μL each.
4. Eject the solidified plugs into 1.2 mL of CPES (from **Step 2**) in a 1.5 mL microcentrifuge tube (making sure all plugs are submerged) and incubate at 37°C for 1 h 15 min, mixing periodically by inversion or by placing on a rotator.
5. Remove the CPES with a flame-polished Pasteur pipette carefully, so as not to tear the plugs. Add approximately 1.2 mL of Lysis Buffer with Proteinase K (1 mg/mL) until the plugs are completely submerged. Mix by inversion. Incubate plugs at 50°C overnight, inverting the tubes after 30 min to mix.
6. The next day, remove Lysis Buffer and add fresh Lysis Buffer with Proteinase K (1 mg/mL). Repeat the 50°C incubation overnight, inverting the tubes after 30 min to mix.

7. Remove the Lysis Buffer and add about 1.2 mL of TE with PMSF (1 mM) and incubate plugs at room temperature for 1–2 h, mixing periodically by inversion. The PMSF is needed to inactivate any residual Proteinase K.
8. Remove the TE-PMSF and add TE until the plugs are covered completely. Mix on a rotator at room temperature for an hour. Repeat this step two more times.
9. After the final TE wash, store the plugs at 4°C for further use. The plugs and the DNA are stable for at least two years, though the freshest preparations are ideal.
10. Flow cytometry can be used to verify that replication, and therefore the initiation of meiosis, has occurred. Spin the cell samples stored in 70% ethanol from **Step 1** for 1 min at 13,000 rpm (16,000*g*), wash once in 1 mL of 50 mM trisodium citrate, spin again for 1 min, and resuspend the cells in 0.5 mL of 50 mM trisodium citrate. Add RNase A to 10 μg/mL and incubate at least 2 h at 37°C. Pipette cells into a 5 mL round bottom tube and add 0.5 mL of propidium iodide (4 μg/mL). Vortex and store at 4°C. Sonicate the cells at the lowest output setting with a microtip for 15 s to break any cell clumps just prior to flow cytometry.

3.3. Gel Electrophoresis of Recombination Intermediates

3.3.1. Pulsed-field Gel Electrophoresis for DSB Analysis

3.3.1.1. Whole Chromosomes

1. To perform DSB analysis on whole chromosomes, take an agarose plug with DNA from each time point of a meiotic induction and soak in 250–500 μL of 1 × TAE in a 1.5 mL microcentrifuge tube at room temperature for 1 h, mixing periodically to aid diffusion. Melt 0.8% Chromosomal Grade Agarose in 1 × TAE and cool to the touch (about 40°C). With the gel comb lying flat, place one plug on each comb tooth, in order from 0 to 6 h. *S. pombe* whole chromosome standards (BioRad #170-3633) can also be used if desired. Wick away any excess buffer from around the plugs, place the comb in the gel mold, and pour gel around comb and plugs, which should remain adhered to the comb. If a plug starts to pull away, use a spatula to replace it. Alternatively, the gel can be poured and solidified first, and the plugs then placed into the empty wells, making sure that the plug and gel are in complete contact.
2. Place the gel in a PFGE box in 2 L of cold 1 × TAE, and electrophorese for 48 h at 2 V/cm, 100° angle, with a 30 min switch time (initial and final) at 14°C.
3. Stain the gel for 30 min with EtBr (0.5 μg/mL), destain in water for 15 min, and photograph under UV light (**Fig. 15.1A**).

3.3.1.2. 20 kb–1 Mb DNA Fragments

1. The digestion of the chromosomes into DNA fragments allows one to visualize sites of DSBs. First, select a restriction enzyme that will give fragments of the desired

size range. In order to prevent shearing of the DNA, the digestion is performed on agarose-embedded DNA. Use an agarose plug with DNA from each time point in the induction and soak in 250 μL of 1 × digestion buffer in a 1.5 mL microcentrifuge tube for 30 min at 4°C, mixing periodically to aid diffusion. Repeat this step once. Remove the buffer and add 250 μL of fresh 1 × buffer with 10–50 units of restriction enzyme (*see* **Note 7**). Incubate at 4°C for 4–6 h to allow the enzyme to diffuse into the plug (*see* **Note 7**), and then incubate at 37°C overnight.

2. The next day remove the buffer and soak the plugs in 0.5 × TBE at room temperature for 1 h. Melt 1.0% Chromosomal Grade Agarose in 0.5 × TBE and cool to the touch (about 40°C). Load the gel as above in **Section 3.3.1.1, Step 1**, and use appropriate sized PFGE markers.
3. Place the gel in a PFGE box in 2 L of cold 0.5 × TBE and electrophorese at 6 V/cm, 120° angle at 14°C. The length of run and the switch interval depends on the size of the restriction fragments; the longer each parameter is, the larger the fragments that can be resolved (consult BioRad with questions). If using a BioRad CHEF Mapper, use the Auto Algorithm function and enter the upper and lower limits of the desired separation.
4. Stain the gel for 30 min with EtBr (0.5 μg/mL), destain in water for 15 min, and photograph under UV light (**Figs. 15.1B and C**). Proceed to the Southern blotting procedure (**Section 3.4**) below.

3.3.2. Electrophoresis of Crossover DNA Products

1. In order to observe the physical (DNA) products of meiotic recombination, a proper arrangement of markers around a very intense DNA hotspot, one with approximately 10% DSBs or greater, is needed. Heterozygous changes that create or destroy restriction enzyme cleavage sites (*see* **Note 8**) have to be made in the DNA flanking the hotspot on each side. Additional homozygous restriction sites, one to each side, are also required. A diploid containing the heterozygous restriction sites is induced (*see* **Note 1**). Crossover DNA begins to appear at about 4 h and accumulates to its maximum at 6 h. The construct using the meiotic break site *mbs1* is shown in **Fig. 15.2A**, which requires three restriction sites for analysis; use enzymes appropriate for the construct being analyzed. The restriction sites should be approximately, but not exactly, equally spaced and span about 2–15 kb; this will make the two parental fragments, and consequently the two intersister JMs, have different masses and hence allow them to separate upon gel electrophoresis.

2. Using plugs from an induction of a diploid described above, soak each plug in 250 μL of 1 × digestion buffer for 30 min at room temperature, mixing periodically to aid diffusion. Repeat this step. Remove buffer and melt the plugs at 65°C for 1–2 min, until the agarose is completely liquefied. Add one unit of β-agarase and 10–20 units of each of the restriction enzymes and digest at 37°C for 2 h. Concentrate the DNA to about 50 μL in a vacuum oven at 40–50°C (about 1 h).
3. Melt 0.8% agarose in 1 × TAE and pour a gel with wells 1 cm wide (*see* **Note 9**) and let it solidify. Add loading buffer directly to the DNA samples and load the gel with appropriate markers, radioactively labeled when possible. Run the gel at 1 V/cm at room temperature until the desired separation is achieved (48 h for a 22 cm gel). Buffer circulation is highly recommended.
4. Stain the gel for 30 min with EtBr (0.5 μg/mL), destain in water for 15 min, and photograph under UV light. Proceed to the Southern blotting procedure (**Section 3.4**) below (*see* **Fig. 15.2B** for an example of a finished blot).

3.3.3. 2D Electrophoresis of Branched DNA Intermediates

3.3.3.1. Meiotic Time Course for JM Formation and Resolution

1. Select one or two restriction enzymes that generate a DNA fragment containing the DSB hotspot. The DNA fragment can be 5–15 kb long and should have the DSBs sufficiently far from the end of the fragment so that branch migration, and consequent loss, of the junction is not an issue. Using plugs from a meiotic induction (*see* **Note 10**), soak a plug from each time point in 250 μL of 1 × digestion buffer in a 1.5 mL microcentrifuge tube for 30 min at 4°C, mixing periodically to aid diffusion. Repeat this step. Remove the buffer and add 250 μL of fresh 1 × buffer with 10–50 units of enzyme (*see* **Note 7**). Incubate at 4°C for 4–6 h to allow the enzyme to diffuse into the plug (*see* **Note 7**), and then incubate at 37°C overnight.
2. The next day, remove the buffer and heat the plugs at 65°C for 1–2 min, until the agarose is completely liquefied. Add one unit of β-agarase and incubate at 37°C for 1 h. Add one-tenth volume of 3.0 M sodium acetate pH 5.2, put samples on ice for 15 min, and spin at 13,000 rpm (16,000*g*) for 15 min to pellet any debris. From the supernatant, precipitate the DNA with ethanol. Resuspend the DNA pellet in 30 μL of TE and immediately load (*see* **Note 11**) into a 0.4% agarose gel (10 × 8.5 cm works well) in 1 × TAE with wells 0.5 cm wide, and electrophorese in 1 × TAE at 1 V/cm for 22 h at room temperature.
3. Stain the gel for 30 min with EtBr (0.5 μg/mL) and UV photograph. Carefully cut out each lane from the gel and place each along the top of a new gel mold, rotating the lane 90° counterclockwise so the DNA will run perpendicular

from the gel slice during the second dimension of electrophoresis. Pour a 1.0% agarose gel in 1 × TAE with EtBr (0.5 μg/mL) around the gel slice and electrophorese in 1 × TAE at 4 V/cm for 14.5 h at 4°C (*see* **Note 12**).

4. Proceed to the Southern blotting procedure (**Section 3.4**) below (*see* **Fig. 15.3A** for an example of a finished blot).

3.3.3.2. Identifying Intersister and Interhomolog JMs

1. Heterozygous restriction sites (**Section 3.3.2. Step 1** and **Fig. 15.2A**) are needed to observe the presence of intersister and interhomolog JMs via 2D gel electrophoresis. The two IS species must have different masses; the IH species will have a mass intermediate between those of the IS JMs. DNA from 4.5 or 5 h usually has the maximum JMs; this is therefore the best time point to use for this analysis (*see* **Note 10**). Take two or three plugs from 0 h (as a control) and 4.5 h, and digest with restriction enzymes and β-agarase. Precipitate the DNA with ethanol as described above in **Section 3.3.3.1 Steps 1 and 2**, pooling the DNA from each time point upon precipitation.
2. Dissolve the DNA in TE and immediately load the DNA (*see* **Note 12**) into a 0.35% agarose gel in 1 × TBE with wells 0.5 cm wide. Electrophorese at 0.75 V/cm for 44–48 h at room temperature. It is very important to get good DNA separation, so conditions are altered slightly *(12)* from those in **Section 3.3.3.1 Steps 2 and 3**.
3. Stain the gel for 30 min with EtBr (0.3 μg/mL) and UV photograph. Carefully cut out each lane from the gel and place each in another gel mold at the top, rotating the lane 90° counterclockwise. Pour a 1.0% agarose gel in 1 × TBE with EtBr (0.3 μg/mL) around the gel slice and then electrophorese in 1 × TBE with EtBr (0.3 μg/mL) at 2 V/cm for 44–48 h at 4°C (*see* **Note 13**).
4. Proceed to the Southern blotting procedure (**Section 3.4**) below (*see* **Fig. 15.3B** for an example of a finished blot).

3.4. Southern Blot Hybridization and Analysis of Intermediates

1. The gel can be prepared for Southern blot transfer by either UV nicking (in a Stratagene UV Stratalinker, for example) at 60 mJ/cm^2 (for large DNA from PFGE), or by depurinating in 0.25 N HCl for 15 min at room temperature with gentle shaking.
2. Soak the gel in Alkaline Transfer Solution at room temperature for 15 min with shaking and transfer DNA to BioRad Zeta-probe membrane (or another positively charged nylon membrane) via the standard alkaline Southern blotting method *(13)*.
3. After blotting for 24 h remove the membrane and neutralize by soaking in 0.40 M Tris-HCl pH 7.5 at room temperature for 5 min with shaking, followed by a brief rinse in 2 × SSC.

The DNA becomes bound to positively charged nylon membranes during alkaline transfer, so it is not necessary to bake the blot.

4. Preheat Church Buffer to 65°C. Place the membrane in a hybridization bottle and prehybridize (soak) in 10 mL of Church buffer for 30 min at 65°C with rotation in a hybridization oven.
5. DNA probes for hybridization are generated by PCR and gel purified. For DSB blots, 1 kb probes at either the right end or the left end of the restriction fragment to be probed are desirable for easier mapping. For crossover and JM blots, the 1 kb probe should be specific to the central region of the DNA fragments to detect all of the relevant DNA species. Label the probes by random primer synthesis with the incorporation of [α-^{32}P] dCTP. Start with 25 ng of DNA template, add water to 30 μL, boil for 5 min, and cool for 5 min on ice. Add 10 μL of 5 × Labeling Buffer, 2 μL of 0.5 mM dNTPs (minus dCTP), 2 μL of 10 × BSA, five units of Klenow Fragment of DNA polymerase I, and 5 μL of 0.01 mCi/μL [α-^{32}P] dCTP (3,000 Ci/mmol). Incubate at room temperature for 1 h, and then purify the probe by spinning through a Microspin S-200 HR column at 3,000 rpm (800*g*), and measure the specific radioactivity with a scintillation counter.
6. Remove the prehybridization buffer from the bottle (*see* **Step 4** above) and add 10 mL of Church Buffer and the labeled probe to 10^6 cpm/mL of buffer. The labeled probe should be placed in a boiling water bath for 2 min to denature it just prior to adding it to the membrane.
7. Incubate the membrane at 65°C for 16–22 h with rotation in a hybridization oven.
8. Remove the hybridization solution and dispose of it according to the lab's radioactive waste disposal policy. Briefly rinse the bottle with 2 × SSC, 0.1% SDS, and properly dispose. Wash the membrane as follows, each wash being for 15 min: twice in 2 × SSC, 0.1% SDS at 40°C; twice in 2 × SSC, 0.1% SDS at 52°C; and twice in 1 × SSC, 0.1% SDS at 65°C. A higher stringency wash of 0.1 × SSC, 0.1% SDS at 65°C can be done if needed.
9. Wrap the blot in plastic (Saran) wrap and place in a film cassette. Expose on BioMax MS film for 20–24 h at −80°C. Develop in a film processor (*see* **Note 14**).
10. A digital image can be made using a Typhoon Trio variable mode imager (or one similar). Expose the blot on a recently erased storage phosphor screen overnight at room temperature (*see* **Note 15**). Then place the screen in the imager and scan the appropriate area on the storage phosphor setting

(under the "Acquisition Mode" when using the Typhoon) with the desired pixel size. Quantitate the relative radioactive DNA signals from the gel image using Image Quant (or similar) software and then import the data points into an Excel spreadsheet.

11. The amount of DSBs at each site can be measured as the ratio of the DSB signal to the total DNA signal for the entire lane, after an appropriate background has been subtracted for both. First, choose a time point at which DSBs have reached the maximum (3.5 or 4 h in *rad50*$^+$; 5 or 6 h in *rad50S*) and an early time point without DSBs (0 or 1 h, to establish the background) and subtract from all data points the lowest value present in either lane to establish the zero baseline. Next, sum the data points in each lane and adjust for the different DNA contents (due to replication and differential recovery) by dividing the total counts in the early lane by the total counts in the late DSB lane to get a normalization ratio, and multiply every data point in the early lane by this ratio. Calculate the signal associated with the unbroken DNA fragment by subtracting the background established from the local minimum on each side of the peak signal, and then sum the values across the peak. For all data points below the unbroken fragment in the late DSB lane subtract the signal from the early lane to get the signal that is above background. Sum the values across each DSB peak [e.g. *mbs1* (*) in **Fig. 15.1B and 15.1C**] to calculate the associated signal. Divide the corrected signal from each individual band below the unbroken fragment by the total DNA (the unbroken DNA plus all DSB signal above background) to calculate the DSB frequency at each site.
12. The amount of JMs for a time point (4.5 or 5 h is maximal) can be measured as the ratio of the JM signal (for each branched DNA species) to the total DNA (branched and linear), after subtracting the background locally for each DNA spot. Use an area that is the same size as the DNA spot – JM and parental DNA – but is in a "blank" region (no DNA) as the local background. Sum all distinct DNA species (**Fig. 15.3**) for the total DNA, and divide each individual nonlinear JM species by this total to calculate JM frequency.

4. Notes

1. A completely prototrophic strain or a strain with only one auxotrophy is ideal. Strains with more than one auxotrophy can fail to induce completely or at all. Also, when inducing stable h^+/h^+ or h^-/h^- diploids selected and maintained by

heterozygous auxotrophies, always grow in minimal medium to maintain diploidy. The complementing alleles *ade6-M210* and *ade6-M216* are useful for making such diploids.

2. This timing is for healthy strains. Mutant strains with growth defects (*rad51*, *mus81*, *rad32*, etc.) will need to grow an extra 4–6 h to reach the appropriate density. They will still, however, undergo a synchronous meiosis.
3. Whenever possible, maintain h^+/h^- diploids with *ade6-M210/ade6-M216* complementation for easy selection and high stability. Mitotic recombination between these markers, which could produce a prototrophic haploid, is rare. Streak such a strain onto a YEA plate without added adenine and select a white colony (Ade$^+$) to be sure that diploidy is maintained. Auxotrophic segregants make red colonies. It is also advantageous to keep any additional auxotrophic markers heterozygous to select for diploidy
4. The necessary dilution will vary depending on the growth of the strains. For a *rad50S* strain 1:100 is the appropriate dilution; for *rad50*$^+$ strains, use 1:200.
5. The shift to 34°C is necessary only for strains with the *rad50S* mutation, since it is a temperature-sensitive mutation *(14)*. It is not necessary for meiotic induction. Vigorous shaking of the cultures helps more of the cell population to proceed through meiosis. The use of baffled flasks to increase aeration is also recommended.
6. If psoralen cross-linking is needed for further manipulations, e.g., electron microscopy, do not wash the cells with 50 mM EDTA at this step. Instead, resuspend the cells in 1.5 mL of working psoralen solution (freshly prepared). Pipette the cell suspension into a plastic Petri dish and place on a long-wave UV box and irradiate with 360 nM UV light for 10 min (~50 mJ/cm^2), swirling the plate to mix the cells every few minutes. Pipette the cells back into the same tube and rinse the Petri dish once with 50 mM Tris-HCl pH 8.0, 50 mM EDTA pH 8.0 to collect the remainder of the cells *(15)*. Centrifuge at 5,000 rpm (3,000*g*) at 4°C, wash once in 50 mM EDTA pH 8.0, decant and put on ice. Continue as above.
7. The amount of restriction enzyme needed for digestion of agarose-embedded DNA varies from enzyme to enzyme. Consult the manufacturer for this information. The 4°C preincubation keeps the enzyme stable during its diffusion into the agarose. A 4–6 h preincubation is sufficient for most enzymes.
8. Insertions and deletions have been used for similar constructions, but we prefer to use single bp mutations to minimize unexpected changes in DSB patterns and recombination.

Such changes, which can be dramatic and affect events tens of kb from the DNA change, appear to result from unexplained changes in chromatin structure.

9. Wide gel wells prevent overcrowding of DNA samples and give sharper bands more readily quantitated after radioactive hybridization.
10. Analysis of HJs is greatly enhanced in a *mus81* deletion strain, in which HJs accumulate. Without HJ accumulation it is more difficult (but not impossible) to recover enough DNA intermediates to get a strong signal after Southern blot hybridization, especially for IS and IH JMs.
11. Psoralen cross-linking is not necessary for preserving JMs as long as the cells and DNA remain embedded in agarose, in which branch migration of the junctions is inhibited. However, the agarose needs to be melted and treated with β-agarase for gel analysis; directly loading agarose plugs with DNA into a 2D gel gives poor resolution. After melting the agarose, Mg^{2+} ions inhibit branch migration of the junctions. Load the DNA as soon as it is dissolved in TE to limit the potential migration of the junctions in the absence of Mg^{2+}.
12. If analysis by electron microscopy is to be performed, psoralen cross-linked DNA must be used. Digest the DNA and electrophorese in the first dimension as described. Keep the DNA ladder from this run. In the second dimension, use 1% low melting point agarose and electrophorese as described. Place the gel on a long-wave UV box and cut out an agarose slice, about 1×0.5 cm, starting just above the linear arc in the size range needed, using the first dimension DNA ladder as a guide. Extract the DNA by soaking the slice in $1 \times$ β-agarase buffer, twice for 30 min at 4°C, and then remove the buffer and melt the agarose at 65°C. Precipitate the DNA with ethanol and resuspend in TE.
13. These separation conditions work best with a 22 cm gel. The electrophoresis can also be done in $1 \times$ TAE buffer with higher voltage and shorter run times, but the separation will not be as good.
14. Exposure times will vary based on level of radioactivity. This film step can be skipped and a digital image can be created directly if preferred.
15. It is important not to expose the blot for too long. The signal can become saturated if overexposed (more than 10^5 counts per pixel), which will result in skewed quantitation data. 12–16 h is usually sufficient, depending on the level of radioactivity of the blot.

Acknowledgments

We are grateful to Sue Amundsen and Gareth Cromie for helpful comments on the manuscript. Our laboratory is supported by research grants GM031693 and GM032194 from the National Institutes of Health.

References

1. Bähler, J., Wyler, T., Loidl, J., and Kohli, J. (1993) Unusual nuclear structures in meiotic prophase of fission yeast: a cytological analysis *J. Cell Biol.* **121**, 241–256.
2. Iino, Y., and Yamamoto, M. (1985) Mutants of *Schizosaccharomyces pombe* which sporulate in the haploid state. *Mol. Gen. Genet.* **198**, 416–421.
3. Egel, R. (Ed.) (2004) The molecular biology of *Schizosaccharomyces pombe*, Springer, Berlin.
4. Cromie, G. A., Hyppa, R. W., Cam, H. P., Farah, J. A., Grewal, S. I., Smith, G. R. (2007) A discrete class of intergenic DNA dictates meiotic DNA break hotspots in fission yeast. *PLoS Genetics* **3,** e141.
5. Beach D., Rodgers L., Gould J. (1985) ran1 + controls the transition from mitotic division to meiosis in fission yeast. *Curr. Genet.* **10,** 297–311.
6. Brewer, B. J., and Fangman, W. L. (1987) The localization of replication origins on ARS plasmids in *S. cerevisiae*. *Cell* **51**, 463–471.
7. Keeney, S. (2001) Mechanism and control of meiotic recombination initiation. *Curr. Top. Dev. Biol.* **52**, 1–53.
8. Hyppa, R. W., Cromie G. A., Smith G. R., (2008) Indistinguishable landscapes of meioctic DNA breaks in rad50+ and rad50S strains of fission yeast revealed by a novel rad50+ recombination intermediate. *PLoS Genetics* **4,** e1000267.
9. Cervantes, M. D., Farah, J. A., and Smith, G. R. (2000) Meiotic DNA breaks associated with recombination in *S. pombe*. *Mol. Cell.* **5**, 883–888.
10. Young, J. A., Schreckhise, R. W., Steiner, W. W., and Smith, G. R. (2002) Meiotic recombination remote from prominent DNA break sites in *S. pombe*. *Mol. Cell.* **9**, 253–263.
11. Cromie, G. A., Hyppa, R. W., Taylor, A. F., Zakharyevich, K., Hunter, N., and Smith, G. R. (2006) Single Holliday junctions are intermediates of meiotic recombination. *Cell* **127**, 1167–1178.
12. Krysan, P. J., and Calos, M. P. (1991) Replication initiates at multiple locations on an autonomously replicating plasmid in human cells. *Mol. Cell. Biol.* **11**, 1464–1472.
13. Sambrook, J., Fritsch, E. F., and Maniatis, T. (1989) Molecular Cloning: A Laboratory Manual, Cold Spring Harbor Laboratory Press, Cold Spring Harbor, NY.
14. Farah, J. A., Hartsuiker, E., Mizuno, K., Ohta, K., Smith, G. R. (2002) A 160-bp palindrome is a Rad50-Rad32-dependent mitotic recombination hotspot in Schizosaccharomyces pombe. *Genetics* **161,** 461–468.
15. Schwacha, A., and Kleckner, N. (1994) Identification of joint molecules that form frequently between homologs but rarely between sister chromatids during yeast meiosis. *Cell* **76**, 51–63.

Chapter 16

Analysis of Chromatin Structure at Meiotic DSB Sites in Yeasts

Kouji Hirota, Tomoyuki Fukuda, Takatomi Yamada, and Kunihiro Ohta

Abstract

One of the major features of meiosis is a high frequency of homologous recombination that not only confers genetic diversity to a successive generation but also ensures proper segregation of chromosomes. Meiotic recombination is initiated by DNA double-strand breaks that require many proteins including the catalytic core, Spo11. In this regard, like transcription and repair, etc., recombination is hindered by a compacted chromatin structure because *trans*-acting factors cannot easily access the DNA. Such inhibitory effects must be alleviated prior to recombination initiation. Indeed, a number of groups showed that chromatin around recombination hotspots is less condensed, by using nucleases as a probe to assess local DNA accessibility. Here we describe a method to analyze chromatin structure of a recombination hotspot in the yeasts *Saccharomyces cerevisiae* and *Schizosaccharomyces pombe*. This method, combining micrococcal nuclease (MNase) digestion of chromatin DNA and subsequent Southern blotting, is expected to provide information as to chromatin context around a hotspot. Moreover, by virtue of MNase preferentially targeting linker DNA, positions of several nucleosomes surrounding a hotspot can also be determined. Our protocol is a very powerful way to analyze several-kb regions of interest and can be applied to other purposes.

Key words: Meiotic recombination, recombination hotspot, chromatin structure, nucleosome, linker DNA, micrococcal nuclease (MNase), yeast, indirect end labeling.

1. Introduction

1.1. Chromatin Structure

Eukaryotic DNA is associated with many proteins to form a complex termed chromatin. The fundamental unit of chromatin is the nucleosome, where ~146 bp of DNA tightly wraps around histone proteins, and arrays of nucleosomes are further folded into a highly condensed structure. Thus, chromatin structure is favorable for accommodating the immense size of DNA in a tiny nucleus.

Scott Keeney (ed.), *Meiosis, Volume 1, Molecular and Genetic Methods, vol. 557*

DOI 10.1007/978-1-59745-527-5_16 Springerprotocols.com

However, it intrinsically suppresses various chromosomal events such as transcription or recombination by sterically preventing access of *trans*-acting factors to DNA. Therefore, how this inhibition is alleviated is an essential question to understand mechanisms of such processes *(1)*.

Recent work has showed that several systems including chromatin remodeling and histone modifications decondense chromatin prior to DNA-templated processes [reviewed in *(2, 3)*]. Many of these studies exploit nucleases to assess the compaction states of chromatin: open chromatin regions are more sensitive to nucleases than closed ones are *(4)*. Among various nucleases, micrococcal nuclease (MNase) is particularly of use because it preferentially cleaves linker DNA and its digestion patterns are indicative of nucleosome positions.

1.2. Regulation of Chromatin Structure at Meiotic Recombination Hotspots

Meiotic recombination is also inhibited by chromatin structure. Since recombination is initiated by a programmed DNA double-strand break (DSB) catalyzed by complexes containing Spo11 and other factors *(5)*, one important step would be to modify local chromatin to render residing DNA competent for break formation. This can be accomplished by functions of chromatin remodelers or histone modifiers, and/or by assembly of recombination machineries per se. Supporting these notions, several reports including ours showed chromatin structure alteration is related to recombination initiation *(6–8)*. Explained in the following paragraph is a summary from our findings on this point.

We have been working on meiotic recombination in yeasts and analyzing chromatin structure around recombination hotspots, specific sites of elevated recombination frequency. In budding yeast *Saccharomyces cerevisiae*, recombination hotspots usually coincide with MNase hypersensitive sites, which suggests that hotspots are located in open chromatin regions. Importantly, their sensitivities increase prior to recombination initiation *(7)*. This may indicate that chromatin around hotspots is arranged to allow efficient initiation of recombination. Another plausible interpretation is that assembly of DSB-introducing complex(es) at hotspots might loosen local chromatin structure. We did similar experiments on a meiotic recombination hotspot *ade6-M26* of another but evolutionarily distant yeast *Schizosaccharomyces pombe (8)*. In this case, although an MNase hypersensitive site is not found around *ade6-M26* during mitosis, it does appear after meiosis induction. In addition, an MNase digestion pattern drastically changes over a several kb region spanning *M26*. This means that chromatin structure around *ade6-M26* undergoes dramatic alteration in early meiosis, and altered structure most likely reflects a recombination-competent state. These findings collectively suggest that opened chromatin would serve as a target for DSB formation and hence underscore the importance of regulating chromatin in meiotic recombination. Here, we describe an MNase-based method for analyzing chromatin structures around a recombination hotspot.

1.3. Overview of the Procedure

The protocol consists of three parts: (1) meiosis induction, (2) preparation of crude chromatin fractions and MNase digestion, and (3) restriction enzyme treatment and Southern blotting. (1) In yeast, meiosis is easily induced by changing culture media, which gives sufficient synchrony for further analyses. (2) Crude chromatin fractions isolated by Zymolyase treatment are partially digested with MNase. This step is to fragment chromatin DNA, roughly according to positions of nucleosomes (**Fig. 16.1A**). (3) After deproteinization (**Fig. 16.1B**), DNA is cut with appropriate restriction enzyme(s) to completion to produce DNA fragments around the hotspot spanning a range of sizes (**Fig. 16.1C**). The resultant DNA is analyzed by agarose gel electrophoresis followed by Southern blotting (**Fig. 16.1D**). By using a hybridization

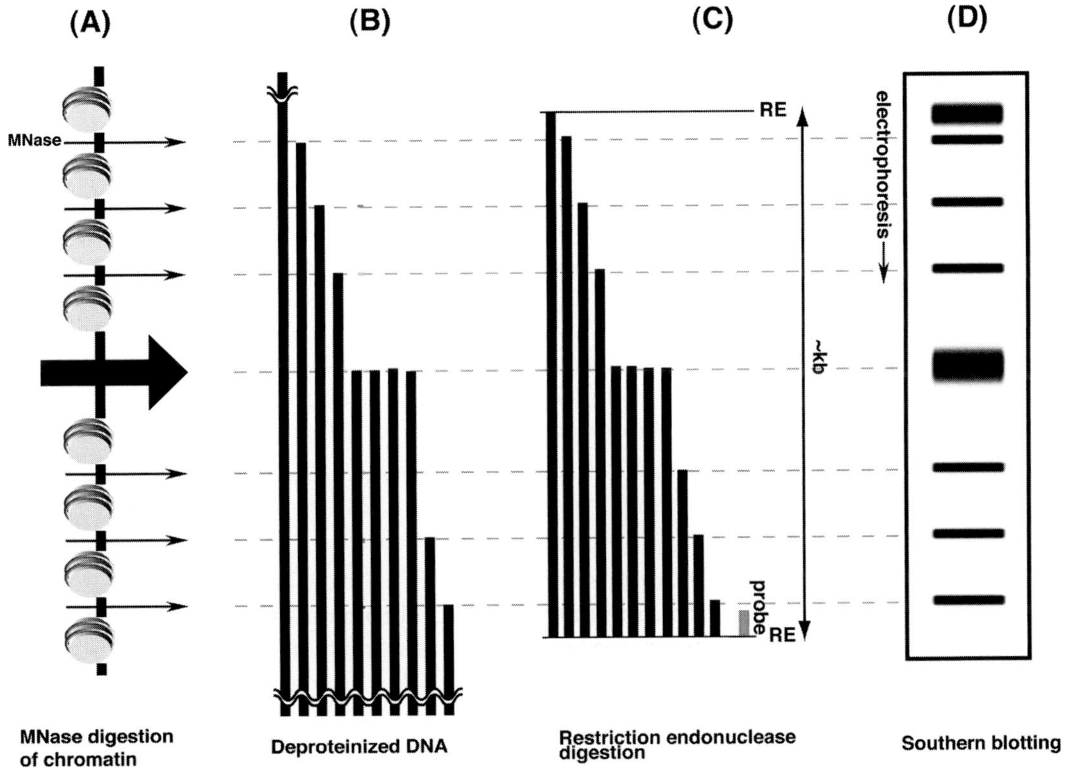

Fig. 16.1. A schematic drawing of procedures for the chromatin analysis. (**A**) MNase digestion of chromatin fractions. The *horizontal arrows* indicate MNase hypersensitive sites. As shown by the *thick arrow*, "Open" chromatin regions are more sensitive to MNase. (**B**) Purified DNA. Note DNA molecules at this stage vary in length due to partial digestion with MNase. (**C**) Complete digestion by appropriate restriction enzymes to yield hotspot-containing DNA fragments shorter than several kbp (shown by the *double-headed arrow*). (**D**) Southern blot analysis using a probe that recognizes one end of the restriction site. Note the MNase hypersensitive site shown in (**A**) gives a stronger signal. The *vertical arrow* indicates the direction of electrophoresis.

probe recognizing either end of the restriction fragment (so called "indirect end labeling"), a series of signals are detected. These signals are derived from DNA molecules that are cut by the restriction enzyme on one end, and by MNase on the other. Accordingly, setting the restriction site (which is shared by all the signals) as a reference, positions of signals indicate positions of linker DNA. In other words, a blank between adjoining signals represents DNA that is incorporated into a nucleosome.

The methods for both yeasts (*S. cerevisiae* and *S. pombe*) are very similar and the protocol for steps after restriction enzyme digestion is exactly the same (*see* **Sections 2.5 and 3.5–3.7**). There are, however, important points that differ between the two yeasts. These include culture media and conditions (*see* **Sections 2.1 and 3.1** for *S. cerevisiae*, and **Sections 2.3 and 3.3** for *S. pombe*), and procedures for preparing the chromatin fraction (*see* **Sections 2.2 and 3.2** for *S. cerevisiae*, and **Sections 2.4 and 3.4** for *S. pombe*).

2. Materials

2.1. Cell Culture for S. cerevisiae

1. YPG plate: 1% (w/v) Bacto-yeast extract (Becton, Dickinson and Company, MD, USA), 2% (w/v) Bacto-peptone (Becton, Dickinson and Company, MD, USA), 3% (v/v) glycerol, 2% (w/v) Bacto-agar (Becton, Dickinson and Company, MD, USA) in distilled water. Sterilize the mixture by autoclaving at 121°C for 15 min. One liter of medium is sufficient for approximately 40 standard plates (90 mm dia.). Allow the plates to dry at room temperature for 2–3 days after pouring. Store at room temperature in sealed plastic bags.
2. YPD plate: 1% (w/v) Bacto-yeast extract, 2% (w/v) Bacto-peptone, 2% (w/v) dextrose, 2% (w/v) Bacto-agar in distilled water. Prepare as YPG plates.
3. SPS (pre-sporulation medium): 0.5% (w/v) Bacto-yeast extract, 1% (w/v) Bacto Peptone, 0.17% (w/v) yeast nitrogen base without ammonium sulfate (Becton, Dickinson and Company, MD, USA), 0.05 M potassium hydrogen phthalate, 1% (w/v) potassium acetate, 0.5% (w/v) ammonium sulfate, 0.01% (v/v) silicone oil KM72 (Shinetsu Kagaku Kogyo, Tokyo, Japan), appropriate nutritional supplements in distilled water. First, dissolve potassium hydrogen phthalate in distilled water, adjust pH with NaOH to 5.0, and then dissolve the other components. Prepare the medium in cotton-plugged 5-L flasks with baffles, with each flask containing less than 1 L of medium, and sterilize by autoclaving at 121°C for 15 min. Store at room temperature.

4. SPM (Sporulation medium): 1% (w/v) potassium acetate, 0.001% (v/v) polypropylene glycol (#2,000) (Nacalai Tesque, Kyoto, Japan), and appropriate nutritional supplements at one fifth of the standard concentration in distilled water. Prepare the medium in cotton-plugged 5- or 2-L flasks with baffles, with each flask containing less than 500 or 200 mL of medium, respectively. Sterilize by autoclaving at 121°C for 15 min and store at room temperature.
5. Nutritional supplements (100 ×) (each prepared separately): 2 mg/mL adenine sulfate; 2 mg/mL uracil; 2 mg/mL L-tryptophan; 2 mg/mL L-histidine-HCl; 3 mg/mL L-leucine; 2 mg/mL L-arginine-HCl. Sterilize by autoclaving at 121°C for 15 min. Store at room temperature and add to media according to the genotypes of the strain analyzed.

2.2. Isolation of Chromatin Fraction from S. cerevisiae

1. Preincubation solution: 0.7 M β-mercaptoethanol, 3 mM EDTA, 20 mM Tris-HCl (pH 8.0) in distilled water. Prepare immediately before use (50 mL) and keep the solution on ice until use.
2. 1 M sorbitol in distilled water : Dissolve 182.71 g of sorbitol in 1L of distilled water. Sterilize by autoclaving at 121°C for 15 min and store at room temperature. Place the solution on ice prior to use.
3. Zymolyase solution: 1 mg/mL Zymolyase 100T (Seikagaku Kogyo Corporation, Tokyo, Japan) in 1 M sorbitol. Dissolve 10 mg of Zymolyase 100T in 10 mL of 1 M sorbitol. Prepare immediately before use and keep on ice.
4. Lysis buffer: 18% (w/v) Ficoll 400, 10 mM KH_2PO_4, 10 mM K_2HPO_4, 1 mM $MgCl_2$, 0.25 mM EGTA, 0.25 mM EDTA in distilled water. Make 50 mL aliquots and store at –20°C. Immediately before use, 1 mM Pefabloc SC (Roche, Mannheim, Germany) may be added.
5. Buffer A: 10 mM Tris-HCl (pH 8.0), 150 mM NaCl, 5 mM KCl, 1 mM EDTA. Sterilize by autoclaving at 121°C for 15 min and store at 4°C until use. Immediately before use, 1 mM Pefabloc SC (Roche, Mannheim, Germany) may be added.

2.3. Cell Culture for S. pombe

1. YE medium *(9)*: 0.5% (w/v) Bacto-yeast extract, 2% (w/v) dextrose, 0.1 mg/mL adenine in distilled water. Sterilize by autoclaving at 121°C for 15 min. Store at room temperature.
2. MM medium *(10)*: 0.3% (w/v) potassium hydrogen phthalate, 0.22% (w/v) Na_2HPO_4, 1% (w/v) dextrose, 20 mL/L Salts stock (see below), 1 mL/L Vitamins stock (see below), 0.1 mL/L Minerals stock (see below) and 0.1 mg/mL

adenine in distilled water. Add other nutritional supplements if needed. Sterilize the mixture by autoclaving at 121°C for 15 min. After autoclaving, add 20 mL of sterilized 25% (w/v) NH_4Cl to 1L of MM medium to make MM+N medium. Store at room temperature. Prepare medium in cotton-plugged 5- or 2-L flasks with baffles, with each flask containing less than 500 or 200 mL of medium, respectively.

3. Salts stock: 5.2 mM $MgCl_2$, 0.1 mM $CaCl_2$, 13.4 mM KCl, and 0.28 mM Na_2SO_4 in distilled water. Sterilize by autoclaving at 121°C for 15 min. Store at 4°C.
4. Vitamins stock: 1% (w/v) nicotinic acid, 1% (w/v) inositol, 0.001% (w/v) biotin, 0.1% (w/v) pantothenic acid in distilled water. Filter-sterilize and store at 4°C.
5. Minerals stock: 8.1 μM H_3BO_3, 2.37 μM $MnSO_4$, 1.39 μM $ZnSO_4$, 0.74 μM $FeCl_3$, 0.25 μM Na_2MoO_4, 0.6 μM KI, 0.16 μM $CuSO_4$, 1% (w/v) citric acid in distilled water. Filter-sterilize and store at 4°C.

2.4. Isolation of Chromatin Fraction from S. pombe *Cells and Micrococcal Nuclease (MNase) Treatment*

1. Preincubation solution: 0.7 M β-mercaptoethanol, 3 mM EDTA, 20 mM Tris-HCl (pH 8.0) in distilled water. Mix 0.25 mL of β-mercaptoethanol, 30 μL of 0.5 M EDTA (pH 8.0), and 0.1 mL of Tris-HCl (pH 8.0) in distilled water to make 5 mL of preincubation solution. Prepare immediately before use.
2. Zymolyase solution: 0.75 M sorbitol, 37.5 mM Tris-HCl, 1.25% (w/v) dextrose, 6.25 mM EDTA, 2.08 mg/mL Zymolyase 100T (Seikagaku Kogyo Corporation, Tokyo, Japan) in distilled water. Prepare 12 mL of this solution immediately before use and keep solution on ice. Also prepare 12 mL of the same solution without Zymolyase.
3. Lysis buffer: Prepare as described in **Section 2.2.**
4. Buffer A: Prepare as described in **Section 2.2.**
5. Sorbitol buffer: 1 M sorbitol, 10 mM EDTA in distilled water. Sterilize by autoclaving at 121°C for 15 min. Store at room temperature.
6. Micrococcal Nuclease (MNase) stock: Dissolve MNase (GE Healthcare) at 10 U/μL in sterilized 50% (v/v) glycerol. Store at –30°C.
7. $CaCl_2$ stock solution: 1 M $CaCl_2$ in distilled water. Sterilize by autoclaving at 121°C for 15 min. Store at room temperature.
8. 0.5 M EDTA: 0.5 M EDTA (pH 8.0) in distilled water. Sterilize by autoclaving at 121°C for 15 min. Store at room temperature.

2.5. Gel-Electrophoresis and Southern Blotting Analysis

1. 50 × TAE: Dissolve 242 g of Tris base, 57.1 mL of acetic acid, and 100 mL of 0.5 M EDTA (pH 8.0) in distilled water to make 1 L of the solution. Store at room temperature.
2. Denaturation buffer: 1.5 M NaCl, 0.5 M NaOH in distilled water. Store at room temperature.
3. Transfer buffer: 1.5 M NaCl, 0.25 M NaOH in distilled water. Store at room temperature.
4. Pre-wet buffer: 0.125 M sodium phosphate (pH 7.4) in distilled water. Dissolve Na_2HPO_4 to 0.125 M in distilled water and adjust pH to 7.4 with phosphoric acid. Store at room temperature.
5. Hybridization buffer: 1% (w/v) bovine serum albumin (BSA), 7% (w/v) sodium dodecyl sulfate (SDS), 1 mM EDTA, 0.25 M Na_2HPO_4 (pH 7.4) in distilled water. Dissolve Na_2HPO_4 in distilled water and adjust pH with phosphoric acid to 7.4 and then dissolve remaining compounds. Make 50 mL aliquots and store at –20°C.
6. Wash buffer: 1% SDS, 1 mM EDTA, 0.1 M Na_2HPO_4 (pH 7.4) in distilled water. Dissolve $Na_2HPO_4 \cdot 12H_2O$ in distilled water and adjust pH with phosphoric acid to 7.4 and then dissolve remaining compounds. Store at room temperature.
7. 2 × SSC buffer: 0.3 M NaCl, 33.7 mM sodium citrate (pH 7.0) in distilled water. Dissolve NaCl and sodium citrate in distilled water and adjust pH to 7.0 with NaOH. Store at room temperature.
8. TE buffer: 10 mM Tris-HCl (pH 8.0), 1 mM EDTA (pH 8.0) in distilled water. Sterilize by autoclaving at 121°C for 15 min and store at room temperature.

3. Methods

Elevated sensitivity of chromatin to MNase is found at meiosis-specific hotspots in both budding and fission yeasts *(6–8)*. To obtain reliable and reproducible results, it is important to prepare spheroplasts carefully using the same conditions. Therefore, the same lot of Zymolyase 100T should be used in this procedure in every experiment. When the lot of Zymolyase is changed, the amounts of Zymolyase used in this procedure may need to be optimized. If mutants are analyzed, it is also important to achieve equal levels of spheroplasting among the mutants examined. For example, we use half the amounts of Zymolyase to analyze *atf1*Δ and *pcr1*Δ that we use for wild-type strains, because these mutants are sensitive to Zymolyase *(11)*. Here we show the method of the analysis for chromatin configuration at meiotic DSB sites using budding yeast (*S. cerevisiae*) and fission yeast (*S. pombe*) (*see* **Note 1**).

3.1. Synchronous Sporulation of S. cerevisiae

1. These instructions assume the use of wild-type diploid strains of SK1 background, which show high sporulation efficiency. The instructions could be applied to other strains, with appropriate alteration in incubation time, temperature, and so on.
2. Streak cells from a –80°C glycerol stock to a YPG plate and incubate at 30°C overnight.
3. Restreak cells from the YPG plate to a YPD plate and incubate at 30°C for two days.
4. Inoculate a single colony from the YPD plate into 10 mL of SPS medium in a 50 mL tube. Grow 24 h at 30°C with vigorous shaking.
5. Inoculate a small amount of the culture into 500–750 mL of SPS in a 5-L flask with baffles and incubate at 30°C with vigorous shaking (125–140 rpm, 10 cm stroke) to a density of 2–5 × 10^7 cells/mL.
6. Collect cells by centrifugation and resuspend in 50 mL of sterile water.
7. Remove half of the cells, pellet in a 50 mL tube and freeze in liquid nitrogen after weighing the pellet (wet weight; *see* **Note 2**). Store at –80°C as a t = 0 h (premeiotic) sample.
8. Transfer the other half of the cells to 500–200 mL of prewarmed SPM and incubate at 30°C with vigorous shaking (125–140 rpm, 10 cm stroke). Remove cells at appropriate time points as meiotic samples. Typical chromatin transition at hotspots is observed after 3–4 h incubation in SPM. Wash the cells in 50 mL of sterile water, pellet in a 50 mL tube, freeze in liquid nitrogen after weighing the pellet, and store at –80°C .

3.2. Preparation of Crude Chromatin from S. cerevisiae Cells and MNase Digestion

1. All procedures are performed at 4°C or on ice unless otherwise stated.
2. Resuspend the frozen cell pellets in 2 mL of preincubation solution per one gram of cells. Take a fraction of the suspension corresponding to 1 g of cells and transfer to a fresh tube to start with a fixed amount of cells. Incubate the samples at 30°C for 10 min with shaking.
3. Collect the cells by centrifugation at 3,500 rpm (2,330*g*) for 5 min. Resuspend in 5 mL of 1 M sorbitol and collect by centrifugation at 3,500 rpm (2,330*g*) for 5 min. Carefully remove the supernatant and resuspend in 4.5 mL of 1 M sorbitol.

4. Add 0.5 mL of Zymolyase solution to each sample and mix well. Incubate with occasional gentle shaking for 5 min (*see* **Note 3**). Collect the resultant spheroplasts by centrifugation at 3,500 rpm (2,330*g*) for 5 min. Carefully remove the supernatant and resuspend in 5 mL of 1 M sorbitol (*see* **Note 4**).
5. Collect the spheroplasts by centrifugation at 3,500 rpm (2,330*g*) for 5 min. Carefully remove the supernatant and resuspend thoroughly in 7 mL of lysis buffer.
6. Collect the crude nuclear pellet by centrifugation at 14,000 rpm (20,406*g*) for 30 min. Carefully remove the supernatant and resuspend in 4 mL of buffer A. Mix 1 mL aliquots of the crude nuclear suspension with 5 μL of 1 M $CaCl_2$ in a 1.5 mL tube.
7. Add MNase at an appropriate concentration (0, 5, 10, and 20 U/mL, for example) to each sample and incubate at 37°C for 5 min.
8. Terminate the reaction by adding 50 μL of 0.5 M EDTA (pH 8.0). Add 0.1 mL of 10% (w/v) SDS and 3 μL of 20 mg/mL Proteinase K, then incubate each sample at 55°C for 2 h to overnight.
9. Add 0.4 mL of 5 M potassium acetate and incubate for 15 min.
10. After centrifugation to remove cell debris, subject the supernatant to phenol-chloroform extraction and isopropanol precipitation for DNA purification. Dissolve the pellet in 0.1 mL of TE buffer.

3.3. Cell Culture for S. pombe

1. Inoculate diploid *S. pombe* cells in YE medium, and culture at 30°C overnight.
2. Inoculate a part of the overnight culture into 500 mL of MM+N medium in a 5-L flask with baffles and incubate at 30°C with vigorous shaking (125–140 rpm, 10 cm stroke) for at least 12 h to a density of 0.5–1 $\times$ 10^7 cells/mL.
3. Harvest cells and freeze half of them in liquid nitrogen after weighing the pellet (wet weight) and store at –80°C. Wash the remainder of the cells with distilled water twice, transfer to MM medium lacking a nitrogen source (NH_4Cl) and culture with vigorous shaking (125–140 rpm, 10 cm stroke) to induce meiosis.
4. Harvest the cells at appropriate time points after medium shift. Typical chromatin remodeling at *ade6-M26* is observed in 3~4 h after medium shift in which premeiotic DNA synthesis is occurring *(8)*. Freeze the harvested cells in liquid nitrogen after weighing the pellet (wet weight) and store at –80°C until the analysis.

3.4. Isolation of Chromatin Fraction from S. pombe Cells and MNase Digestion

1. The samples of chromatin are prepared from a fixed amount of cells (0.5 g wet weight). Suspend these cells in 1 mL of preincubation solution and incubate at 30°C for 10 min. After the incubation, collect the cells by centrifugation at 3,500 rpm (2,330*g*) at 4°C for 5 min and resuspend in 5 mL of ice-cold sorbitol buffer.
2. Collect the cells by centrifugation at 3,500 rpm (2,330*g*) at 4°C for 5 min and resuspend in 2.5 mL of ice-cold Zymolyase solution that contains no Zymolyase. Add 2.5 mL of freshly prepared Zymolyase solution containing Zymolyase 100T and mix well. Incubate the cells with gentle agitation at 30°C for 5 min and collect resultant spheroplast by centrifugation at 3,500 rpm (2,330*g*) at 4°C for 5 min and resuspend in 5 mL of ice-cold sorbitol buffer (*see* **Note 4**).
3. Collect the spheroplasts by centrifugation at 3,500 rpm (2,330*g*) at 4°C for 5 min and resuspend well by pipetting in 7 mL of ice-cold lysis buffer and centrifuge at 14,000 rpm (20,406*g*) at 4°C for 30 min. Remove the supernatant carefully.
4. Resuspend the crude nuclear pellet in 4 mL of ice-cold buffer A. Mix 1 mL aliquots of crude nuclear suspension with 5 μL of 1 M $CaCl_2$ and keep the aliquots on ice.
5. Digest 1 mL aliquots of crude nuclear suspension with different amounts of MNase (0, 10, 20, and 50 U/mL) at 37°C for 5 min.
6. Terminate the reaction by adding 50 μL of 0.5 M EDTA, and incubate with 1% SDS and 20 μg of proteinase K at 55°C for 16 h.
7. After centrifugation, purify the DNA from supernatant by phenol-chloroform extraction (three times) and ethanol precipitation. Resuspend resultant purified DNA in 0.1 mL of TE buffer.

3.5. Gel Electrophoresis and Transfer to Nylon Membrane

1. To analyze chromatin structure by an indirect end-labeling Southern analysis, the choice of restriction enzymes to digest the DNA derived from MNase-treated chromatin is important. **Figure 16.2** depicts restriction enzymes and the location of the probes to analyze chromatin structure around well-known DSB sites in *S. cerevisiae* and *S. pombe*. Generally, the restriction enzymes should generate a 1.5–5 kb fragment in which the sites of interest (DSB sites) are located about 0.8–2 kb away from either end.
2. Digest the DNA (10–20 μL) derived from MNase-treated chromatin with appropriate restriction enzymes, and separate using agarose gel electrophoresis (1–1.5% [w/v] agarose, 40 cm long) in TAE buffer at 40–60 volts for 16–22 h.

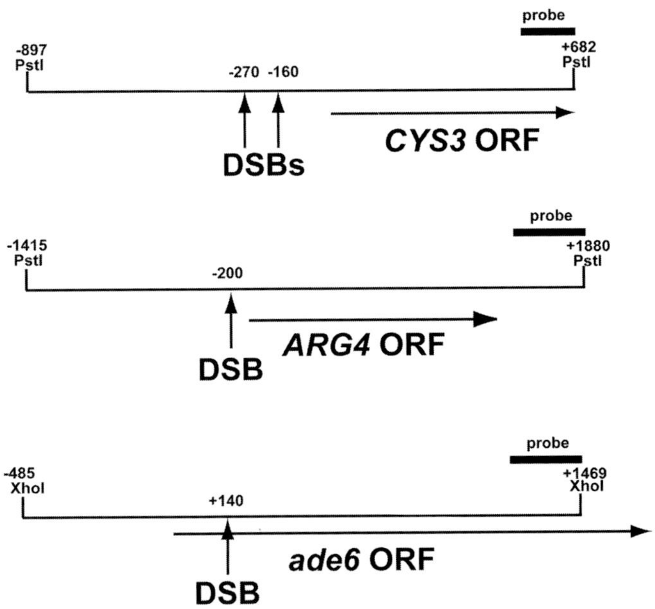

Fig. 16.2. Illustration of the position of the probe and restriction site to analyze typical DSB hotspots in *S. cerevisiae* and *S. pombe*. The *vertical* and the *horizontal arrows* indicate the positions of DSB sites and the open reading frames (ORFs), respectively. The numbers indicate the relative positions from the first ATG of each ORF.

3. After electrophoresis, transfer the separated DNA fragments by alkaline transfer to charged Nylon membranes (Biodyne B membrane; Pall Corporation, FL, USA) using a vacuum blotter (GE Healthcare) as follows. First, cut the membrane to the appropriate size and place on vacuum blotting apparatus, then place the gel containing DNA fragments on the membrane. Then, turn on the vacuum pump at 50 mbar, decant 50 mL of denaturation buffer on the gel, and allow buffer to absorb for 30 min. After that, remove the remaining denaturation buffer from the gel and decant 50 mL of transfer buffer onto the gel. During a couple of hours of vacuum blotting, additional transfer buffer might be decanted so that the surface of the gel does not dry out. After vacuum blotting, wash the membrane with 2 × SSC buffer. Dry the membrane at room temperature and store at 4°C until subsequent use.

3.6. Probe Preparation

1. The probe used for the indirect end labeling is prepared from a PCR-amplified DNA fragment. PCR-amplify an approximately 200-bp DNA fragment that covers one end of the region generated by restriction enzymes. The site of interest should be located at about ~0.8–2 kbp from the end (illustrated in **Fig. 16.2**).
2. Label the DNA fragments with ^{32}P using a random-priming kit (GE Healthcare).

3. Purify the ^{32}P-labeled probe using G50 micro column (GE Healthcare) and denature by boiling followed by quick cooling on ice prior to use.

3.7. Hybridization and Detection

1. Place the membrane in a hybridization bottle, and soak the membrane in 50 mL of prewet buffer. After discarding the buffer, pour 25 mL of preheated hybridization buffer into the bottle. Rotate the bottle containing the membrane at 30 rpm at 62°C for 30 min.
2. Discard the hybridization buffer and pour 25 mL of fresh hybridization buffer into the bottle. Then add the prepared ^{32}P-labeled probe to the hybridization buffer in the bottle, being careful not to add the probe directly onto the membrane.

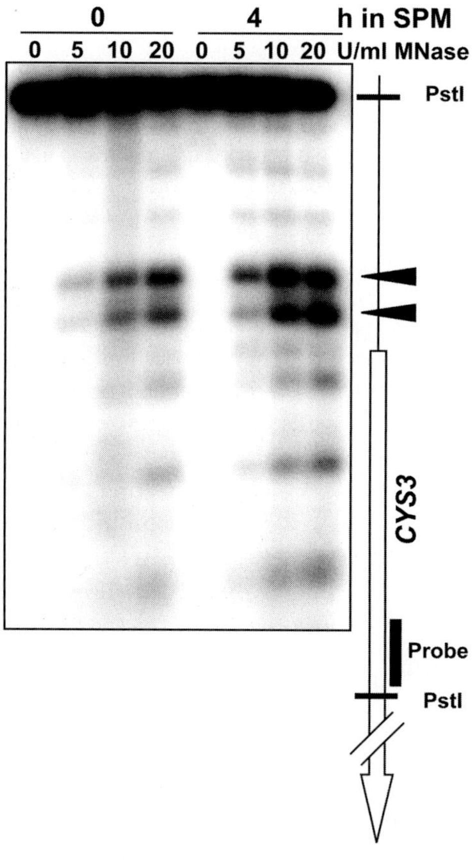

Fig. 16.3. An example of chromatin analyses: chromatin structure around the *S. cerevisiae CYS3* DSB hotspot. Crude chromatin prepared from premeiotic cells (lanes marked 0 h in SPM) and meiotic cells (4 h in SPM) was treated with varying amounts of MNase. After deproteinization, DNA was digested with *Pst* I and analyzed by Southern blotting. Note signal strength at the hotspot (*filled triangles*) increases in meiotic cells. The *open arrow*, the *filled rectangle*, and *horizontal thick lines* indicate positions of *CYS3* ORF, the hybridization probe, and *Pst* I sites, respectively.

3. After incubation with rotating at 30 rpm at 62°C for overnight, remove the hybridization buffer and rinse the membrane with 50 mL of wash buffer.
4. Wash the membrane with 50 mL of wash buffer at 62°C for 5 min and repeat the wash procedure three times or more.
5. Wrap the washed membrane with clear-plastic wrap, and analyze using densitometry apparatus Fuji BAS 2000 (Fuji Film, Tokyo, Japan). **Figure 16.3** shows typical data from chromatin analysis of the *CYS3* DSB hotspot in *S. cerevisiae*.

4. Notes

1. This method is also useful to analyze chromatin structure in promoter regions, as we have demonstrated the chromatin remodeling around promoters of stress inducible genes *(12–14)*.
2. For reproducible analysis, at least 1 g (wet weight) of cells is required. 5×10^9 cells generate roughly 1.5 g of wet pellet.
3. Zymolyase-treatment critically affects the results. Concentration, reaction temperature, and incubation time may need to be varied according to the lot number of Zymolyase to avoid partial cell lysis, nucleosomal rearrangements, and a substantial decrease in meiotically induced MNase sensitivity at DSB sites, which can be caused by excess Zymolyase-treatment.
4. After Zymolyase treatment, because resultant spheroplasts are very fragile, it is important to mix the spheroplasts pellet with buffers gently by pipetting up and down.

Acknowledgments

We thank Dr. Hajime Murakami for the original image of **Fig. 16.1**.

References

1. Wolffe, A. (1997) in: *Chromatin: Structure and function, 3rd edn*, Academic Press, San Diego, USA.
2. Nightingale, K.P., O'Neill, L.P. and Turner, B.M. (2006) Histone modifications: signalling receptors and potential elements of a heritable epigenetic code. *Curr. Opin. Genet. Dev.* **16**, 125–36.
3. Eberharter, A. and Becker, P.B. (2004) ATP-dependent nucleosome remodelling: factors and functions. *J. Cell Sci.* **117**, 3707–11.

4. Felsenfeld, G., Boyes, J., Chung, J., Clark, D. and Studitsky, V. (1996) Chromatin structure and gene expression. *Proc. Natl. Acad. Sci. U. S. A.* **93**, 9384–8.
5. Keeney, S. and Neale, M.J. (2006) Initiation of meiotic recombination by formation of DNA double-strand breaks: mechanism and regulation. *Biochem. Soc. Trans.* **34**, 523–5.
6. Wu, T.C. and Lichten, M. (1994) Meiosis-induced double-strand break sites determined by yeast chromatin structure. *Science* **263**, 515–8.
7. Ohta, K., Shibata, T. and Nicolas, A. (1994) Changes in chromatin structure at recombination initiation sites during yeast meiosis. *EMBO J.* **13**, 5754–63.
8. Mizuno, K., Emura, Y., Baur, M., Kohli, J., Ohta, K. and Shibata, T. (1997) The meiotic recombination hot spot created by the single-base substitution *ade6-M26* results in remodeling of chromatin structure in fission yeast. *Genes Dev.* **11**, 876–86.
9. Moreno, S., Klar, A. and Nurse, P. (1991) Molecular genetic analysis of fission yeast *Schizosaccharomyces pombe. Methods Enzymol.* **194**, 795–823.
10. Isshiki, T., Mochizuki, N., Maeda, T. and Yamamoto, M. (1992) Characterization of a fission yeast gene, *gpa2*, that encodes a G alpha subunit involved in the monitoring of nutrition. *Genes Dev.* **6**, 2455–62.
11. Hirota, K., Steiner, W.W., Shibata, T. and Ohta, K. (2007) Multiple modes of chromatin configuration at natural meiotic recombination hotspots in fission yeast. *Eukaryot. Cell* **6**, 2072–80.
12. Hirota, K., Hasemi, T., Yamada, T., Mizuno, K.I., Hoffman, C.S., Shibata, T. and Ohta, K. (2004) Fission yeast global repressors regulate the specificity of chromatin alteration in response to distinct environmental stresses. *Nucleic Acids Res.* **32**, 855–62.
13. Hirota, K., Hoffman, C.S. and Ohta, K. (2006) Reciprocal nuclear shuttling of two antagonizing Zn finger proteins modulates Tup family corepressor function to repress chromatin remodeling. *Eukaryot Cell.* **5**, 1980–9.
14. Hirota, K., Hoffman, C.S., Shibata, T. and Ohta, K. (2003) Fission yeast tup1-like repressors repress chromatin remodeling at the *fbp1*(+) promoter and the *ade6-M26* recombination hotspot. *Genetics* **165**, 505–15.

Chapter 17

Analysis of Protein–DNA Interactions During Meiosis by Quantitative Chromatin Immunoprecipitation (qChIP)

Marco Antonio Mendoza, Silvia Panizza, and Franz Klein

Abstract

During meiotic prophase a number of important events require recombination between maternal and paternal chromosomes, which is initiated through the introduction of DNA double-strand breaks (DSBs). The majority of DSBs, which mostly occur at so-called hotspots, have been located between cohesin binding sites. qChIP (chromatin immunoprecipitation quantified by real-time PCR) is a sensitive, accurate, and cost-efficient alternative to ChIP-on-Chip for the analysis of noncovalent protein–DNA interactions at defined binding sites in vivo. Here we use qChIP to study Mre11 binding to three chromosomal loci during meiosis. We show that Mre11 interacts with a known hotspot region (*YCR048*) in the R-band of chromosome III, but not with a cold region in the G-band (*YCR011*). Interestingly Mre11 binds to a cohesin binding site (*YCR067*), 20 kb distal to *YCR048*, with similar intensity as to the hotspot, despite the absence of DSBs in this region.

Key words: ChIP, qChIP, real-time PCR, qPCR, DNA binding, Mre11, meiotic recombination, double-strand break repair.

1. Introduction

1.1. When to Use qChIP?

qChIP is ChIP (immunoprecipitation of in vivo cross-linked and sonicated protein–DNA complexes) followed by qPCR (quantitative, real-time PCR). ChIP is used to analyze the in vivo interaction of a query protein with chromosomal target sequences. Several techniques using ChIP differ only in the analysis of the precipitate. In earlier protocols *(1)*, multiplex PCR was used to obtain a semi-quantitative answer as to whether or not a protein was associated with a target DNA sequence. This technique is useful to answer the following question: Does protein X interact

Scott Keeney (ed.), *Meiosis, Volume 1, Molecular and Genetic Methods, vol. 557*

DOI 10.1007/978-1-59745-527-5_17 Springerprotocols.com

more with sequence A or with sequence B in vivo in a cell population at a certain moment? It is not well suited to answer questions requiring quantification of such associations, and therefore also not to compare association of protein X with sequence A under different conditions (time points, treatments, mutants, tissues, etc.). In ChIP on Chip (ChIPChip), precipitated DNA is amplified by a random amplification procedure, labeled, and hybridized to a tiling microarray *(2)*. In a way, ChIPChip is the superior technique because of the relatively unbiased, genome-wide information that can be obtained. However, ChIPChip also has disadvantages, foremost the high price per DNA tiling array, making it unsuitable for large measurement series. Second, because DNA is usually amplified beyond the linear range, current ChIPChip protocols yield little information about the absolute amount of binding. As with multiplex PCR, signals within one and the same experiment can be compared with high sensitivity. Third, ChIPChip results suffer from noise (limiting spatial resolution) and from possible artifacts introduced by the random amplification and/or by the hybridization steps. When a genome-wide profile of protein–DNA interactions has been obtained, it may be desirable to confirm binding at several key binding sites, and to quantify it with a method independent of random amplification and hybridization. This is an indication to perform qChIP. Furthermore, at these sites a number of conditions can be tested and compared (time points, treatments, mutants, tissues, etc.), because the costs for qChIP are only a fraction of those for ChIPChip and in addition information about the absolute yield can be obtained.

1.2. How Quantitative is qChip?

qPCR is linear over a large dynamic range. The comparison between different target sites within the same precipitate depends only on qPCR, and can therefore be regarded as highly exact. In the case study examined here, this corresponds to the values for hotspot, cohesin site, and cold G-band region for one strain (Mre11-HA) at a single time point.

As an internal standard it would be useful to include a constantly occupied reference point in the genome in each measurement, e.g., a protein X carrying the tag of interest, binding strongly and exclusively to a single target site. Unfortunately, such a reference construct is generally not available. Technical difficulties include the search for a promoter that keeps a constant intracellular level of protein X-tag under all experimental conditions, because the yield of an IP will depend on the ratio of standard to query protein and of DNA bound and unbound fractions. It is also essential that protein X-tag does not interfere with the DNA-binding ability of query proteins, or binds at other places than its designated target site. Preliminary trials using as such an internal standard a Myc-tagged repressor under the control of the *URA3* promoter and a "unique" repressor binding site failed, most likely because its expression fluctuated, but also because we detected some binding in addition to the "unique" binding site in other regions of the genome (unpublished observations).

In the absence of an internal standard, comparisons between different IPs are influenced by the efficiency of the IP (i.e., the reproducible handling) and the expression of the bait protein. While the expression of the bait can be followed by Western blotting if required, experimental conditions can be kept relatively constant, giving rise to meaningful time course studies *(3)*. The influence of the yield of a particular extraction is partially eliminated by expressing the IP always relative to the whole cell extract (WCE) of each data point. The outlined protocol is derived from ChIP *(1)* and ChIPChip *(4)* protocols, combined with qPCR.

1.3. Mre11-HA Interaction with Meiotic Chromosomes

Mre11, a key protein in meiotic double-strand break (DSB) repair *(5)*, was tagged with six copies of the HA-tag on the C-terminus using PCR-mediated recombination, tested and determined to be functional in meiotic recombination. Before choosing the target sites for analysis used in this study, we performed ChIPChip on Mre11-HA and obtained a profile similar to one previously published *(6)*, with the exception that we also observe strong signals not corresponding to DSB sites (data not shown). We therefore wished to investigate whether these sites might be artifacts, either of the ChIPChip technique or of our bioinformatic evaluation. We chose chromosome III, where DSB sites have been mapped *(7)*, to test Mre11-HA interactions with three loci: one at a prominent DSB site (*YCR048*), one in the cold G-band region (*YCR011*) and one, (*YCR067*), identified as a cohesin-binding site (data not shown), where we had seen a strong Mre11 signal by ChIPChip. Our results confirmed that Mre11 indeed interacts with the cohesin site, despite the absence of DSBs, with similar intensity and similar kinetics as with a strong DSB site. On the basis of the signal strength and the timing, we hypothesize that this interaction may be of similar importance to meiotic recombination, as the interaction with the DSB site.

2. Materials

2.1. Yeast Culture and Sample Collection

1. Pre-sporulation medium (SPS): 0.5% Yeast extract, 1% Peptone, 0.17% Yeast nitrogen base (without amino acids and without ammonium sulfate), 1% potassium acetate, 1% ammonium sulfate, 0.05 M potassium biphtalate, pH 5.5. Adjust pH with potassium hydroxide, autoclave.

2. Sporulation medium (SPM): 1% potassium acetate pH 7.0. Autoclaved. Supplement with amino acids (320 μL amino acid complementation medium per 100 mL of SPM) and PPG (100 μL 1% PPG per 100 mL of SPM).
3. 1% Polypropylene glycol (PPG).
4. Amino acid complementation medium: 1.5% lysine, 2% histidine, 2% arginine, 1% leucine, 0.2% uracyl, 1% tryptophan. Filter sterilize.
5. 37% Formaldehyde.
6. 2.5 M Glycine (autoclaved).
7. TBS buffer: 20 mM Tris-HCl, pH 7.5, 150 mM NaCl.
8. 0.2 μg/mL DAPI in water.

2.2. Preparation of Cell Extracts

1. Lysis buffer: 50 mM Hepes-KOH pH 7.5, 140 mM NaCl, 1 mM EDTA, 1% Triton X-100, 0.1% Na-deoxycholate (filter sterilized).
2. Lysis buffer complete: Immediately before use, add PMSF (1 mM final concentration), 250 μL of Aprotinin (1.4 mg/mL), and one protease inhibitor tablet (Roche, Complete Protease Inhibitor Cocktail) per 50 mL of lysis buffer.
3. Glass beads, diameter 0.40–0.60 mm (Sartorius, BBI-8541701).

2.3. Immunoprecipitation

1. Dynabeads: Magnetic Dynabeads (Dynal Biotech, Pan Mouse IgG).
2. PBS: 137 mM NaCl, 2.7 mM KCl, 4.3 mM Na_2HPO_4, 1.47 mM KH_2PO_4, pH 7.4.
3. PBS/ BSA: PBS with 5 mg/mL bovine serum albumin.
4. Washing buffer: 10 mM Tris-HCl pH 8, 250 mM LiCl, 0.5% Na-deoxycholate, 1 mM EDTA.
5. TE buffer: 10 mM Tris-HCl pH 8, 1 mM EDTA.
6. Elution buffer: 50 mM Tris-HCl pH 8, 10 mM EDTA, 1% SDS.

2.4. DNA Purification

1. Proteinase K (20 mg/mL) (Roche)
2. Glycogen (10 mg/mL)
3. Phenol/ chloroform / isoamylalcohol: 25/24/1 ratio.
4. 5 M NaCl.
5. DNase-free RNase (500 μg/mL) (Roche).
6. 96% Ethanol.
7. 3 M Sodium acetate.

2.5. Real-Time Quantitative PCR Analysis

1. IQ SYBR® Green Supermix (Biorad #170-8885).
2. Oligonucleotides: 10 pmol/μL.
3. 96-Well iQ Optical PCR plates (Biorad 223-9441).
4. Microseal 'B' Seal (Biorad MSB-1001).
5. iQ5 real-time PCR detection system.

3. Methods

3.1. Synchronized Yeast Sporulation

1. Inoculate the diploid yeast cells in pre-sporulation medium (SPS) at a final concentration of about 2×10^6 cells/mL (*see* **Note 1**). Set up the cultures in 3-L Erlenmeyer baffled flasks. Prepare enough culture to allow 50 mL for each sample to be collected, but do not exceed 15% of the total flask volume to ensure optimal oxygenation.
2. Grow cells under vigorous shaking (250 rpm) for 12–16 h at 30°C until the density reaches 4×10^7 cells/mL. Collect cells by centrifugation (4 min, 3,000*g*) and resuspend at 4×10^7 cells/mL in SPM supplemented with amino acids and PPG (time point t = 0 h). Shake 250 rpm for the whole time-course. Note that the SPM culture should not exceed 10% of the flask capacity (maximum 300 mL in a 3 L flask). Follow the progression of the time-course by regularly drawing samples for DAPI staining: take aliquots (100 μL of culture into 500 μL ethanol) at appropriate intervals (typically 1 h intervals) and store at 4°C until the end of the time-course (*see* **Step 5**).
3. At the desired times, collect 50-mL aliquots for ChIP. Incubate each sample with formaldehyde (1% final concentration) for 30 min (*see* **Note 2**) at room temperature shaking under a fume hood. Stop the cross-linking process by adding glycine to a final concentration of 131 mM and shaking for 5 more min.
4. Finally, wash the samples twice with ice-cold TBS (use a centrifuge, pre-cooled to 4°C), transfer cells to a 1.5 mL Eppendorf tube and freeze in liquid nitrogen. Samples can be stored at –80°C.
5. Before proceeding with the preparation of the cell extracts, examine the progression through meiotic divisions for sufficient synchrony. Pellet the samples collected in ethanol (from **Step 2**) for 10 s at the highest speed in a microcentrifuge and resuspend the cells in 50 μL of 0.2 μg/mL DAPI in water.

6. To facilitate counting, sonicate the DAPI samples for 1 s using a rod-based sonicator at medium strength. Place 5 μL of cell suspension on a glass slide and evaluate the percentage of mononucleate, binucleate, and tetranucleate cells for each time point. In our hands, synchronous cultures of a wild-type SK1 strain typically reach the highest number of binucleate cells at 5–6 h after transfer to SPM. This should correspond to a peak of at least 25% binucleates for wild-type SK1 cells (*see* **Note 3**).

3.2. Yeast Extract Preparation

1. Resuspend the cell pellets (50 mL = 2×10^9 cells) in 1,600 μL Lysis buffer complete and equally distribute the suspension to four 2-mL screw-cap tubes containing ca. 600 μL of glass beads. For cell breakage, place the tubes in a multibeads shocker (YASUI-KIKAI, Osaka) 2,500 rpm, 28 cycles of 30 s on and 30 s off, at 4°C or into a precooled vibrax unit (14 cycles of 1 min on, 1 min off at 4°C. At this stage, check the breakage efficiency by phase contrast microscopy (*see* **Note 4**).
2. After breakage, poke a small hole into the cap of the screw-cap tubes, and place each inverted into a 15 mL Falcon tube. Collect the cell extract by centrifugation 1 min, 1,700*g*, 4°C. Transfer the extract into a 1.5 mL Eppendorf tube and centrifuge again for 1 min at 4,000*g* to remove air bubbles.
3. Sonicate the cell extracts to shear chromatin to the desired length. We obtain an average fragment length of slightly above 500 bp by sonicating five times at 37% power for 15 s (*see* **Note 5**).

3.3. Chromatin Immunoprecipitation

1. Prepare 50 μL of pan mouse IgG magnetic Dynabeads per sample by washing twice with 1 mL of PBS/BSA: add the PBS/BSA to the beads, vortex, fix beads at the wall of the tube with the magnet, then remove liquid. Add 250 μL primary antibody (*see* **Note 6**) (e.g., anti-Myc antibody for IP against the Myc epitope) in PBS/BSA and incubate for 3 h at 4°C rotating end over end on a wheel (*see* **Note 7**). Wash antibody-bound beads twice with 1 mL PBS/BSA and resuspend in 100 μL PBS/BSA.
2. After sonication (**Section 3.2**, **Step 3**), centrifuge the yeast extracts at 13,300*g* for 5 min to remove cell debris. Transfer the supernatant to a new 1.5 mL Eppendorf tube, and remove a 20 μL aliquot to prepare whole cell extract (WCE); store this aliquot at 4°C. Note that the WCE should be prepared from each sample, because the final result will be expressed as IP/WCE(sample) to correct for different amounts of template in the input.

3. Incubate the remaining extract with the antibody-coated Dynabeads (25 μL of bead suspension for each of the four aliquots from each sample from **Section 3.2**, **Step 1**) for 1–3 h at 4°C with nutation or rocking.
4. Wash beads as follows: two times with 1 mL lysis buffer, two times with 1 mL lysis buffer plus 360 mM NaCl, two times with 1 mL washing buffer, one time with 1 mL TE.
5. Centrifuge the samples at 92*g* for 10 s and remove the supernatant completely. Hold beads at the wall of the tube using the magnet while pipetting.
6. Add 40 μL of elution buffer and incubate at 65°C for 15 min shaking. Centrifuge at 13,300*g* for 1 min and transfer the supernatant into a new Eppendorf tube containing 160 μL of TE/1%SDS.
7. To the 20 μL WCE sample from **Step 2**, add 380 μL of TE/ 1%SDS, then split the sample into two aliquots.
8. Incubate all samples overnight at 65°C to reverse the crosslinks.

3.4. DNA Purification

1. To each of the 200 μL aliquots, add 140 μL TE, 3 μL of 10 mg/mL glycogen, and 7.5 μL of 20 mg/mL proteinase K and incubate at 37°C 2 h.
2. Perform phenol/chloroform/isoamyl alcohol (PCI) extraction twice: add 300 μL PCI, vortex briefly, then centrifuge at 13,300*g* for 1 min and recover the upper (aqueous) phase taking care to avoid any contamination from the interphase.
3. Supplement the recovered aqueous phase with NaCl to 200 mM final concentration and add 2 vol of 96% ethanol. Vortex the samples and incubate them at –20°C for at least 30 min.
4. Centrifuge the samples for 30 min at 4°C. Remove the supernatants and wash the DNA pellets with ice-cold 70% ethanol.
5. Dry the DNA pellets, then resuspend each in 30 μL TE/ RNase solution (4 μg DNase-free RNase per sample) and incubate at 37°C for 1h. At this point, pool the different IP aliquots that were split in four at the beginning (**Section 3.2**, **Step 1**). Also pool the WCE aliquots (from **Section 3.3.**, **Step 7**).
6. Optional: an additional PCI extraction can be performed (*see* **Note 8**), but will result in reduced yield.
7. Precipitate the purified DNA by adding 0.1 vol of 3 M sodium acetate and 2 vol of 96% ethanol, incubating at –20°C for at least 30 min, then centrifuging.

8. Wash the precipitated DNA with ice-cold 70% ethanol, dry the pellet and resuspend it well in 30 μL ddH_2O. Store the samples at –20°C until analysis.

3.5. Real-Time PCR Analysis

Many different protocols and reagents for real-time analysis of the precipitated DNA are available. We use the "SYBR Green" method, because it provides flexibility in choosing different chromosomal positions. Alternatively, if the analysis is performed routinely at the same chromosomal positions, techniques requiring modified oligonucleotides and which offer the possibility of multiplexing may also be economical. We use an iQ5 instrument (Biorad), but the "SYBR Green" method can also be performed on simpler real-time PCR stations. Also "SYBR Green" can be substituted with "MESA Green" (Eurogentec), which offers a larger dynamic range for monitoring the exponential amplification phase. For DNA quantification of a sample, the instrument needs a minimal series of three dilutions of a standard of "known" concentration. All measurements will be expressed relative to this standard. For the purpose of quantifying ChIP we recommend using the DNA purified from the WCE as the standard. Negative controls are also important: to obtain meaningful results, ChIP must be performed in parallel using the same conditions on cells lacking the tag (as negative control for tagged baits), or on cells lacking the bait (if antibodies against the bait are used). Only the difference in signals between tagged and untagged (or plus/minus bait) can be regarded as specific. We perform qPCR as follows:

1. Set up a series of three 10-fold dilutions of the WCE sample (e.g. 1/30, 1/300, 1/3,000).
2. PCR mix per sample:

• 12.5 μL IQ SYBR Green Supermix	
• 0.5 μL Primer oligo mix	(10 pmol/μL stock)
• 9 μL ddH_2O	
• 3 μL DNA sample	(IP sample or WCE dilution)
25 μL total reaction volume	

3. Prepare one master mix containing the Supermix, primers and water for all samples for each primer pair (i.e., per locus analyzed). Measure all samples in duplicate (*see* **Note 9**). Note that for each primer pair, six wells are required for the WCE dilution series, plus two wells for each sample. Thus, it is economical to group samples analyzed with compatible primer pairs together in a run.

4. First place the 22 μL of the master mix into the wells of a 96-well optical plate and then add the DNA samples (3 μL of either IP or WCE dilution) (*see* **Note 10**).
5. Place a sealing film over the 96-well plate to avoid evaporation. Centrifuge the plate for 1 min at 300*g* to remove air bubbles and place the plate into the real-time thermocycler.
6. Run the qPCR with the following setting (*see* **Note 11**):

Cycle 1: (1 ×)		
Step 1:	94.0°C	for 3 min
Cycle 2: (40 ×)		
Step 1:	94.0°C	for 15 s
Step 2:	60.0°C	for 1 min
Data collection and real-time analysis enabled.		
Step 3:	72.0°C	for 1 min
Cycle 3: (1 ×)		
Step 1:	94.0°C	for 1 min
Cycle 4: (1 ×)		
Step 1:	60.0°C	for 1 min
Cycle 5: (71 ×)		
Step 1:	60.0°C–95.0°C	for 30 s
Increase set point temperature after cycle 2 by 0.5°C		
Melting curve data collection and analysis enabled.		

7. During the qPCR run, the amplification can be observed online. This allows one to stop the run manually after all samples have reached the plateau. (If this is done, the melting curve analysis has to be initiated manually.)
8. Data interpretation: Even though the instrument's software package may offer programs for fully automated evaluation, it is useful to understand the underlying algorithms to be able to judge the quality and robustness of the obtained results. We can evaluate the amount of chromatin immunoprecipitated relative to the standard curve established for the WCE samples, and compare the fold enrichment obtained for the region of interest and the cold-spot region.

3.6. Interpretation of qChIP Results

1. Typical qPCR results are shown in **Fig. 17.1** (only a single "unknown" is shown for simplicity). The semi-logarithmic plot (**Fig. 17.1A**) transforms the exponential

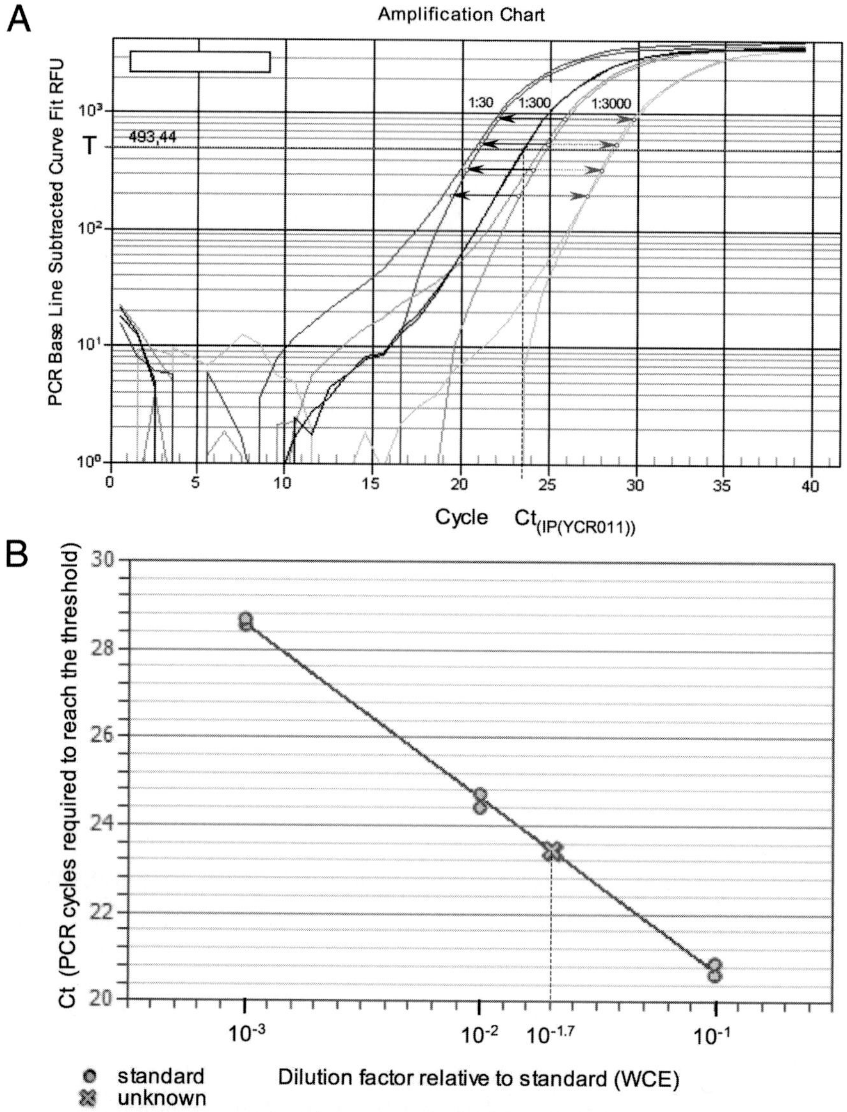

Fig. 17.1. (**A**) Real-time PCR amplification chart. 3 μL of a 1:30, 1:300, and 1:3,000 dilution of WCE were used as standard. PCR amplification is shown in a semi-logarithmic plot, with RFU (Relative Fluorescence Units) plotted against the corresponding number of PCR cycles. RFU values are "base-line corrected," i.e., the constant value of the 20 nM fluorescein has been subtracted. While the fluorescein may help correct uneven reaction volumes, it likely shortens the "linear range." In the end, inexact pipetting cannot be fully corrected for, thus the spiked-in fluorescein may be only of limited use. T (threshold) is an arbitrary RFU value within the linear range of all amplification curves that are to be compared. The slope of all these curves should be nearly the same and is ideally one full duplication per cycle. Primer pairs that differ in their efficiency (i.e., that show a different slope in the linear range) need to be compared to "their own standard" (i.e. WCE amplified with the same primer pair). Usually the results are robust against moving the threshold value up and down within the linear range (*Arrows* demonstrate similar ratio between dilutions over almost an order of magnitude). "MESA green" instead of SYBR green significantly increased the linear range in our hands. Ct (threshold cycle) is the number of cycles of a given reaction required to reach the threshold. (**B**) Ct vs. dilution plot. Here the Ct values determined in **Fig. 17.1A** are plotted against the standard (entered by the user). 10^{-1} stands for the template amount in 0.1 μL WCE (3 μL of 1:30 dilution). Ct values of the standard are correlated with dilution factors by linear regression. From this correlation the dilution factors of the "unknown" samples are estimated (*dashed line*).

amplification into a straight line and the program suggests a threshold (T), at which all curves should be in the linear range. The threshold can be manually edited. Within the linear range such changes will have only very small effects, if all curves are parallel. T should not be chosen from cycles that are too early, where curves tend to deviate from linear. Much of this nonlinearity may be caused by the FITC (spiked-in for "volume" control) and by the attempt of the system to subtract this signal (baseline subtraction).

2. From the number of PCR cycles required to reach the threshold, the original amount of template is estimated (**Fig. 17.1B**) based on the assumption that amplification in the early "invisible" cycles occurred with the same efficiency as in the "visible" ones. The results are expressed relative to the standard (for which we use WCE, treated under similar conditions as the IP samples). Relative comparisons, such as time point x to time point zero (for meiosis-specific interactions), or cells containing the antigen (or tag) versus cells not containing it, are most intuitive and can be done right away.

3. In order to refer to the efficiency of the technique, it is useful to refer to TT, the "total template present in the input", corresponding to 240 μL of the sample, because 1/10 of an IP is used per qPCR reaction. The way we have defined the standard in **Fig. 17.1B**, a value of 1 corresponds to the amount of template present in 1 μL of the WCE (30 μL of WCE are derived from 20 μL of sample, therefore containing 20/240, or 1/12 of TT). A unit of 1 in the standard curve corresponding to 1 μL WCE therefore is $1/(12 \times 30)$ of TT, thus division by a factor of 360 converts IP values from WCE to TT units. (For example, the value of 6% of WCE, reached by Mre11-HA at the hotspot at t=4 h in **Fig. 17.2**, corresponds to 0.0167% of TT). This demonstrates that ChIP recovers only a small fraction of the total template present. This may be due to the low efficiencies of the IP and artificial cross-linking. In addition, Mre11 may not occupy all templates. Nevertheless, the dynamic range between Mre11-HA and untagged is sufficiently large to identify even very weak binding sites. (The cold G-band at YCR011 is close to the weakest binding sites of Mre11 on chromosome 3, yet it is still significantly different from untagged, *see* **Fig. 17.2**).

4. The generation of data can be observed in "real time" during the PCR run. After all dilutions have reached the plateau, a melting analysis should be carried out. Since the melting

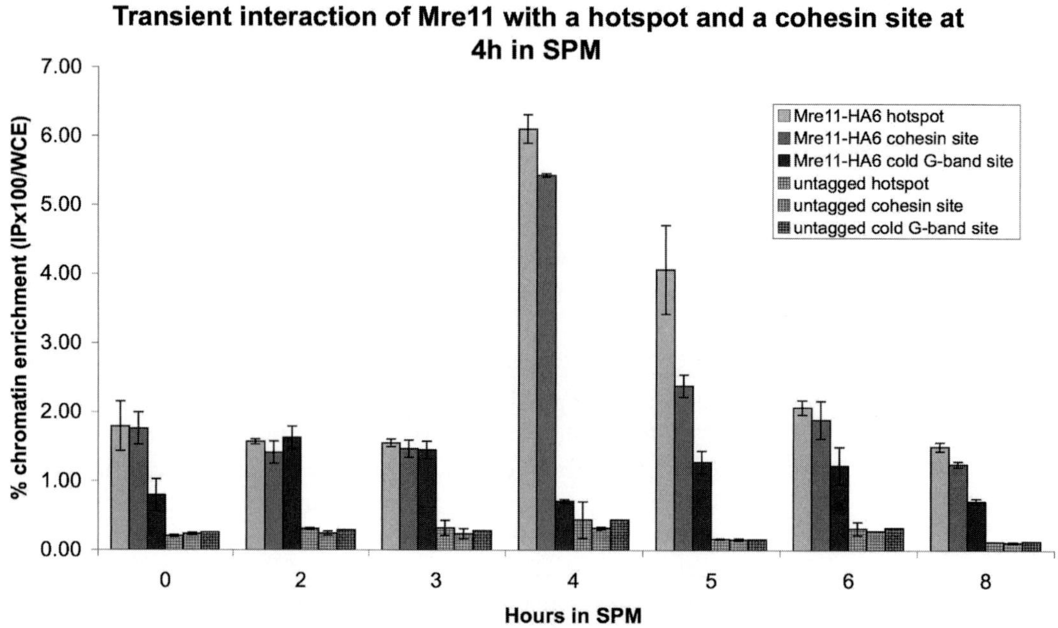

Fig. 17.2 Mre11 interacts transiently with a hotspot and a cohesin binding site in meiosis. ChIP was carried out in a time course experiment in two strains in parallel, one containing Mre11-HA6 and one lacking the tag (untagged). Immunoprecipitated DNA was analyzed with three primer pairs at three different loci. The "hotspot" corresponds to *YCR048* at position 212,000; the cohesin site to *YCR067* at position 233,600; and the cold G-band site to *YCR011* at position 136,500, all on chromosome III. Chromatin enrichment is expressed as IP/WCE, where WCE was determined for each of the 42 samples. Error bars correspond to the duplicate measurement results.

temperature strongly depends on the product length, this is a convenient method to detect and, if necessary, quantify primer dimer products, which are template-independent artifacts that could severely distort the result. For an example, *see* **Fig. 17.3**. Primer-dimer formation occurs rarely if the "hotstart" strategy is used.

5. What influences the qChIP result? Probably the most important parameter is the fraction of template occupied by the bait during crosslinking. However, also the number of baits per template will significantly influence the qPCR value, as long as the efficiency of the IP is not close to 100%. Third but not least, the ability to be cross-linked has a drastic effect. This includes the availability of amino acid side chains to interact with formaldehyde and DNA, but also whether the bait interacts directly with the DNA or via another protein. Finally and perhaps most importantly, solubility after crosslinking may be limiting.

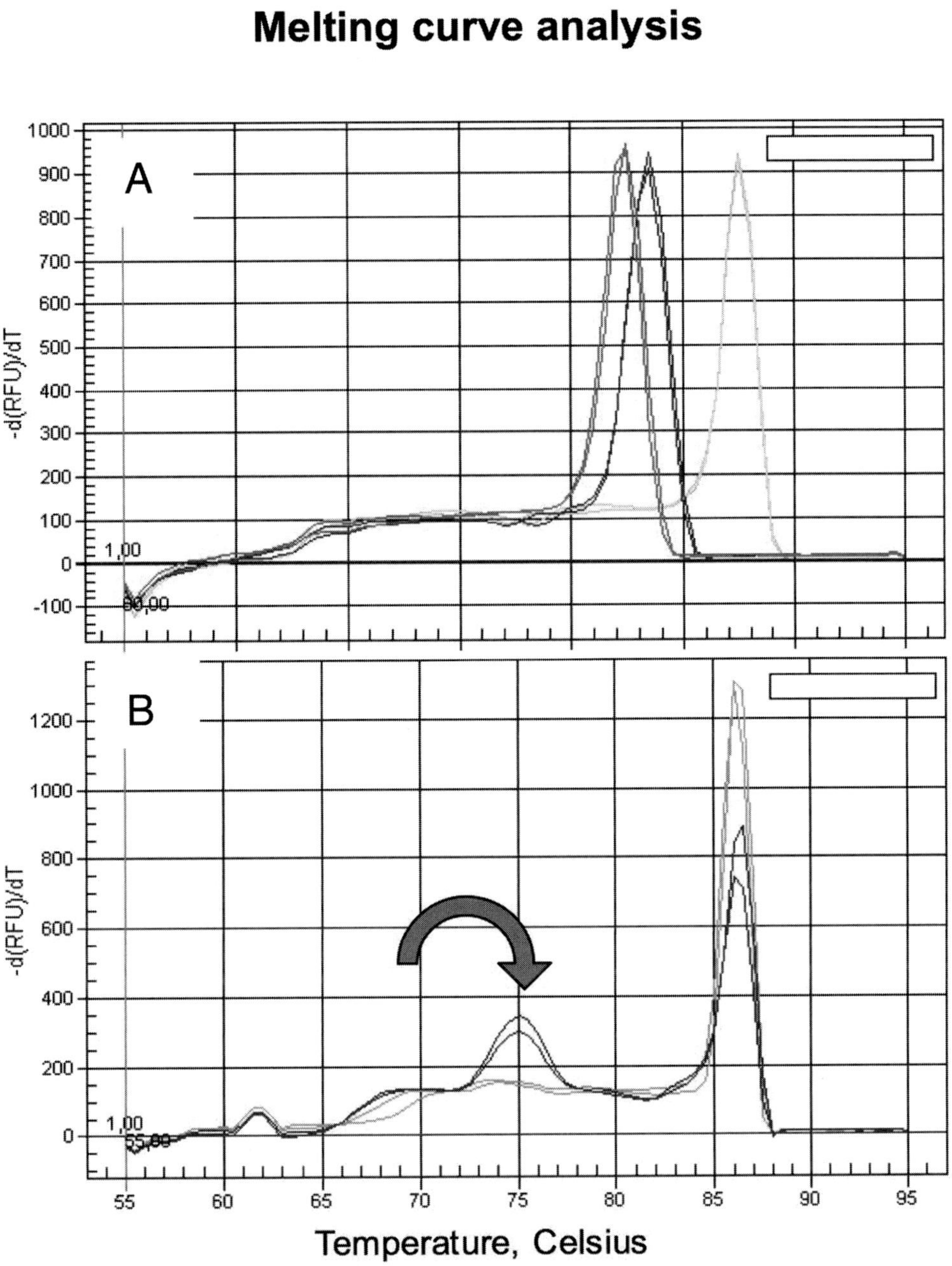

Fig. 17.3 (**A**) Melt curve analysis for the primer pairs for hotspot, cohesin site, and cold G-band used in **Fig. 17.2** and **Fig. 17.1**. The observed peaks correspond to the PCR products, not to primer dimers. (**B**) Melt curve analysis example provided by the Biorad company. The peaks indicated by the *arrow* are from primer dimers, as identified by their melting at 75°C. The area under the peaks is proportional to the amount of product.

4. Notes

1. To monitor cell growth in the pre-sporulation medium (SPS), it is recommended to establish a table of correlation between the number of yeast cells per mL determined with a hemocytometer and the corresponding optical density (OD_{660}). Note that this correlation curve may differ between different strain backgrounds, cells grown in different media, or for different spectrophotometers. This is a critical step to obtain a synchronous meiotic time-course. For instance, cell cultures that were overgrown (more than 5×10^7 cells/mL) may exhibit poor synchrony, and in such a case it is recommended to restart the procedure from the beginning. In the case that a cell culture did not reach the concentration of 4×10^7 cells/mL, we recommend to wait until enough cells have accumulated.
2. The time of incubation with formaldehyde should be optimized for each protein of interest. In our hands, a 30-min incubation works well for cohesin and recombination proteins. Overnight incubation at 4°C was reported for cases where the DNA-protein interaction is expected to be particularly weak. A balance between too long fixation times, which will increase the background, and too short incubation times, which result in insufficient chromatin enrichment, has to be determined.
3. Conditions described here refer to the SK1 background, which undergoes meiosis relatively fast, synchronously and efficiently. It should be considered that some mutant backgrounds may exhibit different kinetics of meiotic divisions than wild-type cells. For instance, cells unable to form DSBs (such as *spo11Δ*) may show binucleate cells earlier than wild type, while those defective in repair may spend a certain time arrested by the DNA damage checkpoint and will display a corresponding delay. For strains expected to sporulate with wild-type efficiency, we only use experiments with "good" synchrony, because the intensity of transient signals depends on synchrony. We evaluate synchrony using the signal intensity of the transient signal "binucleate cells," which should reach at least 25% (*see* **Fig. 17.4**). Of course, this criterion cannot be used for mutants that arrest or delay in meiotic prophase.
4. We have also obtained satisfactory results without splitting the samples, at least for proteins yielding relatively strong signals. Also, the split samples can be combined at different steps after extract preparation to keep the manipulations

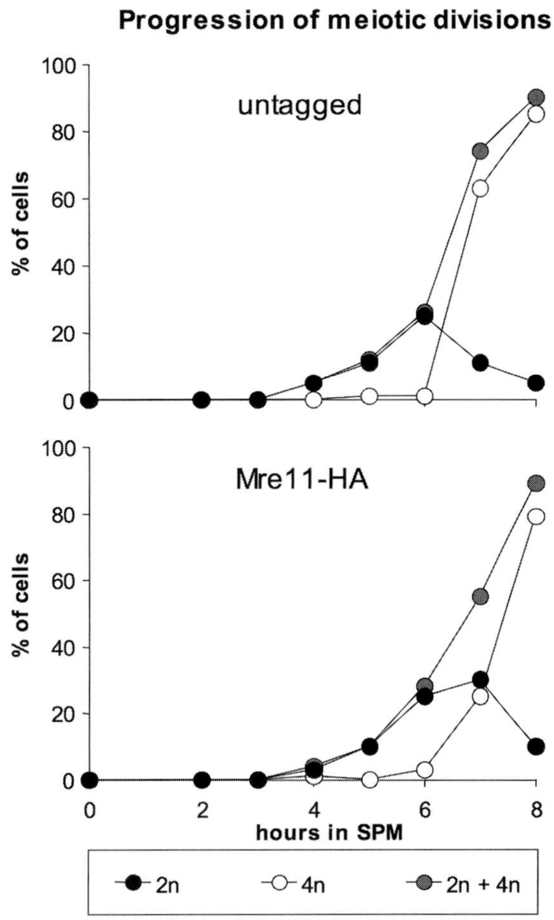

Fig. 17.4. Progression of meiotic divisions. Meiotic progression is followed by counting cells containing one, two, or four nuclei (1n, 2n, 4n) after staining with DAPI. Here meiotic progression is almost identical between the untagged strain and the strain carrying Mre11-HA. Both strains undergo synchronous meiotic divisions with high efficiency (90% past the first meiotic division at t = 8 h). The 7 h time point reveals that Mre11-HA is slightly less synchronous than the untagged strain. This is not seen regularly in time courses of this particular strain.

manageable, albeit with the risk that manipulations on the combined samples might not be as efficient as on the split samples. After cell opening, check breakage efficiency in a microscope. More than 90% of the cells should be opened. In our hands, the optimal diameter of the glass beads is 0.40–0.60 mm. Much longer crushing times may cause uncontrolled shearing of DNA. The multibead shocker can be used for optimal temperature control during opening, but is very expensive. A regular Vibrax controlled by an electronic timer (which can be set to the minimum of 1 min intervals) has also given satisfactory results.

5. The sonication step needs to be optimized. Sonicate for a certain number of cycles (e.g. five times) for about 15 s interrupted by 15 s breaks. During sonication and the breaks, samples should be cooled in ice-cold water. When optimizing sonication conditions, use formaldehyde cross-linked samples, keep sample volume constant and before analysis of the DNA length, remove crosslinks and extract with phenol-chloroform-isoamyl alcohol (**Section 3.3**, **Step 7 and Section 3.4**). Determine the average fragment size by agarose gel electrophoresis.
6. How much antibody is required to load the beads needs to be determined empirically. This can be done by using the ChIP signal as read out, or—if little is known about the expected signals—by determining the ratio of antibody to extract required to cause detectable depletion of the antigen from the extract.
7. Beads can be chosen from a wide range of products. We recommend the use of Dynabeads over regular agarose beads, because of better recovery rates and lower background (i.e., non-specific precipitation of target DNA by the beads). Protein A or Protein G coated beads are versatile because they interact with a large range of different antibodies. To reduce nonspecific interactions, a pre-clearing step is recommended by some protocols when using Protein A/G beads. We prefer to use pan mouse IgG or pan rabbit IgG beads (M-280, Dynabeads). Finally, we observed that adding salmon sperm DNA (2 mg/mL) to the PBS/BSA blocking solution can further increase the signal/background ratio, but will also, as a side effect, decrease the absolute signal intensity.
8. After protein and RNA removal, earlier protocols recommended an additional PCI extraction. However, we found that it can be safely omitted.
9. Duplicates of PCR reactions typically differ by less than 10%. The SYBR GREEN Supermix contains 20 nM fluorescein, which allows the iQ5 to correct for small volume differences caused by pipetting errors.
10. If significant differences between duplicates occur, it is possible to spin the samples briefly directly before loading to avoid incompletely dissolved DNA. One may also consider increasing the volume of the loaded DNA from 3 to 6 μL, depending on the availability of high-precision pipettes.
11. We design all primer pairs to work with the same PCR program. To achieve this goal, our primers are 20 nucleotides long, have a GC content between 50 and 55%, and a Tm >55°C and <61°C. The product length can be between 100 and 500 nucleotides (ideally below 200 to insure rapid and

complete amplification). Before a new primer pair is accepted, it is tested for efficient amplification and for the absence of primer dimers under our exact qPCR conditions.

Acknowledgments

We thank Marie Therese Hauser, Kim Nasmyth, and Katsuhiko Shirahige for helping us to set up, or improve parts of the technique and former lab member Silvia Prieler for tagging Mre11. This work was supported by FWF grant P18847 to F.K.

References

1. Tanaka, T., Knapp, D., and Nasmyth, K. (1997) Loading of an Mcm protein onto DNA replication origins is regulated by Cdc6p and CDKs. *Cell* **90**, 649–660.
2. Katou, Y., Kanoh, Y., Bando, M., Noguchi, H., Tanaka, H., Ashikari, T., Sugimoto, K., and Shirahige, K. (2003) S-phase checkpoint proteins Tofl and Mrc1 form a stable replication-pausing complex. *Nature* **424**, 1078–1083.
3. Prieler, S., Penkner, A., Borde, V., and Klein, F. (2005) The control of Spo11's interaction with meiotic recombination hotspots. *Genes Dev.* **19**, 255–269.
4. Katou, Y., Kaneshiro, K., Aburatani, H., and Shirahige, K. (2006) Genomic approach for the understanding of dynamic aspect of chromosome behavior. *Methods Enzymol.* **409**, 389–410.
5. Nairz, K. and Klein, F. (1997) *mre11S* - a yeast mutation that blocks double-strand-break processing and permits nonhomologous synapsis in meiosis. *Genes Dev.* **11**, 2272–2290.
6. Borde, V., Lin, W., Novikov, E., Petrini, J. H., Lichten, M., and Nicolas, A. (2004) Association of Mre11p with double-strand break sites during yeast meiosis. *Mol. Cell* **13**, 389–401.
7. Baudat, F. and Nicolas, A. (1997) Clustering of meiotic double-strand breaks on yeast chromosome III. *Proc. Natl. Acad. Sci. U. S. A.* **94**, 5213–5218.

Chapter 18

Genome-Wide High-Resolution Chromatin Immunoprecipitation of Meiotic Chromosomal Proteins in *Saccharomyces cerevisiae*

Kazuto Kugou and Kunihiro Ohta

Abstract

Cooperative actions of chromosomal proteins play critical roles in the dynamics, structural transition, segregation, and maintenance of meiotic chromosomes. A high-resolution genome-tiling array combined with a chromatin immunoprecipitation assay (ChIP-chip) is a powerful tool for uncovering the precise and dynamic distributions of various proteins along meiotic chromosomes. In this chapter, we describe a method to map the binding sites of meiotic chromosomal proteins such as Spo11, Mre11, and Rec8 using the high-resolution ChIP-chip technology. This system provides us with informative knowledge on the interplay of meiotic chromosomal proteins.

Key words: Recombination, meiosis, ChIP-chip, genome-tiling array, DNA–protein interaction.

1. Introduction

Dynamics of meiotic chromosomes are choreographed by the actions of various proteins. DNA–protein and protein–protein interactions govern the dynamic features of meiotic chromosomes that are important for premeiotic DNA replication, monopolar kinetochore organization, remodeling of chromatin structure, recombination, pairing of homologous chromosomes, formation of the synaptonemal complex, and chromosome segregation. These processes are regulated interdependently and consist of systematic regulatory pathways.

Scott Keeney (ed.), *Meiosis, Volume 1, Molecular and Genetic Methods, vol. 557*

DOI 10.1007/978-1-59745-527-5_18 Springerprotocols.com

To understand the molecular interplay of meiotic chromosomal proteins, it is essential to study their precise localizations along the meiotic chromosomes. For this purpose, cytological methods have been successfully employed. In addition, chromatin immunoprecipitation (ChIP) analyses can provide information on the interaction of meiotic proteins with chromosomes. Recently, ChIP combined with DNA microarray technology has been developed to investigate the genome-wide and comprehensive status of chromosomal proteins. The ChIP-chip technology was originally developed by Ren et al. *(1)*. The high-resolution oligonucleotide tiling array was introduced by Katou et al. *(2)*. Now the dynamics of many chromosomal proteins have been uncovered by the high-resolution ChIP-chip technology *(2, 3)*. In this section, we describe a standard protocol for the high-resolution ChIP-chip technology and its application to the investigation of meiotic recombination proteins in *S. cerevisiae*.

2. Materials

2.1. Cell Culture and In Vivo Crosslinking

1. YPG medium: 1% (w/v) Bacto yeast extract (Becton, Dickinson and Company, Sparks, MD), 2% (w/v) Bacto peptone, 3% (v/v) glycerol, 2% (w/v) Bacto agar.
2. YPD medium: 1% (w/v) Bacto yeast extract, 2% (w/v) Bacto peptone, 2% (w/v) glucose, 2% (w/v) Bacto agar.
3. SPS medium: Mixture of 1 volume of 0.1 M potassium phthalate, pH 5.0 and 1 volume of medium containing 1% (w/v) Bacto yeast extract, 2% (w/v) Bacto peptone, 0.34% (w/v) Difco yeast nitrogen base without amino acids and ammonium sulfate (Becton, Dickinson and Company), 2% (w/v) potassium acetate, 1% (w/v) ammonium sulfate, supplemented with amino acids (e.g., 0.79 g of CSM (Complete Synthetic Medium powder, MP Biomedicals, Solon, OH) for 1 L of SPS medium) and a defoaming agent (*see* **Note 1**). Autoclave at 121°C for 15 min.
4. Sporulation medium (SPM): 1% (w/v) potassium acetate, 1/5 of usual concentration of amino acids (e.g. 0.158 g/L CSM), 0.001% (v/v) polypropylene glycol. Autoclave at 121°C for 15 min.
5. Crosslink reagent: 36–38% formaldehyde.
6. Crosslink stop reagent: 2.5 M glycine. Autoclave at 121°C for 10 min, and store at room temperature.
7. Tris-buffered saline (TBS) (10 ×): 200 mM Tris-HCl, 1.5 M NaCl, pH 7.6. Autoclave at 121°C for 15 min. Chill on ice before use.

2.2. Cell Lysis and Fragmentation of Genomic DNA

1. 1 M HEPES-KOH, pH 7.5. Sterilize by passing through a 0.22-μm cellulose nitrate (CN) filter. Store at 4°C.
2. 0.5 M EDTA, pH 8.0. Autoclave at 121°C for 10 min, and store at room temperature.
3. Lysis buffer: 140 mM NaCl, 0.1% (w/v) sodium deoxycholate, 1 mM EDTA, 50 mM HEPES-KOH, pH 7.5, 1% (v/v) Triton X-100. Sterilize by passing through a 0.22-μm CN filter. Store at 4°C.
4. Protease inhibitor cocktail (50 ×): Dissolve one tablet of Complete EDTA-free (Roche Diagnostics GmbH, Mannheim, Germany) in 1 mL of sterilized water just before use.
5. Zirconia beads: Add 1.5 g of zirconia beads (0.5 mm diameter, Yasui Kikai Corporation, Osaka, Japan) to a 2-mL tube with a screw cap plus O-ring (Yasui Kikai Corporation). Autoclave at 121°C for 10 min and air dry at 55°C. Prepare four sets for each sample and chill them on ice before use.
6. Cell disruptor: Multi-beads shocker (Yasui Kikai Corporation).

2.3. Preparation of Antibody-Conjugated Beads

1. Magnetic beads conjugated with recombinant Protein A: Dynabeads Protein A (Invitrogen Corporation, Carlsbad, CA) (*see* **Note2**).
2. ChIP grade antibody: Anti-FLAG M2 monoclonal antibody (Sigma-Aldrich, Inc., St. Louis, MO).
3. Phosphate-buffered saline (PBS) (10 ×): 1.37 M NaCl, 80 mM Na_2HPO_4, 27 mM KCl, 15 mM KH_2PO_4. Autoclave at 121°C for 10 min, and store at room temperature.
4. BSA-PBS: Prepare 1 × PBS with sterilized water. Dissolve 0.5% (w/v) BSA (fraction V bovine serum albumin, Sigma-Aldrich, Inc.) and chill on ice until use.

2.4. Immunoprecipitation (IP)

1. 1 M Tris-HCl, pH 8.0. Autoclave at 121°C for 10 min, and store at room temperature.
2. 20% (w/v) SDS. Store at room temperature.
3. Wash buffer I: 500 mM NaCl, 0.1% (w/v) sodium deoxycholate, 1 mM EDTA, 50 mM HEPES-KOH, pH 7.5, 1% (v/v) Triton X-100. Sterilize by passing through a 0.22-μm CN filter. Store at 4°C.
4. Wash buffer II: 250 mM LiCl, 0.5% (w/v) sodium deoxycholate, 1 mM EDTA, 10 mM Tris-HCl, pH 8.0, 0.5% (v/v) Nonident P-40 (NP40). Sterilize by passing through a 0.22-μm CN filter or autoclaving at 121°C for 10 min.
5. Wash buffer III (1 × TE: Tris EDTA buffer): 10 mM Tris-HCl, 1 mM EDTA, pH 8.0. Autoclave at 121°C for 10 min, and store at room temperature.

6. Elution buffer: 50 mM Tris-HCl, 10 mM EDTA, pH 8.0, 1% (w/v) SDS. Prepare just before use.
7. Secondary elution buffer: 50 mM Tris-HCl, 10 mM EDTA, pH 8.0, 0.67% (w/v) SDS. Prepare just before use.

2.5. Preparation of DNA

1. Proteinase K: Recombinant Proteinase K PCR grade (14–22 mg/mL) in 10 mM Tris-HCl, pH7.5 (Roche Diagnostics GmbH).
2. TE (pH 8.0): 10 mM Tris-HCl, 1 mM EDTA, pH 8.0. Autoclave at 121°C for 10 min, and store at room temperature.
3. Carrier or co-precipitant for DNA purification: 20 mg/mL glycogen, molecular biology grade (Sigma-Aldrich, Inc.).
4. Phenol: chloroform: isoamyl alcohol (25:24:1) saturated with 0.1 M Tris-HCl, pH 8.0 (Invitrogen Corporation).
5. Salt solution for DNA precipitation: 3 M sodium acetate, pH 5.3. Dissolve sodium acetate·$3H_2O$ in distilled water. Adjust the pH to 5.3 with glacial acetic acid. Autoclave at 121°C for 10 min, and store at room temperature.
6. 99.5% Ethanol
7. 75% (v/v) Ethanol
8. TE (pH 7.5): 10 mM Tris-HCl, 0.1 mM EDTA, pH 7.5. Autoclave at 121°C for 10 min, and store at room temperature.
9. 10 mg/mL RNase A in TE (pH 7.5) containing 50% glycerol (EMD Chemicals, Inc., San Diego, CA).

2.6. Amplification of DNA

1. Random primer (primer A): 40 μM, 5′-GTT TCC CAG TCA CGA TCN NNN NNN NN-3′.
2. DNA polymerase for random priming: 13 units/μL Sequenase version 2.0 DNA polymerase (USB Corporation, Cleveland, OH).
3. Sequenase reaction buffer (5 ×) supplied with Sequenase version 2.0 DNA polymerase: 200 mM Tris-HCl, pH 7.5, 100 mM $MgCl_2$, 250 mM NaCl.
4. Sequenase dilution buffer supplied with Sequenase version 2.0 DNA polymerase: 10 mM Tris-HCl, pH 7.5, 5 mM DTT, 0.1 mM EDTA.
5. Mixture A of dNTPs: Dilute 10 mM dNTPs mixture (TOYOBO Co., Ltd, Osaka, Japan) with TE (pH 7.5) to 3 mM each. Store at –20°C.
6. 0.1 M Dithiothreitol (DTT) in distilled water. Sterilize by passing through a 0.22-μm CN filter. Store in aliquots at –20°C.

7. Bovine serum albumin (BSA): Dilute 50 mg/mL BSA (Invitrogen Corporation) with sterilized water to 0.5 mg/mL. Prepare just before use.
8. Primer for PCR (primer B): 100 μM, 5′-GTT TCC CAG TCA CGA TC-3′
9. *Taq* DNA polymerase: 5 units/μL recombinant *Taq* (r*Taq*) DNA plymerase *TaKaRa Taq* (TaKaRa Bio Inc., Shiga, Japan).
10. PCR buffer (10 ×) Mg^{2+} plus: Use reagents supplied with *TaKaRa Taq*.
11. Mixture B of dNTPs: 2.5 mM each. Use reagents supplied with *TaKaRa Taq*.

2.7. Fragmentation and End-Labeling

1. Centrifugal filter devices to concentrate DNA: Microcon YM-10 (Millipore Corporation, Billerica, MA).
2. 1 units/μL DNase I, amplification grade (Invitrogen Corporation).
3. One-phor-all buffer PLUS (10 ×): 100 mM Tris-acetate, 100 mM magnesium acetate, 500 mM potassium acetate, pH 7.5 (GE Healthcare).
4. DNA deoxynucleotidylexotransferase: 400 units/μL terminal transferase, recombinant (Roche Diagnostics GmbH).
5. 25 mM $CoCl_2$. (Supplied with the recombinant terminal transferase.)
6. Biotin-labeled 2′,3′-dideoxynucleotide analog: 1 mM Biotin-11-ddATP (PerkinElmer Inc., Waltham, MA).

2.8. Hybridization

1. Saline-sodium phosphate-EDTA buffer (SSPE) (20 × stock): 3 M NaCl, 0.2 M sodium phosphate buffer, 20 mM EDTA, pH 7.4. Sterilize by passing it through a 0.22-μm CN filter. Store at room temperature.
2. 0.1% (v/v) Triton X-100: Dilute Triton X-100 with sterilized water. Store in aliquots at –20°C.
3. 10 mg/mL Herring sperm DNA in 10 mM NaCl, 1 mM EDTA, 10 mM Tris-HCl, pH 7.5 (Promega Corporation, Madison, WI).
4. Eukaryotic hybridization control kit: Kit containing 20 × Eukaryotic Hybridization Controls and 3 nM Control Oligonucleotide B2 (Affymetrix, Inc., Santa Clara, CA).
5. GeneChip (Affymetrix, Inc.): *Saccharomyces cerevisiae* tiling array (e.g. rikDACF, chromosome VI tiling array and SC3456a520015F, chromosome III-V tiling array).
6. Hybridization oven: GeneChip hybridization oven 640 (Affymetrix, Inc.)

2.9. Washing, Staining, and Scanning

1. 12 × MES buffer: 1.22 M MES (2-(*N*-morpholino)ethanesulfonic acid), 0.89 M [Na^+]. Dissolve 17.6 g of MES free acid monohydrate (Sigma-Aldrich, Inc.) and 48.3 g of MES sodium salt (Sigma-Aldrich, Inc.) in 250 mL of molecular biology-grade water (Invitrogen Corporation). pH between 6.5 and 6.7 is acceptable. Sterilize by passing it through a 0.22-μm CN filter. Store at 4°C. Must be shielded from light. If it turns yellow, do not use.
2. Antifoam stock solution: Dissolve 5% (w/v) antifoam O-30 (Sigma-Aldrich, Inc.) in sterilized water. Autoclave at 121°C for 10 min, and store at room temperature.
3. Wash buffer A (*see* **Note 3**): Nonstringent wash buffer. 6 × SSPE, 0.01% (w/v) Tween-20, 0.005% (w/v) antifoam O-30. Dilute 300 mL of 20 × SSPE stock solution with 700 mL of molecular biology-grade water (Invitrogen Corporation). Add 1 mL of Surfact-Amps 20 (10% (w/v) Tween 20, Thermo Fisher Scientific Inc., Waltham, MA). Sterilize by passing it through a 0.22-μm CN filter. Add 1 mL of 5% (w/v) antifoam O-30. Store at room temperature.
4. Wash buffer B (*see* **Note 3**): Stringent wash buffer. 100 mM MES, 0.1 M [Na^+], 0.01% (w/v) Tween 20. Add 83.3 mL of 12 × MES buffer stock solution and 5.2 mL of 5 M NaCl to 900 mL of molecular biology-grade water. Adjust the volume to 1 L with molecular biology-grade water. Add 1 mL of Surfact-Amps 20. Sterilize by passing it through a 0.22-μm CN filter. Store at 4°C.
5. Stain buffer (2 ×): 200 mM MES, 2 M [Na^+], 0.1% (w/v) Tween 20. Add 8.34 mL of 12 × MES buffer stock solution, 18.5 mL of 5 M NaCl, 0.5 mL of Surfact-Amps 20 to 20 mL of molecular biology-grade water. Adjust the volume to 50 mL with molecular biology-grade water. Sterilize by passing it through a 0.22-μm CN filter. Store at 4°C.
6. 50 mg/mL Acetylated BSA (Invitrogen Corporation).
7. 1 mg/mL Streptavidin, R-phycoerythrin conjugate (SAPE; Invitrogen Corporation). Remove aggregates by microcentrifugation at 17,610*g* for 5 min at 4°C just before use.
8. Washing and staining instrument: GeneChip fluidics station 400 or 450 (Affymetrix, Inc.).
9. Scanner: GeneChip scanner 3000 7G (Affymetrix, Inc.).

3. Methods

The ChIP-chip protocol can be divided into five sections (*see* **Fig. 18.1**): (1) Chromatin immunoprecipitation (ChIP), (2) amplification of DNA, (3) hybridization to tiling array, (4) detection (acquisition of signal intensity), and (5) analysis. In the first ChIP step, meiotic cells are treated with formaldehyde to crosslink proteins to DNA. The crosslinked cells are then harvested and disrupted with a Multi-beads shocker (Yasui Kikai Corporation). After fragmentation of DNA by sonication, DNA–protein complexes of interest are subjected to immunoprecipitation (IP) using a specific antibody conjugated with magnetic beads, and the immunoprecipitated DNA is further processed to reverse the crosslinking. This technique requires ChIP-grade antibodies, which can recognize crosslinked protein with high specificity. For the analysis of yeast proteins, it is possible to use cells expressing epitope-tagged proteins under the control of their original promoters and the commercially available anti-epitope tag antibodies. However, it is necessary to confirm the functionality of the tagged protein in vivo by testing its ability to complement the defects of the mutation for the gene. In addition, it is recommended to verify whether the IP process is specific and efficient enough by conventional Western blotting analysis (*see* **Note 4**). In this report, we describe a few examples of experiments using yeast cells expressing FLAG-tagged proteins for meiosis.

In the second step, DNA samples are prepared from both the immunoprecipitated and control whole-cell extract (WCE) fractions, and they are further amplified by random primed PCR. The WCE sample is used to determine the reference baseline at the analysis step. Each amplified DNA sample is fragmented to an average length of 100 bp, then end-labeled with biotin-conjugated ddATP by a terminal transferase. This fragmentation step is critical to ensure maximal efficiency of hybridization. Thus, the average fragment length of DNA should be checked very carefully. The Affymetrix system adopts a one-colored approach (i.e. R-phycoerythrin, with efficient excitation at 488 nm and emission at 578 nm) rather than a two-colored approach. The biotin-labeled DNA is detected by staining with SAPE (streptavidin conjugated with R- phycoerythrin) at the fourth step.

The third step is the hybridization to oligonucleotide arrays. The biotin-labeled DNA samples for IP and WCE are independently hybridized to the tiling arrays of interest. Affymetrix currently provides two types of *S. cerevisiae* tiling arrays specially designed for genome-wide ChIP experiments (*see* **Fig. 18.2**). In this report, we show the results using the *S. cerevisiae* chromosome

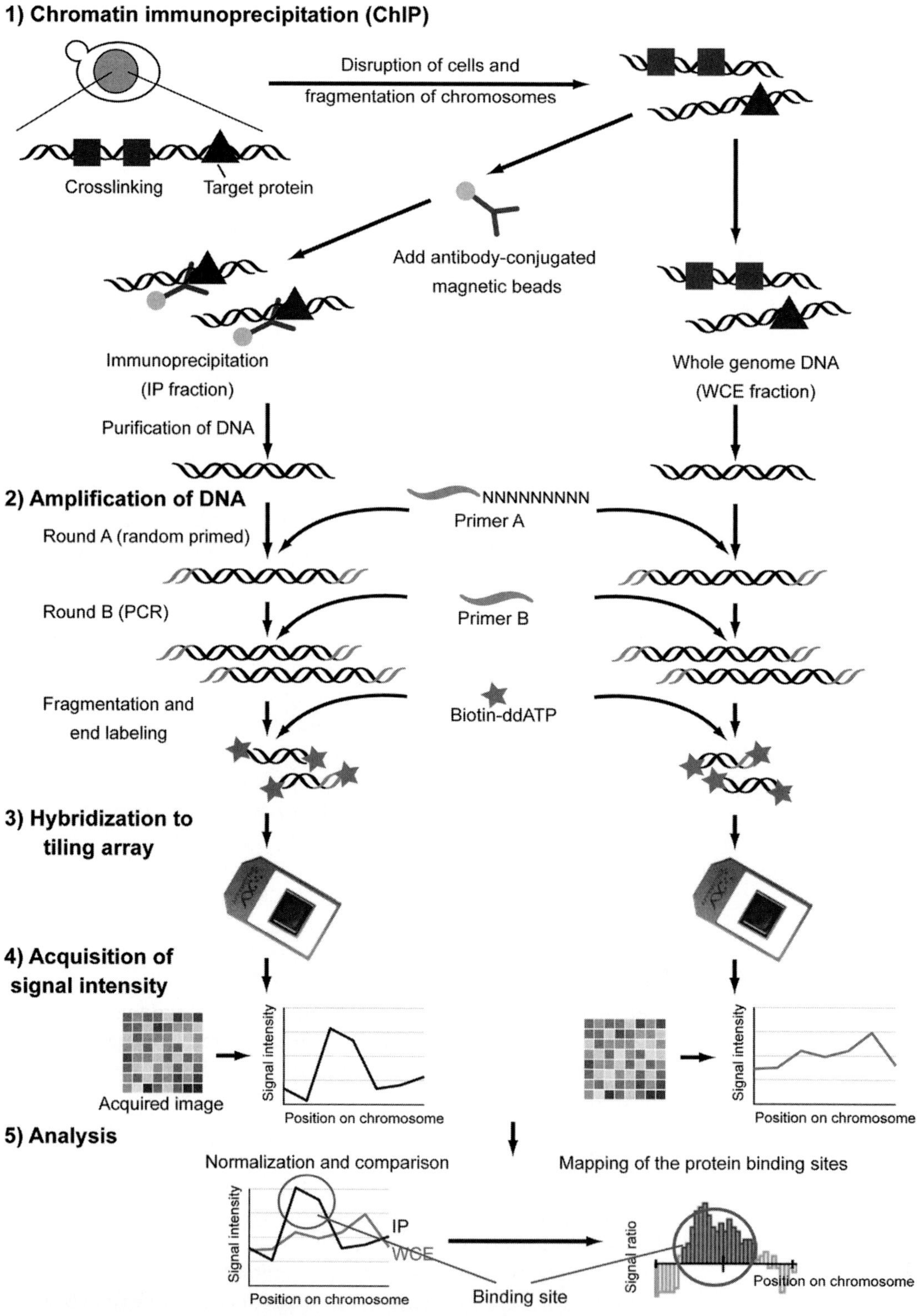

Fig. 18.1. Schematic representation of ChIP-chip protocol. See text for details.

VI tiling array (rikDACF) and chromosome III, IV, V array (SC3456a520015F) (*see* **Fig. 18.2A**). The full-length sequence of entire chromosomes, except some repetitive sequences, are segmented into detection loci of 300 bp or 100 bp. For each detection locus, probes are designed following rigorous criteria established by Affymetrix by taking into account sequence specificity and melting temperatures. Each detection locus contains 16 (for rikDACF array) or 11 (for SC3456a520015F) pairs of 25-mer

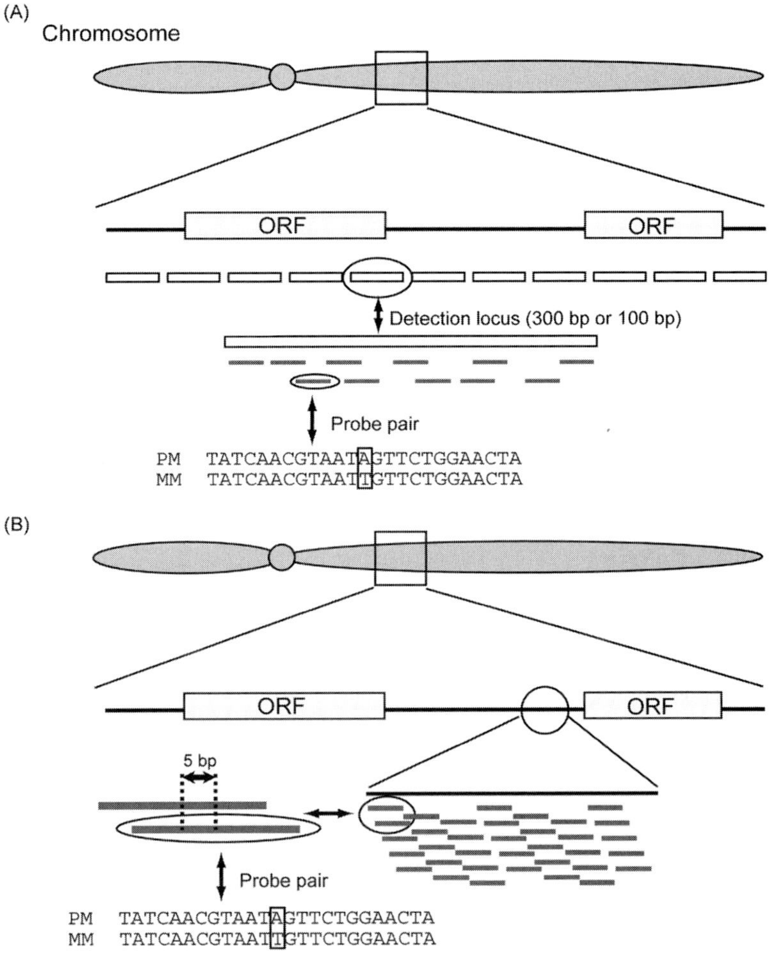

Fig. 18.2. Design of oligonucleotide tiling array. Two types of oligonucleotide tiling arrays are available from Affymetrix. Probe pairs, which are comprised of perfect match (PM) and mismatch (MM) 25-mer oligos, are designed for each type of array in a different way. A horizontal gray bar indicates a probe pair. (**A**) Tiling array, mainly used in this protocol (rikDACF, 1 detection locus = 16 probe pairs; SC3456a520015F, 1 detection locus = 11 probe pairs). Horizontal open box indicates a detection locus. Signal intensity and p-value are calculated for each detection locus. (**B**) The second type of tiling array (*S. cerevisiae* Tiling 1.0R array, *see* **Note 5**). In contrast to the first type of tiling array, the detection loci must be defined by the user. Signal intensity and p-value are calculated for the position of each probe pair.

oligonucleotide probes. Each probe pair includes a perfect match (PM) probe, which is perfectly complementary to a target sequence, and a mismatch (MM) probe, which contains a single base substitution in the center. We can assess the extent of cross-hybridization from the difference in hybridization signals between the PM and MM probe pair. Affymetrix also provides a second type of tiling array (e.g. *S. cerevisiae* Tiling 1.0R array), not fully detailed in this report, for the study of protein distributions at the whole genome level (*see* **Fig. 18.2B**). The overall approach when using these arrays remains the same, except for some modifications at the data analysis step (*see* **Note 5**).

The fourth step is the acquisition of signal intensity. After hybridization, the tiling arrays are washed, stained, and scanned using the Affymetrix GeneChip system. Then, signal intensity for each detection locus is estimated from the acquired image by the single array analysis protocol provided by Affymetrix. Briefly, the signal intensity is calculated using the algorithm which yields a robust weighted mean by taking into account background, noise, and differences between PM and MM. The reliability of the signal intensity is estimated from the detection p-value, which is determined by applying statistical methods to the signal intensity for PM and MM in the locus.

The final step is the comparative analysis. To distinguish the protein binding sites, signal intensity for each locus on two independent arrays is examined by the comparative expression analysis protocol provided by Affymetrix. Before the comparison, normalization is performed to minimize fluctuations due to technical variations between the two arrays. Briefly, the trimmed mean value (the upper and lower 2% of values are eliminated) of the signals for the IP sample is adjusted to be the same as that of the array hybridized with the WCE samples. The comparative expression analysis is then conducted by comparing the signals of the probe pairs (PM and MM) for the IP with their counterpart probe pairs for the WCE. Then a signal $\log_2$ ratio and a change p-value are calculated for each detection locus according to the statistical methods recommended by Affymetrix. Finally, we estimate the binding sites of the protein of interest based on selection criteria as described later (see the representative results shown in **Fig. 18.3**).

Finally, it should be noted that it is necessary to verify that the ChIP signals are actually significant before such analysis, by comparing the signals with the background signals in a mock treated experiment using the nontagged version of the protein. For this purpose, both IP and mock treated samples should be subjected to the same small number of cycles of random primed PCR amplification. If there is only a little difference in the PCR analysis, the ChIP signals are most probably false-positive and not to be used to interpret the protein binding level. In such cases, the protein might not be able to bind at a detectable level. It might be necessary to change the experimental conditions and try other epitope-tags and antibodies.

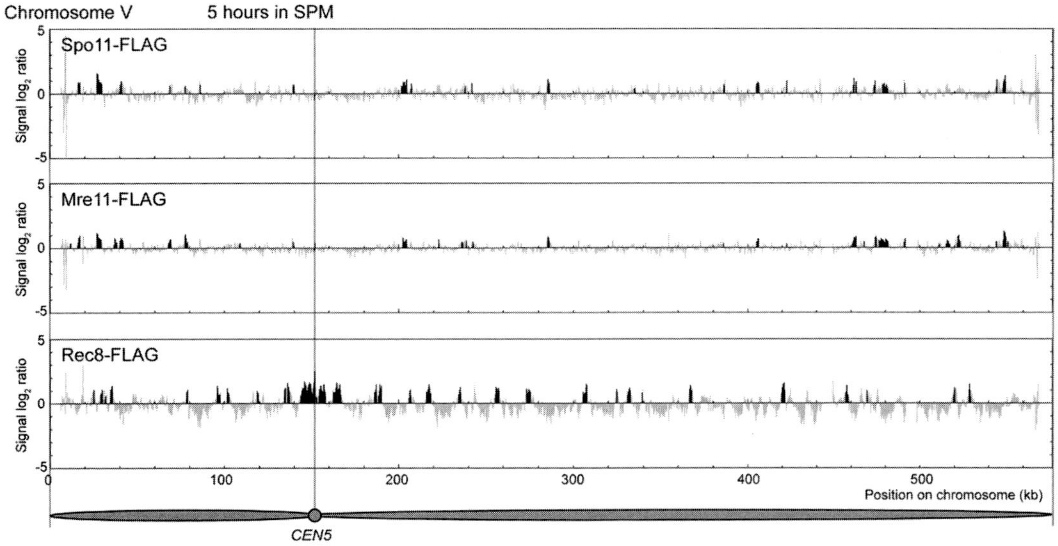

Fig. 18.3. Distribution of Spo11-FLAG, Mre11-FLAG, and Rec8-FLAG during meiosis on chromosome V of budding yeast. The corresponding yeast strains expressing the tagged protein were crosslinked at 5 h after transfer to SPM and analyzed by ChIP-chip technology using the oligonucleotide tiling array (SC3456a520015F). *Upper panel*, Spo11-FLAG. *Middle panel*, Mre11-FLAG. *Lower panel*, Rec8-FLAG. *Vertical bars* represent the protein association level at the analyzed detection loci. The signal $\log_2$ ratio (IP versus WCE fraction) is reported on the vertical axis. The physical position along the chromosome is reported on the horizontal axis. *Bars shaded* in *black* indicate the loci for which the signal $\log_2$ ratio satisfied the two criteria for determination of the binding sites (see text for details).

3.1. Cell Culture and In Vivo Crosslinking

1. In this sub-section, it is essential to obtain samples of good quality from highly synchronized meiotic cultures. For this purpose, prepare SK1 background diploid cells expressing FLAG-tagged proteins under the control of their original promoters. It is important to monitor the progression of meiosis by measuring the recombination frequency using appropriate recombination markers (e.g. *arg4-rv/arg4-bgl* heteroallele) and scoring cell population for completion of meiosis I (e.g., by DNA visualization with Hoechst 33342). If the synchrony of meiosis is not good, discard the sample and retry the sporulation.
2. Grow cells on YPG plates at 30°C for 3 days.
3. Streak the cells from the YPG plates onto YPD plates to isolate a single colony. Incubate at 30°C for 2 days.
4. Pre-culture: Inoculate a single colony into 3 mL of SPS medium and incubate with vigorous shaking at 30°C for 9 h.
5. Transfer the culture to >50 mL of SPS medium and grow the cells to a cell density of 4.0×10^7 cells/mL with vigorous shaking at 30°C.
6. Harvest cells in a centrifuge tube at 4,600*g* for 5 min at room temperature, then resuspend cells in sterilized water and centrifuge at 4,600*g* for 3 min at room temperature.

7. Resuspend and dilute the cells in >100 mL of SPM to a cell density of 2.0×10^7 cells/mL and incubate the suspension with vigorous shaking at 30°C.
8. Take a 100 mL sample from the culture and add 2.8 mL of formaldehyde to a final concentration of 1%. Incubate at room temperature for 10 min (*see* **Note 6**). During the incubation, mix the culture occasionally by inverting three to four times.
9. To stop the crosslinking reaction, add 5.0 mL of 2.5 M glycine to a final concentration of 125 mM, and mix gently. Incubate at room temperature for 5 min.
10. Centrifuge the cells at 1,700*g* for 5 min at 4°C. Suspend the pellet in 40 mL of ice-cold TBS and centrifuge at 1,700*g* for 5 min at 4°C. Repeat this washing step once. Freeze the pellet in liquid nitrogen. Pellets can be stored at this point at –80°C if necessary.

3.2. Cell Lysis and Fragmentation of Genomic DNA

1. These instructions assume the use of a Multi-beads shocker, which offers both a very efficient disruption and a good preservation of proteins from proteolysis under low-temperature conditions. We strongly recommend using this disruptor, but the following procedures are easily adaptable to other instruments such as the BeadBeater (BioSpec Products, Inc., Bartlesville, OK). Note that it is critical to keep the samples at low temperature (4°C) throughout the disruption process.
2. Set the cooling device of the Multi-beads shocker at 0°C 2 h before the experiment.
3. Add 40 μL of freshly prepared 50 × Complete stock solution to 1.8 mL of lysis buffer. Chill the buffer and four sets of 2-mL tubes containing zirconia beads on ice.
4. Suspend and thaw the frozen cells in 1.8 mL of ice-cold lysis buffer.
5. Prepare aliquots of the suspension in the ice-cold 2-mL tubes containing beads (about 500 μL of the suspension/tube).
6. Disrupt the cells with a Multi-beads shocker at 0°C at 2,300 rpm for a total time of 28 min (seven cycles under the following conditions; four sets of 1 min shaking with 1 min pause. Leave a 5 min interval between each cycle). Verify in a microscope that >98% of the cells are broken.
7. Recover the cell extracts as follows. Remove the O-ring from the 2-mL tube. Clean the tube with a paper towel soaked in ice-cold 70% ethanol and then wipe it with a dry paper towel. Puncture the bottom of the tube with a sterilized needle (e.g., 25-gauge) and place the tube into a 1.5-mL microcentrifuge tube (1.5-mL EasiFit cap microtubes, Treff AG, Degersheim,

Switzerland, *see* **Note 7**). Spin at 700*g* for 30 s at 4°C. Loosen the screw cap and centrifuge again for 2 min. Resuspend the cell extracts by pipetting.

8. Fragment the chromosomal DNA to average lengths of ca. 500 bp by sonicating with a handy sonic (Tomy Seiko Co., Ltd., Tokyo, Japan) at a setting of level 7–10 for a total time of 120 s (20 s × 6) (*see* **Note 8**). During the sonication, keep the sample on ice. Chill the sonication probe between each sonication.
9. Centrifuge at 17,610*g* for 5 min at 4°C and transfer the supernatants to a fresh 1.5-mL microcentrifuge tube.
10. Centrifuge the supernatants again for 10 min and then transfer them (whole-cell extract, WCE) to a fresh 1.5-mL microcentrifuge tube. Leave behind >50 μL of the supernatants to avoid aspiration of the precipitates. The procedure can be stopped at this point by freezing the supernatant in liquid-nitrogen and storing at –80°C.

3.3. Preparation of Antibody Binding Beads

1. Thoroughly suspend Dynabeads Protein A in the vial. Transfer 160 μL of the Dynabeads Protein A to a 1.5-mL microcentrifuge tube.
2. Place the tube in a magnetic stand [e.g. Dynal MPC-S (Invitrogen Corporation)] and collect the Dynabeads Protein A. Discard the supernatant carefully.
3. Suspend the Dynabeads Protein A in 1 mL of ice-cold 0.5% BSA-PBS by inverting the tube. Place the tube in a magnetic stand again, and discard the supernatant. Repeat this washing step twice.
4. Resuspend the Dynabeads Protein A in 160 μL of ice-cold 0.5% BSA-PBS and divide into two 80 μL aliquots in 1.5-mL microcentrifuge tubes.
5. Add 8 μL of anti-FLAG M2 antibody (ca. 40 μg) to each tube and mix well by pipetting (*see* **Note 9**).
6. Incubate with gentle rotation for 3 h at 4°C (*see* **Note 10**).
7. Just before use, collect the beads with a magnetic stand, discard the supernatant, and resuspend the beads in 500 μL (each) of ice-cold 0.5% BSA-PBS by inverting the tube several times. Spin down briefly and repeat this washing step once more.

3.4. Immunoprecipitation

1. If starting from frozen WCE, thaw it on ice. Add 800 μL of WCE to each 1.5-mL microcentrifuge tube containing the antibody-conjugated magnetic beads, and suspend by inverting the tube several times.
2. Incubate with a gentle rotation for 6 h at 4°C.

3. Spin down briefly, collect the beads with a magnetic stand, and resuspend the beads in 1 mL of lysis buffer in each tube by inverting several times. Repeat this washing step once.
4. Wash once with 1 mL of wash buffer I in each tube.
5. Wash twice with 1 mL of wash buffer II in each tube.
6. Wash once with 1 mL of wash buffer III in each tube. Spin down briefly, and discard the remaining supernatant carefully to remove the buffer completely.
7. Suspend the beads in 50 μL (each) of elution buffer by vortexing, and incubate for 15 min at 65°C. Mix by vortexing three times for a few seconds during the incubation.
8. Spin the tube and collect the beads with a magnetic stand. Transfer the supernatant (first elution) to a fresh 1.5-mL microcentrifuge tube. Suspend the beads in 75 μL (each) of secondary elution buffer by vortexing and incubate for 5 min at 65°C.
9. Spin the tube and collect the beads with a magnetic stand. Transfer the supernatant (second elution) to the tube containing the first elution. The total volume of the elution (IP fraction) is 250 μL.

3.5. Preparation of DNA

1. Prepare WCE fraction as follows. Mix 8 μL of WCE, 92 μL of lysis buffer, and 400 μL of elution buffer. Incubate for 20 min at 65°C.
2. Add 8.4 μL of Proteinase K to the IP and WCE fractions. Incubate overnight at 37°C.
3. Incubate for 6 h at 65°C to reverse the crosslinks.
4. Add 250 μL of TE (pH 8.0) to the IP fraction.
5. Add 1 μL of 20 mg/mL glycogen and 510 μL of phenol: chloroform: isoamyl alcohol to the IP and WCE fractions, mix well and centrifuge at 20,000*g* for 10 min at room temperature.
6. Transfer 450 μL of aqueous phase to a fresh 1.5-mL tube. Leave >50 μL of aqueous phase to avoid disturbing the interphase.
7. Add 80 μL of TE, pH 8.0 and extract with 530 μL of phenol: chloroform: isoamyl alcohol again.
8. Add 45 μL of 3 M sodium acetate (pH 5.3) and 1 mL of ethanol to the recovered supernatants. Mix by vortexing and incubate for 2 h at –20°C.
9. Centrifuge at 17,610*g* for 30 min at 4°C.
10. Discard the supernatants and then add 1 mL of cold 75% ethanol, mix by inverting the tube, and centrifuge at 17,610*g* for 10 min at 4°C.

11. Discard the supernatants and dry the pellets. Suspend the pellets of the IP fraction in 10 μL of TE (pH 7.5) containing 0.1 mg/mL RNase A and the pellets of the WCE fraction in 100 μL of TE (pH 7.5) containing 0.1 mg/mL RNase A. Incubate for 30 min at 37°C and store at –20°C until amplification of DNA.

3.6. Amplification of DNA

1. Set up the thermocycler program for round A as follows: Two cycles of 94°C for 2 min, 10°C for 5 min, ramping to 37°C over 9 min, 37°C for 8 min.
2. Prepare the round A (random priming step) sample mixture in a 0.2-mL PCR tube as follows for each sample. Mix 7 μL of DNA sample, 2 μL of 5 × Sequenase reaction buffer, and 1 μL of 40 μM primer A.
3. Prepare the round A enzyme premix in a 0.2-mL PCR tube as follows. Mix 1 μL of 5 × Sequenase reaction buffer, 1.5 μL of 0.5 mg/mL BSA, 0.75 μL of 0.1 M DTT, 1.5 μL of 3 mM dNTP, and 0.3 μL of 13 units/μL Sequenase. Mix by pipetting and spin down briefly. Note that the amounts indicated here are for a single reaction.
4. Insert the tubes with the round A sample mixture in the thermocycler and start the round A program. Pause the program at the onset of the incubation step at 10°C of the first cycle. Add 5.05 μL of the round A enzyme premix to the tubes, and mix by pipetting with the sample tubes left in the thermocycler, then resume the program.
5. When the first cycle is completed, prepare an additional enzyme mixture for each tube by mixing 0.9 μL of Sequenase dilution buffer with 0.3 μL of 13 units/μL Sequenase. Mix by pipetting and spin down briefly.
6. Pause the program at the onset of the incubation step at 10°C of the second cycle, add 1.2 μL of the additional enzyme mixture, mix by pipetting, and resume the program.
7. At the end of the round A program, chill the sample on ice, spin briefly, and add 45 μL of sterilized water.
8. Prepare a round B (PCR) mixture in each 0.2-mL PCR tube by mixing 15 μL of the round A product with 85 μL of the *Taq* mixture: 65.5 μL of sterilized water, 10 μL of 10 × PCR buffer, 8 μL of 2.5 mM dNTP, 1 μL of 100 μM primer B, and 0.5 μL of 5 units/μL r*Taq*. Store the remaining round A product on ice or at –20°C.
9. Perform PCR amplification with the following cycling conditions: 94°C for 5 min; 25 cycles of 92°C for 30 s, 40°C for 30 s, 50°C for 30 s, 72°C for 3 min; 72°C for 5 min.

10. Verify the DNA amplification by running 1 μL of the PCR products in 1% agarose gel. A smear should be detected between 300–1,000 bp.

3.7. Fragmentation and End-labeling

1. Add 400 μL of sterilized water to the round B products and concentrate the amplified DNA down to <40 μL using a Microcon YM-10 following the instructions of the manufacturer. After the concentration, transfer the sample to a 0.2-mL PCR tube and adjust the volume to 40 μL with sterilized water.
2. Prepare a DNase I solution in a 0.2-mL PCR tube as follows. Mix 1 μL of 10× One-phor-all buffer PLUS, 7.4 μL of sterilized water, 0.6 μL of 25 mM $CoCl_2$, and 1 μL of 1 units/μL DNase I. This recipe is for six reactions.
3. Add 2.9 μL of 25 mM $CoCl_2$, 4.85 μL of 10× One-phor-all buffer PLUS, and 1.5 μL of the 0.1 units/μL DNase I solution to the sample. Mix by pipetting.
4. Incubate at 37°C until the DNA is fragmented to a length of about 100 bp, and inactivate the DNase I by heating at 99°C for 15 min. In most cases, the incubation step takes 70 s. After the reaction, confirm that an average length of the DNA is ca. 100 bp by agarose gel electrophoresis. If the DNA size is still longer, add 1.5 μL of 0.1 units/μL DNase I solution and incubate again briefly.
5. Add 1 μL of 400 units/μL terminal transferase and 1 μL of 1 mM biotin-11-ddATP, mix by pipetting, and incubate for 1 h at 37°C.

3.8. Hybridization

1. Equilibrate the GeneChips at room temperature and preheat the hybridization oven at 42°C.
2. Prepare a prehybridization buffer as follows. Mix 120 μL of 20× SSPE, 260 μL of sterilized water, and 20 μL of 0.1% (v/v) Triton X-100. This recipe is for a single GeneChip analysis.
3. Prepare a hybridization mixture as follows for each hybridization. Incubate 15 μL of 20× Eukaryotic Hybridization Controls in a 1.5-mL tube at 65°C for 5 min, and spin down briefly. Add 122 μL of sterilized water, 90 μL of 20× SSPE, 5 μL of 3 nM Control Oligonucleotide B2, 3 μL of 10 mg/mL Herring Sperm DNA, and 15 μL of 0.1% (v/v) Triton X-100, mix by vortexing, and spin down briefly.
4. Mix 250 μL of the hybridization mixture with 50 μL of the sample by vortexing, and spin down.

5. Incubate at 99°C for 10 min and then immediately transfer on ice. During this incubation, fill the GeneChip hybridization chamber with the prehybridization buffer as described in the Affymetrix Technical Manual, and incubate at 42°C for more than 10 min in the hybridization oven with rotation at 60 rpm.
6. Spin down the hybridization sample at 17,610*g* for 5 min at 4°C to precipitate any insoluble materials.
7. Remove the prehybridization buffer from the GeneChip and apply the appropriate volume of the hybridization sample according to the Affymetrix Technical Manual (e.g. 200 μL for SC3456a520015F, 150 μL for rikDACF). Avoid contamination with precipitated materials at the bottom of the tube. Store the remaining samples at –20°C.
8. Place the GeneChip into the hybridization oven and rotate at 60 rpm for 16 h at 42°C.

3.9. Washing, Staining, and Scanning

1. Prepare the staining buffer in a 1.5-mL tube as follows for each GeneChip. Vortex 1 mg/mL SAPE and centrifuge at 17,610*g* for 3 min at 4°C to remove aggregation of SAPE. Mix 600 μL of 2 × stain buffer, 540 μL of sterilized water, 48 μL of 50 mg/mL acetylated BSA, and 12 μL of 1 mg/mL SAPE. The staining buffer must be shielded from light by wrapping the tube in aluminum foil and storing at 4°C.
2. Warm up the GeneChip scanner.
3. Set up the GeneChip fluidics station by running the priming protocol supplied by Affymetrix.
4. Remove the hybridization samples from the GeneChip. Store the samples at −80°C for another hybridization, if necessary.
5. Fill the GeneChip with wash buffer A.
6. Place the GeneChip into the GeneChip fluidics station and perform washing and staining with the staining buffer by running the appropriate protocol provided by Affymetrix (e.g. EukGE-WS1v3 for SC3456a520015F, Mini-Euk1v2 for rikDACF). When the program is finished, shield the GeneChip from light until scanning. If the GeneChip is not filled with buffer, fill it with wash buffer A manually.
7. Scan the GeneChip with the GeneChip scanner.
8. Make sure that the main grid is aligned properly. If auto-alignment fails, adjust the alignment manually.
9. Perform single-array analysis with Microarray Suite 5.0 (MAS5) or GeneChip Operating Software (GCOS) using the default settings to obtain the ".chp" file containing the numeric data of all detection loci (e.g. signal, detection p-value).

10. Generate and save the expression report (".rtp" file). Then make sure that the "Number Present" is more than 95%. If not, the data can not be used for the next comparative analysis.
11. Perform comparative analysis with MAS5 or GCOS using the default settings to obtain the signal $\log_2$ ratio and change p-value (".chp" file). Compare signals of the ChIP fraction with those of the WCE fraction. Set the ".chp" file of the WCE fraction to the baseline.
12. Determine the loci which display significant signals indicating the binding of the protein to chromosomes, according to the following criteria: (1) The statistic reliability of the signal intensity (detection p-value) and of the "IP versus WCE" signal ratio (change p-value) should both be less than 0.025. (2) If more than three adjacent loci at a maximal distance of 300 bp from each other (calculated from their center) satisfy the above reliability criterion, the central locus is defined as a binding site.

4. Notes

1. Add 6 μL of emulsion-type defoaming agent Grade KM-72 (Shin-Etsu chemical Co., Ltd., Tokyo, Japan) to 100 mL of SPS medium.
2. Beads coupled with Protein G (e.g. Dynabeads Protein G) are also available, however, we recommend beads coupled with Protein A, because Protein G beads tend to give high background due to contamination of DNA.
3. Wash buffer A and wash buffer B can be purchased from Affymetrix. Affymetrix provides the GeneChip hybridization, wash, and stain kit including all reagents for hybridization, washing, and staining. However, we have not tried them.
4. Before addition of Proteinase K and reversal of crosslinks, take an appropriate volume of sample (e.g. 5 μL) from the IP and WCE fractions, mix with an equal volume of 2 × SDS sample buffer, and incubate at 95°C for 30 min to perform a Western blotting analysis. For this purpose, elute the DNA–protein conjugates from antibody-beads using a small volume of elution buffer (e.g. 50 μL).
5. Since the whole-genome tiling array [*S. cerevisiae* Tiling 1.0R array (*see* **Fig. 18.2B**)], is designed by a special algorithm, specialized software (i.e. GCOS, TAS and IGB, available from Affymetrix) and setting criteria are needed to identify the protein binding sites. Affymetrix-designed probes are selected

at defined intervals (5 bp) throughout the whole genome sequence. At the primary analysis (single and comparative analysis), data are normalized using the "robust quantile normalization", then for each 25-mer probe position, an estimated signal and its corresponding p-value taking into account the neighboring probes are calculated using the TAS software. These values are determined in function of the raw signal data of all the probes located within a region of user-defined length around the considered position (recommended bandwidth: half of the ChIP fragment size, i.e. 250 bp). Subsequently, the detection loci displaying significant positive signals are identified by user-defined selection criteria based on signal and p-value thresholds.

6. In some cases, crosslinking conditions should be optimized. For example, incubate for 10–30 min at room temperature after addition of formaldehyde and then incubate overnight at 4°C.
7. In this step, it is recommended to use the 1.5-mL microcentrifuge tubes. The 2-mL disruption tube might not fit into other types of microcentrifuge tubes. Alternatively, use autoclaved 5-mL Röhren tubes (75 × 13 mm, Sarstedt AG & Co., Nümbrecht, Germany) or 15-mL Falcon tubes (Becton, Dickinson and Company). Recover the flow-through in the tubes and transfer them into 1.5-mL tubes.
8. Optimize the conditions for sonication at the beginning of the experiment. Sonicate 500 μL of whole-cell extracts at various conditions (by changing the power and time of sonication), and prepare DNA as described in "preparation of DNA". Then suspend DNA into a small volume of TE containing RNase A. Verify the smear ranging between 200–1,000 bp by agarose gel electrophoresis.
9. Affinity of Protein A to mouse IgG1 is known to be weaker than that of Protein G. However, we obtained similar efficiency even when we used Dynabeads Protein A instead of Dynabeads Protein G in our ChIP condition.
10. Attention should be paid so that the suspension remains at the bottom of the tube. If necessary, transfer the suspension to a fresh tube or spin down at maximum speed and resuspend the beads again.

Acknowledgments

We would like to thank all the laboratory members of Ohta laboratory and Shibata distinguished senior scientist laboratory for helpful discussion. We also thank Dr. W. Lin for critically reading this manuscript and Ms. Y. Sakuma for technical assistance. This work was supported by grants from the following sources: basic research from the Bio-oriented Technology Research Advancement Institution (to T. Shibata and K. Ohta); grants-in-aid for scientific research on priority areas from the Ministry of Education, Science, Culture, & Sports, Japan (to K. Ohta).

References

1. Ren, B., Robert, F., Wyrick, J. J., Aparicio, O., Jennings, E. G., Simon, I., et al. (2000) Genome-Wide location and function of DNA binding proteins. *Science* **290**, 2306–2309.
2. Katou, Y., Kanoh, Y., Bando, M., Noguchi, H., Tanaka, H., Ashikari, T., et al. (2003) S-phase checkpoint proteins Tofl and Mrc1 form a stable replication-pausing complex. *Nature* **424**, 1078–1083.
3. Lengronne, A., Katou, Y., Mori, S., Yokobayashi, S., Kelly, G. K., Ito, T., et al. (2004) Cohesin relocation from sites of chromosomal loading to places of convergent transcription. *Nature* **430**, 573–578.

Chapter 19

Parallel Detection of Crossovers and Noncrossovers in Mouse Germ Cells

Frédéric Baudat and Bernard de Massy

Abstract

The recombination between homologous chromosomes during the prophase of the first meiotic division plays an essential role in the formation of euploid gametes, as well as contributing to genetic diversity through the generation of new allele combinations. Two types of products are formed, crossovers (CO) and gene conversions not associated with a crossover, also called noncrossovers (NCO). They result from the repair through differentiated pathways of a common initiating event, DNA double-strand breaks, formation of which is programmed during early meiotic prophase. In contrast to CO, little is known on the frequency, distribution, and regulation of NCO in mammals, mostly due to the technical challenge represented by their detection. However, the development of batch sperm-typing methods allowed for the detection of meiotic NCO products at meiotic recombination hotspots in human and mice. Several recent studies based on this technique demonstrated that mammalian CO hotspots are sites of recombination initiation and that the formation of CO and NCO has different genetic requirements. Here, we describe a method for detecting and characterizing CO and NCO, which has been applied to the analysis of the mouse *Psmb9* recombination hotspot.

Key words: Meiosis, recombination, crossover, gene conversion, sperm typing, allele-specific PCR.

1. Introduction

During meiosis, the programmed induction of DNA double-stranded breaks (DSB) leads to the formation of two types of products, crossovers (CO) and gene conversions not associated with a crossover, also called noncrossovers (NCO). CO are at the origin of chiasmata, which are essential for the segregation of homologous chromosomes at the first meiotic division, whereas the function of NCO remains elusive. Studies in

Scott Keeney (ed.), *Meiosis, Volume 1, Molecular and Genetic Methods, vol. 557*

DOI 10.1007/978-1-59745-527-5_19 Springerprotocols.com

budding yeast have shown that both CO and NCO originate from a common initiating event, programmed genomic DNA double-strand breaks (DSBs) catalyzed by the Spo11 protein. Most DSBs are clustered at numerous preferential sites, giving rise to recombination hotspots, which are short intervals surrounding DSB sites that concentrate recombination events (CO and NCO) *(1)*. The formation of CO and NCO results from the repair of the initiating DSBs through differentiated pathways *(2, 3)*. Analyzing both CO and NCO is therefore essential for understanding the mechanisms and regulation of meiotic recombination.

In eukaryotes other than yeasts, including mammals, there is an enormous difference between the relatively large amount of knowledge on CO distribution and the small amount of data on NCO *(4)*. This is essentially explained by the fact that CO lead to changes of the associations between alleles along large chromosomal fragments, whereas NCO cause a localized transfer of genetic information that often includes no or only a few polymorphisms. Detection of NCO thus requires specific methodologies. The distribution of CO in mammals has been analyzed at various scales by several methods *(4, 5)*. In human, in part similar to *S. cerevisiae*, CO tend to be clustered in short intervals called CO hotspots. A technique of sperm typing, based on allele-specific PCR amplification, has allowed for detecting and analyzing large numbers of CO molecules at individual CO hotspots *(6)*. These studies have shown that in human, CO hotspots are 1–2 kb wide, with CO densities higher at their center and decreasing progressively on both sides *(5, 7)*.

Recently, the power of the sperm-typing methods for detecting rare recombinant molecules at CO hotspots has been applied for developing related assays aimed at detecting and characterizing NCO products at mammalian CO hotspots. These approaches succeeded in detecting NCO at the center of CO hotspots in mouse and human sperm, demonstrating that mammalian CO hotspots are sites of meiotic recombination initiation *(8, 9)*. In addition, the parallel analysis of CO and NCO provides insights into the mechanisms regulating their formation. For instance, the analysis of several human hotspots has shown that the relative proportion of CO and NCO varies widely from one hotspot to another *(10)*. Here, we describe one method which we have used for analyzing in parallel CO and NCO at the mouse *Psmb9* hotspot. Our method has been adapted for detecting these recombinant molecules in several tissues (i.e., testes and ovaries), allowing us to analyze the formation of mature recombinant products in both male and female germ lines, as well as in sterile mouse mutants in which no gametes are produced *(11, 12)*.

2. Materials

2.1. Preparation of Ovary Samples

1. 10 × PBS: 1.37 M NaCl, 27 mM KCl, 101.4 mM Na_2HPO_4, 17.6 mM KH_2PO_4, pH 7.4.
2. PBS-EDTA: PBS supplemented with 5 mM EDTA.
3. PBS-BSA: PBS supplemented with 0.05% bovine serum albumin.
4. Fixative: 2% paraformaldehyde, 0.05% Triton-X100 in PBS. Prepare fresh for each experiment.
5. Polylysine-coated slides (Menzel).
6. Primary antibodies: anti-GCNA1 (monoclonal rat IgM 10D9G11), gift from G. Enders (University of Kansas). Anti-SYCP3, guinea pig serum (home made against a mouse SYCP3 peptide).
7. Antibody dilution buffer (ADB), 10 × : 10% normal donkey serum (Jackson Immunoresearch), 3% BSA, 0.05% Triton-X100 in PBS. Dilute in PBS to get 1 × ADB.
8. Secondary antibodies: Cy2-conjugated goat anti-rat IgM, μ chain-specific (Jackson Immunoresearch); Cy3-conjugated donkey anti-guinea pig IgG(H+L) (Jackson Immunoresearch).
9. DAPI (4′,6-diamidine-2′-phenylindole dihydrochloride), 1 mg/mL in water, stored at –20°C.
10. Mounting medium: ProLong antifade or SlowFade antifade (Invitrogen).

2.2. Preparation of the DNA

1. Mouse strains: C57BL/10 J (the Jackson laboratory, Bar Harbor, Maine, USA), hereafter abbreviated as B10, and B10.A(R209) (from T. Shiroishi, National Institute of Genetics, Mishima, Japan), hereafter abbreviated as R209 *(13)*.
2. Lysis buffer: 0.1 M EDTA, 0.5% SDS, 1 mg/mL Proteinase K, 50 mM Tris-HCl pH 8.0. Store in aliquots at –20°C. For extracting the genomic DNA from sperm, supplement before use with 0.7 M of β-mercaptoethanol.
3. Phenol, equilibrated to pH 8.0 and stored at –20°C, and chloroform. Make a 1:1 mix.
4. SST Vacutainer tubes (Becton Dickinson).
5. Isopropanol, 70% v/v ethanol and pure ethanol.
6. 5 mM Tris-HCl, pH 8.0.
7. Tris-Carrier DNA: Denatured herring testes DNA, 10 mg/mL (Clontech). Dilute to 5 μg/mL in 5 mM Tris-HCl, pH 8.0.

2.3. Reaction Mixes for the Nested PCR

1. PCR buffer, 10 × *(14)*: 450 mM Tris-HCl pH 8.8, 110 mM $(NH_4)_2SO_4$, 45 mM $MgCl_2$, 67 mM β-mercaptoethanol, 44 μM EDTA, 10 mM dATP, dCTP, dTTP, and dGTP, 1.13 mg/mL BSA (MP Biomedicals). Store in aliquots at –20°C (*see* **Note 1**).
2. HPSF purified oligonucleotides, dissolved at 100 μM in water and stored at –20°C. Prepare a 10 μM working solution, stored at –20°C.
3. Taq DNA polymerase.
4. Pfu polymerase.
5. Ethidium bromide or SYBRGold (Invitrogen).

3. Methods

In this article, we describe how to detect, measure the frequency, and characterize CO and NCO products in hybrid mice at the *Psmb9* recombination hotspot. Guidelines for setting up an assay at another locus are given (*see* **Note 2**). To obtain a reliable measurement of the frequency of meiotic recombination products, one should first determine the fraction of meiocytes among the cells from which the DNA is extracted (especially for oocytes). Then, the principle of the method consists of determining in a preparation of genomic DNA the total concentration of amplifiable molecules on one side, and the concentration of recombinant molecules (CO and NCO) on the other side (**Fig. 19.1**). The ratio between the second and the first measures gives an estimate of the frequency of each type of recombinant product (CO and NCO, respectively) per haploid genome. In addition, the properties of each type of recombinant molecules are deduced from the characterization of the PCR products. Finally, it should be noted that, in principle, this method could apply for detecting a specific change at a particular nucleotide (either gene conversion or a specific mutation) in any tissue, providing that this change is occurring at a sufficient frequency (see below).

3.1. Preparation of the Ovary Samples

If meiotic recombination is not analyzed in oocytes, proceed directly to **Section 3.2**.

1. Dissect ovaries from new-born mice in PBS and clean them up as much as possible. Pool all ovaries from 1L for further processing (*see* **Note 3**).
2. Incubate the whole ovaries in PBS-EDTA for 30 min at room temperature. Release the cells by pricking the gonads with fine needles in PBS-BSA. Collect the cells (not the debris of tissue) in 500 μL of PBS-BSA.

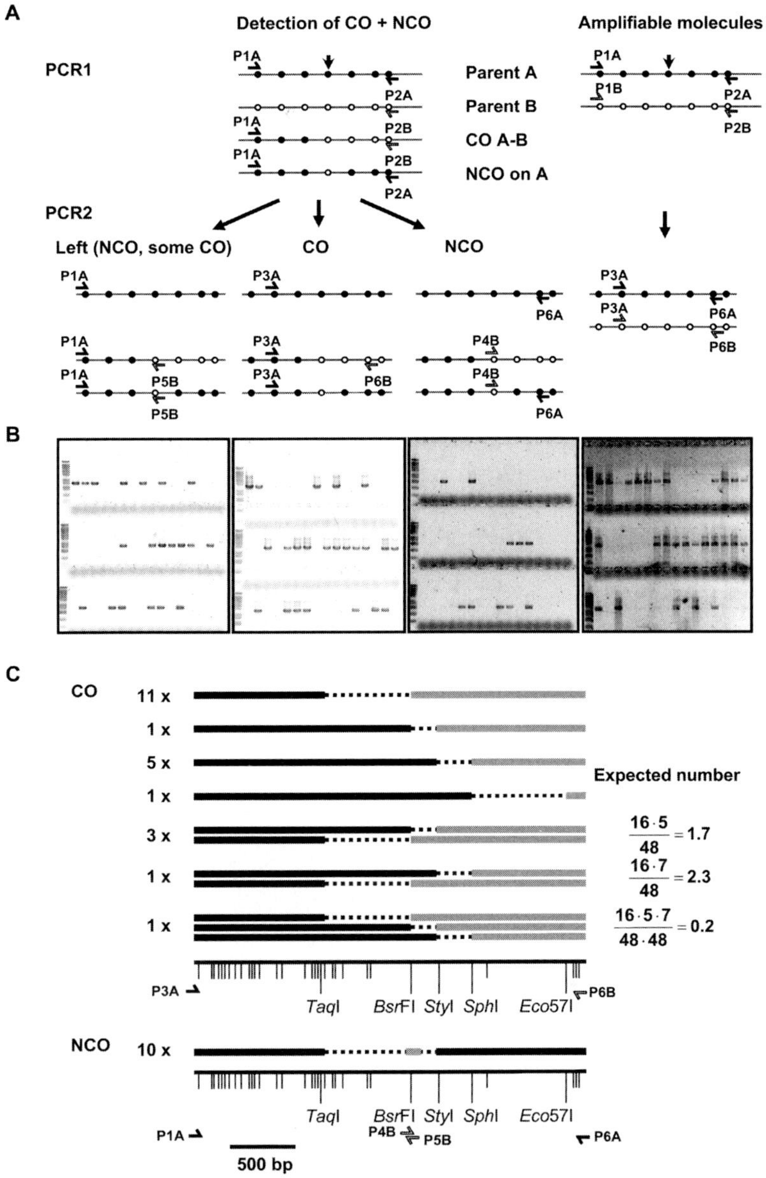

Fig. 19.1. Parallel detection of CO and NCO at the *Psmb9* hotspot in B10 × R209 ovaries. **A** – Strategy for the detection of CO and NCO molecules. For the detection of CO and NCO, pools of 100–1,000 haploid genomes are placed in each PCR1 reaction. For estimating the concentration of amplifiable molecules, one to three haploid genomes are placed in each PCR1 reaction. **B** – Detection of CO and NCO in ovary DNA from B10 (Parent A) × R209 (Parent B) F_1 hybrid mice. Forty-eight pools of 250 haploid genomes (CO + NCO) or 40 pools of 2.5 haploid genomes (Amplifiable molecules) have been amplified as depicted above. The primer sequences are listed in **Table 19.2**. **C** – Mapping of recombination breakpoints for the CO and NCO detected in B. In this experiment, the alleles of the recombinant molecules have been determined at the restriction site polymorphisms only (*Taq* I, *BsrF* I, *Sty* I, *Sph* I, and *Eco*57 I). *Black line*, B10 allele. *Gray line*, R209 allele. *Dotted line*, interval where the breakpoint is located. The number of pools containing each type of recombinant product(s) is indicated to the *left*. The expected number of pools containing more than one type of CO is indicated for each type of mix to the *right*.

3. Take a 25 μL aliquot of each cell suspension. Add 5 μL of this aliquot to 10 μL of PBS, place the cells on a Malassez cell and count the cells. The oocytes are larger and more irregularly shaped than the somatic cells. Usually, they represent 10–20% of the cells.
4. Add the remaining 20 μL of the cell suspension aliquot (from **Step 3**) to 100 μL fixative and incubate 5 min at room temperature. Rinse the cells once with 100 μL PBS, pellet and resuspend them in 100 μL PBS and place 50 μL of the suspension on each of two polylysine-coated microscope slides. Let them dry overnight at 4°C.
5. Rinse the slides once in PBS, then incubate them with the primary antibodies (anti-GCNA1, undiluted supernatant; anti-SYCP3 serum, 1/1,000) for 3 h at room temperature. Wash the slides twice in PBS for 5 min, incubate them with the secondary antibodies (Cy2 goat anti-rat IgM, 1/100; Cy3 donkey anti-guinea pig IgG, 1/250) in 1 × ADB for 1 h at 37°C, wash three times 5 min in PBS, incubate in DAPI (1 μg/mL in PBS) for 15 min at room temperature, rinse once in H_2O and mount the slides with coverslips. The oocytes, identified by the nucleus which is stained with both antibodies, usually represent 10–15% of the cells (**Fig. 19.2**).

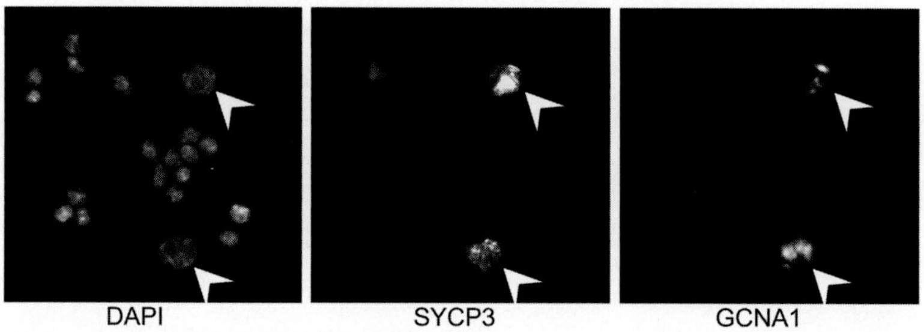

Fig. 19.2. Cells from a new-born ovary stained with DAPI and antibodies directed against SYCP3 and GCNA1. *Arrowheads*: Oocytes are identified by the SYCP3 and GCNA1 signals. Note the large size of their nucleus stained with DAPI.

3.2. Preparation of the DNA

1. Dissect out the tissues of interest with clean tools. The processing of ovaries is described in the former section. A somatic tissue (for instance tail tip or liver) should be taken as a negative control for meiotic recombination. For recovering the sperm, dissect out the caudal epidydimus and place in a drop of PBS, then squeeze it gently with tweezers and collect the sperm with a 20 μL pipette into 500 μL of PBS. Remove 5 μL for counting on a Malassez cell (*see* **Note 4**). Centrifuge the remainder (15,000*g* for 30 s) and proceed to **Step 2**.
2. Resuspend the samples in 500 μL of lysis buffer and incubate overnight at 60°C. Specifically for sperm DNA extraction, supplement the lysis buffer with 0.7 M β-mercaptoethanol.

Perform two extractions with 500 μL of phenol:chloroform (use SST Vacutainer tubes for phase separation), and precipitate the DNA with 300 μL of isopropanol. Wash with 70% ethanol and resuspend the DNA in 50–100 μL of 5 mM Tris-HCl, pH 8.0.

3. Determine the DNA concentration by measuring the absorbance at 260 nm, and verify the presence of high molecular weight DNA (>15 kb) on a 0.7% agarose gel.
4. Incubate 4 μg of each DNA sample with *Xho* I for 2–4 h at 37°C (*see* **Note 5**). Precipitate the DNA with ethanol and resuspend in 50 μL 5 mM Tris-HCl, pH 8.0. Verify completion of the digestion on a 0.7% agarose gel, and carefully determine the DNA concentration by absorbance at 260 nm.
5. Make serial dilutions of the digested DNA in Tris-carrier DNA in a volume sufficient to minimize fluctuations (0.5–1.5 mL), down to 3 and 1 haploid genomes/μL (1 mouse haploid genome = 3 pg).

3.3. Detection of Specific Molecules by Allele-Specific Nested PCRs (Fig. 19.3)

1. The reaction mixes for the two nested PCR amplifications (PCR1 and PCR2) are prepared similarly for the detection of amplifiable molecules, the detection of CO, and the parallel detection of CO and NCO. **Steps 2–4** must be performed in conditions designed to reduce contaminations (*see* **Note 6**).
2. For the PCR1, prepare a mix for the number of reactions to be performed (*see* **Note 7**), containing for each 10 μL reaction:

   1 μL of 10 × PCR buffer,

   0.5 μL of 10 μM oligonucleotide, for each primer,

   0.625 unit of Taq DNA polymerase,

   0.0625 unit of Pfu DNA polymerase,

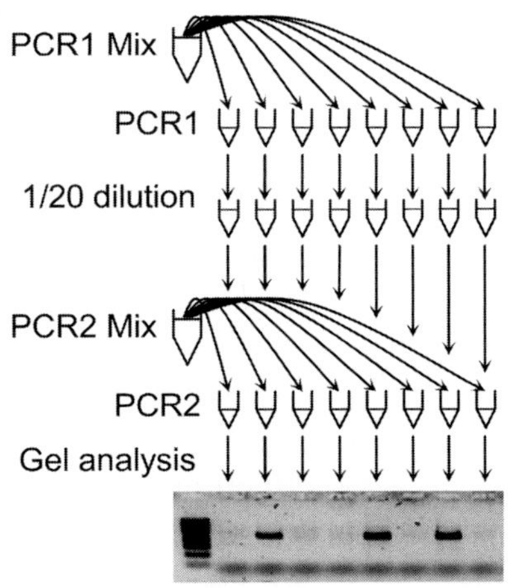

Fig. 19.3. Schematic representation of the nested PCR procedure applied for detecting specific molecules.

1 μL of diluted genomic DNA,

H_2O to 10 μL final volume.

PCR conditions depend upon the primers, and are specified in the appropriate sections below.

3. Prepare tubes containing 20 μL of 5 mM Tris-HCl, pH 8.0 for the dilution of the PCR1 products.
4. Prepare a mix for the PCR2 reactions, containing for each 10 μL reaction:

   1 μL of 10 × PCR buffer,

   0.5 μL of 10 μM oligonucleotide, for each primer,

   0.625 unit of Taq DNA polymerase,

   0.0625 unit of Pfu DNA polymerase,

   H_2O to 9 μL final volume.
5. Outside the dedicated room, make a 1/20 dilution (1 μL in 20 μL) of the PCR1 products in 5 mM Tris-HCl, pH 8.0, in the tubes prepared in **Step 3**.
6. Add 1 μL of the 1/20 dilution of the PCR1 product to each tube of PCR2 mix.

   PCR conditions depend upon the primers, and are specified below in the appropriate sections.
7. Run PCR products on an agarose gel, stain the gel with ethidium bromide or SYBRGold and determine the number of positive and negative pools.

3.4. Determination of the Concentration of Amplifiable Molecules

1. Perform series of PCR1 reactions for each DNA sample to be tested with the primers P1A, P1B, P2A, and P2B, as described in **Section 3.3** (*see* **Note 8**). For each DNA sample, prepare one series of eight PCR1 reactions with three haploid genomes per reaction, and one series with one haploid genome per reaction. PCR1 conditions are 94°C for 2 min, then 24 cycles of 94°C for 10 s, 63°C for 30 s and 68°C for 5 min, followed by 68°C for 7 min.
2. Prepare the mix for the PCR2 with primers P3A, P3B, P6A, and P6B, and put 9 μL of this mix in each tube for the PCR2.
3. Add 1 μL of the 1/20 dilution of the PCR1 product to each PCR2 reaction tube. PCR2 conditions are 94°C for 2 min, then 28 cycles of 94°C for 10 s, 63°C for 30 s and 68°C for 3 min, followed by 68°C for 5 min.
4. Run PCR products on an agarose gel, stain the gel with ethidium bromide or SYBRGold and determine the number of positive and negative pools.
5. For each DNA sample, determine which dilution gives a frequency of negative pools close to 0.5. Using this dilution, perform again **Steps 1–4** with 40–96 reactions instead of eight.

3.5. Determination of the CO Frequency

1. A rough estimate of the CO frequency is obtained by performing a series of eight PCR amplifications on pools containing numbers of genomes expected to yield an average of 0.5–1 CO per pool. For example, for an expected frequency of 0.1% CO per amplifiable molecule, use pools of 5,000, 1,000, and 500 genomes (*see* **Note 9**).

   Prepare a mix for the PCR1 reactions with primers P1A and P2B (*see* **Note 10**). PCR1 conditions are 94°C for 2 min, then 24 cycles of 94°C for 10 s, 63°C for 30 s and 68°C for 5 min, followed by 68°C for 7 min (*see* **Note 11**).
2. Make a 1/20 dilution (1 μL in 20 μL) of the PCR1 products in 5 mM Tris-HCl, pH 8.0 (*see* **Note 12**).
3. Prepare a mix for the PCR2 reactions with primers P3A and P6B (*see* **Note 13**).
4. Add 1 μL of the 1/20 dilution of the PCR1 product to each tube. PCR2 conditions are 94°C for 2 min, then 28 cycles of 94°C for 10 s, 63°C for 30 s and 68°C for 3 min, followed by 68°C for 5 min.
5. Run PCR products on an agarose gel, stain the gel with ethidium bromide or SYBRGold and determine the number of positive and negative pools.

3.6. Parallel Analysis of CO and NCO

1. Perform a series of 40–96 PCR1 reactions for each sample to be tested, with the number of molecules per reaction determined in **Section 3.5** (*see* **Note 14**). Prepare a mix for the PCR1 reactions with primers P1A, P2A, and P2B (*see* **Note 15**). PCR1 conditions: 94°C for 2 min, then 18 cycles of 94°C for 10 s, 63°C for 30 s and 68°C for 5 min, followed by 68°C for 7 min.
2. Make a 1/20 dilution (1 μL in 20 μL) of the PCR1 products in 5 mM Tris-HCl, pH 8.0.
3. Prepare three mixes for PCR2 reactions: one for the detection of CO, one for the detection of NCO, and one allowing the detection of recombinant products to the left side (**Fig. 19.1**) (*see* **Note 16**). For the detection of CO, prepare a mix with primers P3A and P6B. For the detection of NCO, prepare a mix with primers P4B and P6A. For the detection of recombinant product to the left side, prepare a mix with primers P1A and P5B.
4. Make a 1/20 dilution of the PCR1 and add 1 μL of this dilution to each tube of the three series of PCR2 reactions. For the detection of CO, PCR conditions are 94°C for 2 min, then 30 cycles of 94°C for 10 s, 63°C for 30 s and 68°C for 3 min, followed by 68°C for 5 min. For the detection of NCO, PCR conditions are 94°C for 2 min, then 30 cycles of 94°C for 10 s, 63°C for 30 s and 68°C for 1.5 min, followed by 68°C for 5 min. For the detection of recombinant

products to the left side, PCR conditions are 94°C for 2 min, then 30 cycles of 94°C for 10 s, 63°C for 30 s and 68°C for 2 min, followed by 68°C for 5 min.

5. Run PCR products on an agarose gel, stain the gel with ethidium bromide or SYBRGold and determine the number of positive and negative pools.

3.7. Calculation of the Frequencies of Recombinant Products

1. For each series of K pools, determine the number of pools which give an amplification product (K_+ positive pools), and the number which did not (K_0 negative pools). The proportion of negative pools is $P_0 = K_0 / K$.
2. The average number of amplified molecules per pool is estimated as $E = -\ln(P_0) = \ln(K/K_0)$. If $0.2 < P_0 < 0.8$ and $K_0 > 10$, the variance of E can be estimated as $\sigma_E{}^2 = 1/K_0 - 1/K$. E_A, E_{CO}, and E_{NCO} are the numbers of amplifiable molecules per pool in the experiments designed to detect amplifiable molecules, CO and NCO, respectively.
3. The frequency of recombinant products (CO for example) among amplifiable molecules (A) is estimated by $f_{CO} = \frac{n_A}{n_{CO}} \cdot \frac{E_{CO}}{E_A}$, where n_A and n_{CO} are the number of haploid genomes per pool in the series of PCR designed to detect the amplifiable molecule and the CO, respectively. The variance of the frequency of recombinant molecules can be estimated as $\sigma^2_{f_{CO}} = \frac{n^2_A}{n^2_{CO}} \cdot \frac{E^2_{CO}}{E^2_A} \cdot \left[\frac{\sigma^2_{E_{CO}}}{E^2_{CO}} + \frac{\sigma^2_{E_A}}{E^2_A}\right]$.
4. The frequency of recombinant molecules measured in ovary DNA is corrected as follows for estimating the frequency in oocytes (f'_{co}), assuming that the DNA content of prophase I oocytes is twice the content of somatic cells. $f'_{CO} = \frac{1 + p_{oo}}{2 \cdot p_{oo}} \cdot f_{CO}$, where p_{oo} is the proportion of oocytes determined as described in **Section 3.1, Step 5**. The variance of this frequency is estimated as $\sigma_{f'_{co}{}^2} = \left(\frac{1 + p_{oo}}{2p_{oo}}\right)^2 \cdot \sigma_{f_{co}{}^2} + \frac{1 - p_{oo}}{4x_{oo}p^3_{oo}} \cdot f_{co}$, where x_{oo} is the number of cells determined as described in **Section 3.1, Step 5.**
5. When the frequency of recombinant products (either CO or NCO) is too low, such that $P_0 > 0.8$ for any dilution which can be practically used, the frequency of recombinant products is estimated directly as $f_{CO} = \frac{K_{+co}}{K_{co} \cdot n_{CO} \cdot E_A}$, where $K_{co} \cdot n_{co} \cdot E_A$ is the estimate of the total number of amplifiable molecules in K_{co} pools which have been essayed for the presence of CO products. The 95% confidence interval is calculated using the

Poisson approximation for the binomial distribution of the number of recombinant products ($K_{+_{co}}$) over $K_{co} \cdot n_{CO} \cdot E_A$ amplifiable molecules.

6. Examples of experimental data and results of calculations are given in **Table 19.1**.

3.8. Mapping the Recombination Breakpoints

1. Determine the position of exchanges in CO molecules by analyzing the PCR2 products. The parental origin is determined for every polymorphism present along the amplification products. This is done by restriction digestion for the restriction site polymorphisms (**Fig. 19.1C**), and by sequencing the PCR products for the other polymorphisms (*see* **Note 17**). It should be noted that the presence of recombinant products containing a mix of both alleles at some polymorphic positions can in principle be due to either recombinant molecules containing a region of heteroduplex DNA or the presence of more than one recombinant molecule in one single pool of genomic DNA *(11, 15, 16)* (*see* **Note 18**).

Table 19.1
Frequencies of CO and NCO in B10xR209 oocytes

	CO	NCO	Amplifiable molecules
PCR1 primers	P1A / P2A+P2B	P1A / P2A+P2B	P1A+P1B / P2A+P2B
PCR2 primers	P3A / P6B	P4B / P6A	P3A+P3B / P6A+P6B
n	250	250	2.5
K	48	48	40
K_0	25	38	12
E	0.65	0.23	1.20
σ_E	0.14	0.07	0.24
f_{rec}	0.54%	0.19%	–
$\sigma_{f_{rec}}$	0.16%	0.07%	–
x_{oo}	200	200	–
p_{oo}	0.2	0.2	–
f'_{rec}	1.63%	0.58%	–
$\sigma_{f'_{rec}}$	0.51%	0.23%	–

The symbols are described in **Section 3.7**. The data are those from **Fig. 19.1B**.

2. For NCO, the same procedure is followed separately on the PCR2 products resulting from the amplification on one and the other side of the selected polymorphism, allowing for mapping the breakpoints on both ends of the gene conversion tract.

4. Notes

1. When frozen and thawed, a white precipitate often forms in the PCR buffer. The buffer should be mixed thoroughly to homogenize the suspension before use.

   The quality of each batch of PCR buffer should be verified, by comparing it with a former batch in test PCR amplification. (A 30-cycle amplification of 3 ng of mouse genomic DNA with primers P1 and P6 should give a product detectable on a gel stained with ethidium bromide.) Batches displaying efficiency too low or too high (the latter correlating usually with a low allele-specificity) should be discarded. Small differences between batches can be compensated by adding or removing one or two cycles in the first PCR, increasing or decreasing the annealing temperature by 1°C in all PCR, or modifying the dilution ratio of the PCR1 in **Section 3.3**, **Step 5** (between 1/10 and 1/100).
2. Here are a few guidelines for setting up an assay for detecting meiotic CO and NCO at another locus in heterozygous mice. First, both chromosomes (i.e., both parent strains) should have been sequenced in the interval of interest, allowing for the choice of appropriate SNPs where allele-specific primers will be designed. Second, an interval of interest containing a CO hotspot should be limited to a length compatible with PCR amplification (roughly 10 kb). Several methods have been reported for identifying candidate intervals: pedigree analyses *(17)*, the precise mapping of single CO in recombinant inbred strains *(15)*, the systematic analysis of the CO distribution in a given interval by sperm typing *(18)*, the analysis of the genetic diversity between different strains or in a population *(16, 19, 20)*. Third, experiments designed for detecting CO only (described in **Section 3.5**) should be performed first to give information on the distribution of the CO in the interval and on the frequency of recombination events. Indeed, NCO are expected to occur at a significant frequency only at the center of CO hotspots *(9)*. Moreover, the assay for detecting both CO and NCO is less sensitive than the assay designed for detecting CO only. Fourth, allele-specific primers should be designed and tested for efficiency

and specificity. The specificity of the primers is conferred by the 3'-most nucleotide of the primer, which should anneal to a SNP. Additional SNPs in the sequence of the primer could confer a higher specificity. (Some examples are in **Table 19.2**.) Primers can be tested by performing PCR amplification on serial dilutions of a PCR fragment covering the region of interest, from one parent and the other. In our experience, a PCR reaction with a primer specific for a parent A allele should produce at least as much product with a given dilution of DNA from parent A than with a 1,000-fold less diluted DNA solution from parent B. To improve the amplification efficiency of primers with a low GC content, a 5' extension (for instance CCCC) may be added *(19)*. To increase the allele-specificity of primers when another SNP cannot be used instead, the primer sequence can be modified, by adding an additional mismatch or a specific 5' tail [see for example *(21, 22)*].

3. It may happen that mice from a single litter have different genotypes, for instance when homozygous-sterile mutants are analyzed *(11)*. Each pair of ovaries should then be either

Table 19.2
Sequences of the primers used for the detection of recombinant products at the *Psmb9* hotspot in the B10 × R209 hybrid

Name	Sequence
P1A	TTAGCTTCTGAGCCATCTGT
P1B	TCAACTTCTGAGCCGTCTGA
P2A	ATTTGAGAGTGGGAGGGGAT
P2B	ATTTGAGAGTGAGAGGGGAC
P3A	CTTCCTCTTGCCATTACCG
P3B	CCTCCTCCTACCATTACCT
P4A	GTTGTGCAAGATGCCT
P4B	GTTGTGCAAGATGCCG
P5A	GACCTCAGGGCACACA
P5B	GACCTCAGGGCACACC
P6A	AATAAAGGGCTAGCTGTGCA
P6B	AATAAAGGGCTGGCTGTGCC

The SNPs differentiating B10 and R209 are underlined.

frozen for subsequent processing, or processed separately. Samples from mice having the same genotype can be pooled once the mice have been genotyped. To save material, the DNA can be extracted directly from the whole ovaries. However, the exact proportion of oocytes, and therefore the recombination frequency per oocyte, cannot be determined in individual experiments *(11)*.

4. Spermatozoa always represent more than 90% of the cells in the suspension (more than 95% in most experiments). Nevertheless, sperm may be purified further by performing a differential lysis in 1 × SSC supplemented with 0.15% SDS *(23)*.
5. In order to get reliable amounts of DNA in the serial dilutions, the viscosity of the genomic DNA solution is diminished by a treatment with a restriction enzyme. The enzyme is chosen such that no recognition sequence motif is present in the amplified interval.
6. Avoiding contamination by PCR products is a major challenge of the method. Practically, the dissection of the animals, the preparation of the genomic DNA and the preparation of the PCR reactions must be performed in a dedicated room, with dedicated tools and chemicals. The PCR1 mixes are prepared entirely, including the addition of the genomic DNA, in the dedicated room. The tubes containing the Tris solution for the dilution of the PCR1 products (**Section 3.3, Step 3**) and the PCR2 mixes (**Section 3.3, Step 4**) are also prepared in this dedicated room, and are brought out to the lab where the PCR1 products are added. Chemicals, enzymes, and primers should be delivered directly to this room, without passing through a room likely to contain PCR products.
7. Each PCR1 tube should contain a defined volume of genomic DNA. Therefore, the whole volume of the mix should be dispatched into the defined number of PCR tubes, by adjusting slightly the volume dispatched in individual tubes if it appears that there is not enough or too much of the mix. Differently, for the PCR2, it is better to prepare a larger volume of mix and to place the defined volume (i.e., 9 μL) in each tube.
8. In our experience, the number of amplifiable molecules per haploid genome is usually in the range of 0.5–0.9. When several DNA samples are compared, the actual concentration of amplifiable molecules can be determined for only one of these which will be used as a standard. Then the concentration of the genomic DNA will be measured carefully for every sample (make several measures on different dilutions) and the quality of the different samples will be assessed by real-time PCR.

9. The maximum number of amplifiable molecules that can be placed in one pool without inhibiting the amplification of one single recombinant molecule depends upon the locus and the primers used. In our hands, at *Psmb9*, this number is about 10,000 molecules. In order to increase this number up to several thousands of amplifiable molecules, a third PCR with internal allele-specific primers can be performed *(24)*. To increase the sensitivity of the detection, the secondary PCR products can also be analyzed by Southern blot, with an internal probe *(25)*.
10. The reciprocal products, obtained as described in **Sections 3.5** and **3.6** (CO in the reverse orientation and NCO on the homologous chromosome), are detected by replacing the primers "A" by the corresponding primers "B", and vice versa.
11. High extension temperature (70–72°C) might reduce the efficiency of amplification and favor inter-haplotype-jumping PCR artifacts. Therefore, Jeffreys et al. *(19)* performed extension at 61–66°C, and extension at 68°C works well in our assay.
12. On some occasions, PCR artifacts containing partially single-stranded DNA have been observed. These artifacts can be eliminated by digesting the product of the first PCR with S1 nuclease prior to seeding the PCR2 reaction mix, as described in *(6)*.
13. The second PCR can be performed in a real-time PCR machine, circumventing the requirement for gel analysis *(18)*.
14. The number of amplifiable molecules placed in each pool is lower in these experiments, because only the second PCR is specific for the recombination product to be detected. In our hands, the appropriate sample size varies from less than 100 to about 1,000 amplifiable molecules, depending upon the polymorphism selected by the primers P4 and P5. It is important to note that, in contrast to the primers designed for detecting CO, the 3' end of primers designed for detecting NCO (P4 and P5) is fixed by the choice of the SNP where gene conversion events are selected.

    When NCO frequencies are too low (below 10^{-4} to 10^{-3}, depending on the number of amplifiable molecules which can be placed in each reaction), the method of DNA Enrichment by Allele-Specific Hybridization (DEASH) could be used to enrich for a specific allele at a particular SNP *(26)* (Refer to **Chapter 20** in this volume).
15. For the PCR1, the primers P2A and P2B can be replaced by a single "universal" primer amplifying both alleles. If a mix of two allele-specific primers is used (as in this example), it is important to ensure that each primer amplifies its respective

allele with the same efficiency. This could be done by verifying that both parents are equally represented among the products of amplifiable molecules (**Section 3.4**).

16. An alternative method, based on the same principle, is described in *(9)*. In that method, both the first and the second, nested, PCR are performed with one primer specific for one haplotype on one side, and a universal primer on the other side. The secondary PCR products are submitted to dot-blot hybridizations with oligonucleotides specific for the second haplotype, for a series of SNPs along the hotspot.
17. If there is not enough PCR2 product for these analyses, the PCR2 reaction can be performed again in a larger volume and/or with slightly less stringent PCR conditions (more cycles or lower annealing temperature) with an appropriate negative control (a negative pool from the same series of reactions). The recombinant breakpoints can also be mapped by dot-blot hybridization with allele-specific oligonucleotides, as described in *(9, 27)*.
18. The distribution of the number of recombinant molecules per pool follows the Poisson law. Therefore, a certain proportion of the positive pools contain more than one recombinant molecule. The occurrence of two recombinant molecules with breakpoint mapping in different intervals could lead to the erroneous conclusion that some molecules contain heteroduplex DNA. Therefore, the number of pools containing these mixed products should be compared with the expected number of pools containing two recombinant molecules with breakpoint mapping in different intervals, which can be deduced from the number of pools containing a recombinant product with a breakpoint in each of the considered intervals. The expected frequency of pools containing products with breakpoints in two different intervals is the product of the frequencies of pools containing one product with a breakpoint in each interval taken separately (e.g., **Fig. 19.1C**). Alternatively, the number of genomes per pool can be lowered so as to render unlikely the presence of more than one recombinant molecule in a single pool.

Acknowledgments

We thank G. Enders for providing GCNA1 antibody. This work was supported by grants from CNRS, ARC (4626 and 3723), and ANR (ANR-06-BLAN-0160-01) to BdM.

References

1. Petes, T. D. (2001) Meiotic recombination hot spots and cold spots. *Nat. Rev. Genet.* **2**, 360–369.
2. Allers, T., and Lichten, M. (2001) Differential timing and control of noncrossover and crossover recombination during meiosis. *Cell* **106**, 47–57.
3. Hunter, N., and Kleckner, N. (2001) The single-end invasion: an asymmetric intermediate at the double-strand break to double-holliday junction transition of meiotic recombination. *Cell* **106**, 59–70.
4. Kauppi, L., Jeffreys, A. J., and Keeney, S. (2004) Where the crossovers are: recombination distributions in mammals. *Nat. Rev. Genet.* **5**, 413–424.
5. Buard, J., and de Massy, B. (2007) Playing hide and seek with mammalian meiotic crossover hotspots. *Trends Genet.* **23**, 301–309.
6. Jeffreys, A. J., Murray, J., and Neumann, R. (1998) High-resolution mapping of crossovers in human sperm defines a minisatellite-associated recombination hotspot. *Mol. Cell* **2**, 267–273.
7. de Massy, B. (2003) Distribution of meiotic recombination sites. *Trends Genet.* **19**, 514–522.
8. Guillon, H., and de Massy, B. (2002) An initiation site for meiotic crossing-over and gene conversion in the mouse. *Nat. Genet.* **32**, 296–299.
9. Jeffreys, A. J., and May, C. A. (2004) Intense and highly localized gene conversion activity in human meiotic crossover hot spots. *Nat. Genet.* **36**, 151–156.
10. Holloway, K., Lawson, V. E., and Jeffreys, A. J. (2006) Allelic recombination and de novo deletions in sperm in the human {beta}-globin gene region. *Hum. Mol. Genet.* **15**, 1099–1111.
11. Guillon, H., Baudat, F., Grey, C., Liskay, R. M., and de Massy, B. (2005) Crossover and noncrossover pathways in mouse meiosis. *Mol. Cell* **20**, 563–573.
12. Baudat, F., and de Massy, B. (2007) Cis- and Trans-acting elements regulate the mouse Psmb9 meiotic recombination hotspot. *PLoS Genet.* **3**, e100.
13. Shiroishi, T., Sagai, T., Hanzawa, N., Gotoh, H., and Moriwaki, K. (1991) Genetic control of sex-dependent meiotic recombination in the major histocompatibility complex of the mouse. *EMBO J.* **10**, 681–686.
14. Jeffreys, A. J., Neumann, R., and Wilson, V. (1990) Repeat unit sequence variation in minisatellites: a novel source of DNA polymorphism for studying variation and mutation by single molecule analysis. *Cell* **60**, 473–485.
15. Bois, P. R. (2007) A highly polymorphic meiotic recombination mouse hot spot exhibits incomplete repair. *Mol. Cell. Biol.* **27**, 7053–7062.
16. Kauppi, L., Jasin, M., and Keeney, S. (2007) Meiotic crossover hotspots contained in haplotype block boundaries of the mouse genome. *Proc. Natl. Acad. Sci. USA* **104**, 13396–13401.
17. Shifman, S., Bell, J. T., Copley, R. R., Taylor, M. S., Williams, R. W., Mott, R., and Flint, J. (2006) A high-resolution single nucleotide polymorphism genetic map of the mouse genome. *PLoS Biol.* **4**, e395.
18. Tiemann-Boege, I., Calabrese, P., Cochran, D. M., Sokol, R., and Arnheim, N. (2006) High-resolution recombination patterns in a region of human chromosome 21 measured by sperm typing. *PLoS Genet.* **2**, e70.
19. Jeffreys, A. J., Kauppi, L., and Neumann, R. (2001) Intensely punctate meiotic recombination in the class II region of the major histocompatibility complex. *Nat. Genet.* **29**, 217–222.
20. Jeffreys, A. J., Neumann, R., Panayi, M., Myers, S., and Donnelly, P. (2005) Human recombination hot spots hidden in regions of strong marker association. *Nat. Genet.* **37**, 601–606.
21. Hazbon, M. H., and Alland, D. (2004) Hairpin primers for simplified single-nucleotide polymorphism analysis of Mycobacterium tuberculosis and other organisms. *J. Clin. Microbiol.* **42**, 1236–1242.
22. Li, B., Kadura, I., Fu, D. J., and Watson, D. E. (2004) Genotyping with TaqMAMA. *Genomics* **83**, 311–320.
23. Yauk, C. L., Bois, P. R., and Jeffreys, A. J. (2003) High-resolution sperm typing of meiotic recombination in the mouse MHC *Eβ* gene. *EMBO J.* **22**, 1389–1397.
24. Kauppi, L., Stumpf, M. P., and Jeffreys, A. J. (2005) Localized breakdown in linkage disequilibrium does not always predict sperm crossover hot spots in the human MHC class II region. *Genomics* **86**, 13–24.
25. May, C. A., Shone, A. C., Kalaydjieva, L., Sajantila, A., and Jeffreys, A. J. (2002)

Crossover clustering and rapid decay of linkage disequilibrium in the Xp/Yp pseudoautosomal gene *SHOX*. *Nat. Genet.* **31**, 272–275.

26. Jeffreys, A. J., and May, C. A. (2003) DNA enrichment by allele-specific hybridization (DEASH): a novel method for haplotyping and for detecting low-frequency base substitutional variants and recombinant DNA molecules. *Genome Res.* **13**, 2316–2324.
27. Jeffreys, A. J., Ritchie, A., and Neumann, R. (2000) High resolution analysis of haplotype diversity and meiotic crossover in the human *TAP2* recombination hotspot. *Hum. Mol. Genet.* **9**, 725–733.

Chapter 20

Analysis of Meiotic Recombination Products from Human Sperm

Liisa Kauppi, Celia A. May, and Alec J. Jeffreys

Abstract

Traditional methods for surveying meiotic recombination in humans are limited to pedigree and linkage disequilibrium analyses. We have developed assays that allow the direct detection of crossover and gene conversion molecules in batches of sperm DNA. To date, we have characterized 26 recombination hotspots by allele-specific PCR and selectively amplified recombinant DNA molecules from these regions. These analyses have revealed that meiotic crossover hotspots in humans are highly localized and flanked by DNA segments where recombination is suppressed. The centers of crossover hotspots are also active in noncrossover recombination, displaying short conversion tracts.

Key words: Allele-specific PCR, haplotype, sperm DNA, crossover, gene conversion.

1. Introduction

Human DNA diversity arises from germline mutation, creating new haplotypes that are then reshuffled by meiotic recombination. It has become increasingly obvious that meiotic recombination events in mammals are not randomly distributed, but instead cluster into highly localized regions (recombination hotspots).

Several experimental approaches have been used to characterize the frequency and spatial distribution of meiotic recombination events in humans. Examining parent-to-offspring transmission of alleles at several loci (the so-called pedigree approach) has identified localized genomic regions, at the 1–10 kb scale, in which recombination events appear to cluster [for example, **Refs.** *(1–5)*]. However, because of the small size of human families, the number of

Scott Keeney (ed.), *Meiosis, Volume 1, Molecular and Genetic Methods, vol. 557*

DOI 10.1007/978-1-59745-527-5_20 Springerprotocols.com

informative meioses is limited. This approach to measuring meiotic recombination rates thus is not suitable for detecting low frequency events or for defining hotspots with precision.

An alternative method for elucidating the genome-wide distribution of recombination events is to examine patterns of linkage disequilibrium (LD). LD reflects recombination in all of the generations since the rise of the individual polymorphic alleles under study. Various LD measures exist for characterizing the statistical association between alleles at different loci (see, e.g., refs. *6*, *7* for reviews). High-resolution LD analyses of the human genome show extended domains of markers in strong association interrupted by areas of LD breakdown *(8, 9)*.

In principle, if strong LD is found between two loci, little if any historical recombination has occurred between the loci. Conversely, little or no LD (that is, breakdown of LD) implies historical recombination activity. Regions of LD breakdown are therefore putative recombination hotspots, and can be directly tested for crossover and gene conversion activity, as we describe in this chapter. However, patterns of LD can be influenced also by the age of crossover events, selection acting on single or linked alleles, and demographic history *(10, 11)*, highlighting the importance of experimentally verifying recombination activity in a region of interest.

In this chapter, we describe strategies to identify and characterize human recombination hotspots. First, high-resolution LD structure is examined to identify signatures of historical crossing-over (**Fig. 20.1A**). LD profiles can sometimes be complex, so genotype data can also be analyzed by linkage disequilibrium unit mapping to provide a simple summary profile of the rate of LD breakdown across the genotyped region, facilitating the detection of putative hotspots *(12)*. More sophisticated coalescent methods are also available that allow detailed estimation of the historical recombination profile across the region most compatible with contemporary haplotype diversity *(13–16)*. The next step is to test LD block boundaries (putative hotspots) for meiotic recombination hotspot activity using allele-specific PCR to recover crossover molecules directly from sperm DNA (**Fig. 20.1B**). Crossover breakpoint mapping and the resolution of hotspots is limited only by the density of heterozygous markers within the target region; resolution to less than 0.5 kb is usually possible by choosing individuals with appropriate genotypes. Once the precise location of a crossover hotspot is known, assays can be designed to analyze noncrossover recombination events (gene conversions) at the center of the hotspot.

Sperm-typing experiments have demonstrated that areas of LD breakdown frequently correspond with generally good precision to meiotic crossover hotspots *(15, 17–21)*. All hotspots defined share similar widths (1–2 kb wide) but vary considerably in intensity with sperm recombination frequencies varying from

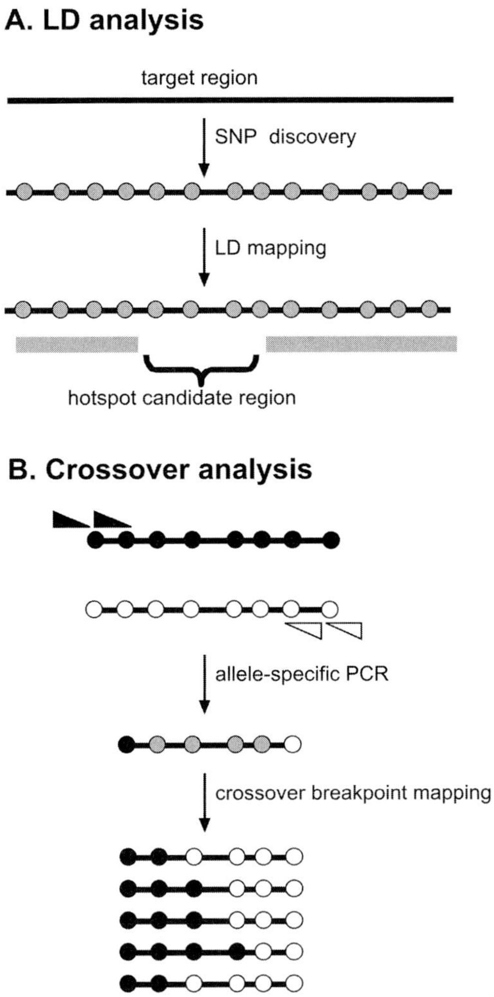

Fig. 20.1. PCR approach for recovering crossover molecules directly from sperm DNA. (**A**) The first step is the discovery of single-nucleotide polymorphisms (SNPs, shown as gray circles) and LD mapping, allowing the identification of LD blocks (gray boxes) and regions where LD breaks down (gap between gray boxes). Such regions of LD breakdown are putative recombination hotspots. (**B**) For putative recombination hotspot regions, two rounds of allele-specific PCR are performed on batches of sperm DNA from an individual with multiple SNP heterozygosities (black and white circles). Allele-specific primers are shown as black and white arrowheads. Recombinant sperm DNA molecules are selectively amplified, and the location of crossover breakpoints can be subsequently mapped by typing internal SNPs.

0.0004% to 1% over currently characterized hotspots. Crossover hotspots often occur in clusters and together account for at least 80% *(13, 16, 22)*, and probably more than 95% *(15, 23)*, of all crossovers in the male germline.

To facilitate the analysis of noncrossover gene conversions at hotspots, a DNA enrichment-based assay has been developed that can capture conversion events involving even single SNP sites *(24)*.

Assays performed at three human crossover hotspots showed that conversions are even more localized, although they peak in activity at the center of the crossover hotspot *(25)*. This is consistent with crossovers and conversions being produced by the same initiation events, presumably DNA double-strand breaks. Conversion tracts are very short (<300 bp on average), and initiation events seem to cluster within a zone of just ~400 bp. However, data from five loci indicate that the relative frequencies of conversion to crossover show major variation between hotspots, ranging from ~10:1 to <1:12 *(25–27)*.

The methods described in this chapter have provided significant insight into the nature of human meiotic recombination. The obvious drawback of sperm PCR techniques is that they can only be used to analyze meiotic recombination in human males. Pedigree data remain essential for analyzing the distribution and rate of female recombination events, though population genetic inferences of historical recombination in females can be gained by LD analysis of the sex-specific region of the X chromosome.

2. Materials

To minimize the risk of contamination with environmental DNA, all reagents should be prepared in a laminar flow hood using dedicated pipettes plus plasticware taken directly from the manufacturers' bags/boxes. Buffers can be further treated by exposure to UV light to eliminate any possible PCR-amplifiable DNA.

2.1. DNA Extractions

1. 20x sodium chloride-sodium citrate (SSC) buffer: 3 M NaCl, 0.3 M sodium citrate-dihydrate; use to make 1x SSC and 0.2x SSC in ddH_2O
2. 10% (w/v) sodium dodecyl sulfate (SDS)
3. 2-mercaptoethanol
4. 20 mg/mL Proteinase K in ddH_2O; prepare fresh, store on ice until use
5. Liquefied phenol, washed in Tris buffer (general purpose grade)
6. 2 M sodium acetate pH 7.0
7. Ethanol, absolute and 80%
8. 5 mM Tris-HCl pH 7.5
9. Screw-top 1.5 mL microcentrifuge tubes (Sarstedt, Nümbrecht, Germany)

2.2. Allele-Specific PCR

1. 11.1x PCR buffer, prepared as indicated in **Table 20.1**. Nonacetylated Ultrapure bovine serum albumin (BSA) is supplied by Ambion (Austin, TX). Freeze in 100 μL aliquots and store at –20°C.
2. *Taq* polymerase (ABgene, Epsom, UK)
3. *Pfu* polymerase
4. 2 M Tris(hydroxymethyl)aminomethane (Ultra grade for molecular biology, Fluka Chemie, Buchs, Switzerland) in ddH_2O
5. S1 nuclease, diluted in S1 nuclease storage buffer (20 mM Tris-HCl pH 7.5, 50 mM NaCl, 0.1 mM zinc acetate, 50% [v/v] glycerol). Store at a concentration of 10 U/μL at –20°C. Stable for >6 months.
6. S1 digestion buffer: 20 mM sodium acetate pH 4.9, 1 mM zinc acetate, 100 mM NaCl. Store at –20°C.
7. S1 dilution buffer: 10 mM Tris-HCl pH 7.5, 5 μg/mL high molecular weight salmon sperm DNA
8. 10x Tris-borate/EDTA (TBE) buffer: 890 mM Tris-Borate, 20 mM EDTA in ddH_2O, pH 8; use to make 0.5x TBE in ddH_2O. Store at room temperature.
9. All-purpose loading dye: 30% (v/v) glycerol in 0.5x TBE plus enough bromophenol blue to give adequate blue color. Store at room temperature.

Table 20.1
Recipe for 11.1x PCR buffer

Component	Concentration of stock solution	Volume (arbitrary units)	Final concentration in PCR
Tris-HCl pH 8.8	2 M	167	45 mM
ammonium sulfate	1 M	83	11 mM
$MgCl_2$	1 M	33.5	4.5 mM
2-mercaptoethanol	100%	3.6	6.7 mM
EDTA pH 8.0	10 mM	3.4	4.4 μM
dATP	100 mM	75	1 mM
dCTP	100 mM	75	1 mM
dGTP	100 mM	75	1 mM
dTTP	100 mM	75	1 mM
BSA	10 mg/mL	85	113 μg/mL
Total Volume		676	

10. Seakem LE agarose
11. Ethidium bromide

2.3. Mapping Crossover Breakpoints by Allele-Specific Oligonucleotide (ASO) Hybridization

1. Denaturing solution: 0.5 M NaOH, 2 M NaCl, 25 mM EDTA. Store at room temperature.
2. 2x SSC and 3x SSC
3. Hybond membrane (Amersham) and 3MM Whatman filter paper
4. Allele-specific oligonucleotides (ASOs), desalted, 18 nucleotides long, with SNP site located at the eighth base from 5′ end. Store at 800 μg/mL concentration in ddH_2O at –20°C. Dilute 1:100 in ddH_2O to make working stock of 8 μg/mL.
5. T4 polynucleotide kinase
6. T4 kinase labeling mix: 700 mM Tris-HCl pH 7.5, 100 mM $MgCl_2$, 50 mM spermidine trichloride (Sigma), 20 mM dithiothreitol. Store at –20°C.
7. 10 mCi/mL [γ-^{32}P]ATP
8. T4 kinase stop mix: 25 mM di-sodium EDTA, 0.1% SDS, 10 μM ATP. Store at –20°C.
9. 50x Denhardt's solution: 5 g Ficoll400, 5 g polyvinylpyrrolidone, 5 g BSA fraction V, dissolved in 500 mL ddH_2O. Store at +4°C.
10. High molecular weight salmon testis DNA or herring sperm DNA, 3 mg/mL in ddH_2O
11. Tetramethyl-ammonium chloride (TMAC) hybridization solution: 3 M TMAC, 0.6% SDS, 10 mM sodium phosphate pH 6.8, 1 mM EDTA, 4 μg/mL yeast (bulk) RNA, in 5x Denhardt's solution. Store at +4°C.
12. TMAC wash solution: 3 M TMAC, 0.6% SDS, 10 mM sodium phosphate pH 6.8, 1 mM EDTA. Store at +4°C.
13. Rotator hybridization oven (e.g. Thermo Hybaid)
14. Pyrex or plastic tray
15. Saran plastic wrap
16. Autoradiography film
17. Stripping solution: 0.1% SDS in ddH_2O

2.4. DNA Enrichment by Allele-Specific Hybridization (DEASH)

1. Restriction enzyme and buffer (*see* **Section 3.5.1**)
2. 0.1 M spermidine trichloride
3. 2 M sodium acetate pH 7.0
4. Ethanol, absolute and 80%
5. 5 mM Tris-HCl pH 7.5

6. 5′ biotinylated ASO purified by HPLC; 18 nucleotides long with allele-specific base as the 11th nucleotide from the 5′ end, store as 5 μM (30 μg/mL) working stock in ddH_2O
7. Nonbiotinylated competitor ASO; corresponding 18mer with opposite allele base at 11nt from the 5′ end, store as 30 μM (180 μg/mL) working stock
8. Dynabeads M-280 Streptavidin (store in the dark at +4°C) and Dynal MPC-E magnetic particle concentrator (Dynal Biotech, Invitrogen)
9. 10x DHB (denaturing/hybridizing/binding buffer); 450 mM Tris-HCl pH 8.8, 110 mM ammonium sulfate, 45 mM $MgCl_2$, 67 mM 2-mercaptoethanol, 44 μM disodium EDTA, 20 μg/mL single-stranded (heat-denatured) high molecular weight herring sperm or salmon testis DNA
10. 500 μg/mL nonacetylated Ultrapure BSA (Ambion)
11. ED (elution buffer): 0.14 x DHB, 4.7 μg/mL single-stranded (heat-denatured) high molecular weight herring or salmon sperm DNA
12. 0.2 mL PCR tubes
13. Siliconized 1.5 mL flip-top microcentrifuge tubes, prepared by soaking open tubes in dimethyldichlorosilane solution (BDH, Lutterworth UK) for 1 min, draining, leaving to air dry completely in a PCR-clean environment (~10 min), then washing thoroughly with dH_2O and air drying again before storing in sealed bag at room temperature. Commercially available siliconized tubes appear to be less reliable and can cause reduced yields due to adsorption of single-stranded DNA.
14. Two waterbaths, one set at temperature to anneal biotinylated ASO to target and other at temperature to elute the bound DNA
15. If Alu competitor is required, 10 μM working stocks of primers Alu-F (5′ GGTGGCTCACGCCTGTAATC 3′) and Alu-R (5′ AGACGGAGTCTCGCTCTGTG 3′)

3. Methods

3.1. DNA Extractions

All DNA extractions should be carried out in a laminar flow hood under conditions that minimize risk of contamination *(28)*. Waste material must be discarded into disinfectant (e.g. solution made with Titan Chlor-Tabs 500, Lever Industrial, Runcorn, Cheshire UK) to destroy any pathogens.

3.1.1. Extracting DNA from Semen

1. Semen donors should be requested to wash the penis prior to ejaculation to minimize the risk of contamination with cellular material from their partner. Ejaculates should be collected in 5 mL 1x SSC and stored at –80°C. Samples so stored are suitable for sperm DNA preparation for at least 10 years.
2. Thaw sample and transfer 1 mL semen/SSC to a screw-top microfuge tube. For a typical ejaculate, this amount should yield ~30 μg sperm DNA. If larger amounts of DNA are required, then process multiple aliquots of semen in parallel. Centrifuge at 13,000 rpm (15,700*g*) in a microcentrifuge for 2 min, discard the supernatant and re-suspend the cell pellet in 1 mL 1x SSC with vigorous shaking until a uniform cell suspension is obtained. Add 20 μL 10% SDS (0.2% final concentration) to lyse any nonsperm cells. Mix contents by inverting vigorously and centrifuge at 13,000 rpm (15,700 rcf) in a microcentrifuge for 2 min. Discard the supernatant and re-suspend the pellet in 1 mL 1x SSC, 0.2% SDS. Centrifuge as before.
3. Discard the supernatant and re-suspend the pellet in 450 μL 0.2x SSC. Lyse sperm heads by the addition of 50 μL 10% SDS (1% final concentration), 35 μL 2-mercaptoethanol (1 M final concentration) and Proteinase K to a final concentration of 200 μg/mL. Mix by gently inverting the tube.
4. Incubate the sample for 30 min to 1 h in a waterbath at 37°C with occasional gentle mixing. Lysis should be evident within 5 min as the lysate becomes viscous. Spin in microcentrifuge at 13,000 rpm (15,700 rcf) for 1 min.
5. Remove any residual proteins by adding 350 μL phenol, followed by gentle mixing by tube inversion to allow emulsification, and spin in a microcentrifuge at 13,000 rpm (15,700 rcf) for 2 min. Transfer the top aqueous layer into a clean microcentrifuge tube. Use a plastic pipette tip with the end cut off to provide a larger bore for pipetting the viscous DNA solution. Be careful not to pipette too close to the inter-phase layer.
6. Re-extract residual DNA from the organic layer by adding 200 μL 1x SSC and 20 μL of 10% SDS to the tube. Mix well by inversion, then centrifuge. Remove the aqueous layer and pool it with the aqueous solution collected from the first phenol step (**Step 5**). Discard organic layer.
7. Subject the pooled aqueous solution to a second round of phenol extraction (i.e. repeat **Steps 5–6**).
8. Divide the pooled aqueous layers (0.9 mL) into three 0.3-mL aliquots in microcentrifuge tubes and precipitate the DNA at room temperature by adding 2 vol (0.6 mL) of absolute ethanol to each tube (*see* **Note 1**). Tap and invert the tube until DNA precipitates as a single, compact, stringy, white lump.

Pellet DNA by centrifugation at 13,000 rpm (15,700 rcf) for 10 s. Remove supernatant and add 800 μL 80% ethanol to wash the pellet. Centrifuge and remove supernatant.

9. Dissolve each DNA pellet in 60 μL ddH_2O per tube by occasional gentle pipetting. The DNA should fully dissolve within 5 min. Pool the DNA solutions from the three tubes and re-precipitate by adding 20 μL 2 M sodium acetate (pH 7.0) and 400 μL absolute ethanol. Invert the tube until DNA precipitates, and spin in a microcentrifuge at 13,000 rpm (15,700 rcf) for 10 s. Discard supernatant and wash pellet with 600 μL 80% ethanol. Pellet DNA by spinning in a microcentrifuge at 13,000 rpm (15,700 rcf) for 10 s. Remove the supernatant. Spin again at 13,000 rpm (15,700 rcf) in microcentrifuge for 10 s and discard any remaining supernatant. Vacuum-dry pellet and dissolve in 50 μL 5 mM Tris-HCl (pH 7.5).

3.1.2. Extracting DNA from Blood

1. Collect venous blood into an equal volume of 1x SSC and store at –80°C. Thaw and remove 1 mL of blood/SSC to use for the DNA extraction. Freezing/thawing lyses almost all erythrocytes.
2. Spin the blood/SSC at 13,000 rpm (15,700 rcf) in a microcentrifuge for 1 min. This step pellets white blood cells, erythrocyte ghosts and any remaining erythrocytes. Remove supernatant and wash the pellet twice with 800 μL 1x SSC, with centrifugation, to remove all hemoglobin.
3. Re-suspend the cell pellet in 450 μL 0.2x SSC by vigorously shaking by hand to produce a homogeneous cell suspension. Lyse the cells by adding 50 μL 10% SDS (1% final concentration), 35 μL 2-mercaptoethanol (1 M final concentration) and Proteinase K to a final concentration of 200 μg/mL. Mix by gently inverting the tube. Lysis will be immediately apparent.
4. Continue with the DNA preparation as described above (**Steps 4–9** in **Section 3.1.1**).

3.1.3. Quantification of Genomic DNA

1. To quantify the DNA, add 0.6, 0.9 and 1.2 μL aliquots of the DNA solution to 1 mL ddH_2O and measure absorbances at 260 nm using a spectrophotometer. Add appropriate volume of 5 mM Tris-HCl (pH 7.5) to make a working stock DNA concentration, typically of 100 to 400 μg/mL. Store DNA at –80°C.
2. Run serial dilutions of genomic DNA of a known concentration (for example, 100 ng, 50 ng, 25 ng and 12 ng per lane) alongside your sample on a 0.8% agarose gel in 0.5x TBE (with 0.2 μg/mL ethidium bromide in both the gel and the buffer) to verify DNA concentration and to check DNA quality.

3.2. Allele-Specific PCR to Detect Crossover Molecules

To minimize contamination, keep reagents and materials used for PCR separate from general laboratory chemicals, and set up PCRs in a laminar flow hood.

3.2.1. Designing Allele-Specific PCR Primers

Crossover assays use nested allele-specific PCR primers and thus require pairs of 5′ and 3′ selector sites (**Fig. 20.1**). These sites can be up to 13 kb apart, the maximum distance that can be efficiently amplified at the single DNA molecule level using our current long PCR protocol.

It is difficult to predict the behavior of an allele-specific primer. When designing them, aim for at least 50% GC content, if possible. Examples of allele-specific primers that we have used successfully *(23)* are shown in **Table 20.2**. The most common length is 18 nucleotides, but very GC-rich allele-specific primers can be as short as 14 nucleotides. Allele-specificity is achieved by virtue of the last 3′ base. The genomic context of the SNP for which the primer is specific mainly dictates the GC content. However, various 5′ extensions can be added to boost the GC content of relatively AT-rich primers and improve primer efficiency, for example C_{4-5} or if necessary a synthetic 19–20 nucleotide sequence which itself can be used as a driver primer during allele-specific PCR (see primers B7RT/Ad and B7RG/Ad in **Table 20.2** for an example).

Table 20.2
Examples of allele-specific PCR primers used in crossover assays. Bases in lowercase are 5′ extensions used to boost GC content, not part of the genomic sequence

Primer name	5′ to 3′ sequence	Length	%GC
F7R2G	GTGGGCCCCAGGAC	14	79
BC4RC	GAGGAAAGCTGGGGGTAG	18	61
A7FG	CGTCCACTGGAAACGTTG	18	56
B7RT/Ad	gtctacgtagtcagctctgGAAATGGTCAAATTTTATGTA	21+Ad	38
B7RG/Ad	gtctacgtagtcagctctgGAAATGGTCAAATTTTATGTC	21+Ad	40
JJ6FT2	GCTCTGGTGGTGTGGT	16	63
JJ6FC	CTGCTCTGGTGGTGTGGC	18	67
J7RC	GCCTAAGAGCAGAGGGAG	18	61
J7RT	GCCTAAGAGCAGAGGGAA	18	56
JJ7FA1	ccccCTTGCTTTGAAATGAGGA	18+cccc	50
JJ7FC2	ccccCTTGCTTTGAAATGAGGC	18+cccc	55

3.2.2. Optimizing Allele-Specific PCR Primers

1. To ensure efficient and specific amplification when using allele-specific primers for crossover analysis, PCR conditions must be tested empirically using long PCR *(29)*. Set up PCRs in 0.2 mL tubes on a MJ Research PTC-225 Tetrad DNA Engine or other suitable thermal cycler, in a total volume of 10 μL with 0.9 μL of 11.1x PCR buffer supplemented with 12 mM Tris base, 0.2 μM allele-specific primer, 0.2 μM universal primer (to give an amplicon 1–6 kb long), 0.03 U/μL *Taq* polymerase and 0.003 U/μL *Pfu* polymerase per reaction (*see* **Note 2**).
2. To determine the optimal annealing temperature of each allele-specific primer, amplify 20 ng of genomic DNA or equivalent whole-genome amplified DNA *(30)* at various annealing temperatures (typically 56, 59, 62, and 65°C). As a template, use DNA from one individual homozygous for the SNP allele at the 3′ end of the allele-specific primer, and from another individual homozygous for the alternative allele in parallel reactions. If you are optimizing PCR conditions for a forward allele-specific primer, use it in combination with a universal reverse primer (22 to 24 nucleotides long) that produces a slightly shorter fragment than the amplicon you are planning to amplify in the crossover assay. If you are optimizing PCR conditions for a reverse allele-specific primer, use it in combination with a universal forward primer. As a positive control, replace the allele-specific primer with a universal primer located near the allele-specific primer site. We routinely denature the template in an initial step of 96°C for 1 min, followed by 34 cycles of denaturing at 96°C for 20 s, 30 s at annealing temperature and extension at 65°C for approximately 90 s/kb (*see* **Note 3**).
3. To each 10 μL PCR, add 2 μL all-purpose loading dye and run 6 μL on a 0.8% agarose gel in 0.5x TBE (with both the gel and the buffer containing 0.5 μg/mL ethidium bromide), with 200 ng of size standard alongside your samples. Photograph the gel, including very long exposures to visualize any very faint PCR products (*see* **Note 4**).

3.2.3. Determining Linkage Phase of Donors to be used in Crossover Assays

1. From your collection of semen donors, identify individuals who are heterozygous for selector SNP sites (F1, F2, R1, R2 in **Fig. 20.2**) and who also carry internal heterozygosities suitable for mapping crossover breakpoints in recombinant molecules.
2. If selector SNP sites F1/F2 and R1/R2 are separated by a putative recombination hotspot, then they will most likely be in free association and their linkage phases will therefore be unknown. Determine linkage phase, i.e. haplotypes, for these SNP sites by allele-specific PCR using allele-specific primer pairs F1-R2, F2-R1, and F2-R2 in all four possible

combinations of each allelic primer (*see* **Fig. 20.2**). If, as is likely, F1 and F2 lie outside a hotspot and are in strong association, then their linkage phase can be inferred from genotype data; the same will probably be true for R1 and R2. Under these circumstances, only one phasing assay needs be done, e.g. F2-R2, to determine complete 4-marker haplotypes. Under no circumstances should phasing be done using the outside primers F1-R1, as this will generate PCR products that could easily contaminate primary PCRs in subsequent crossover analyses.

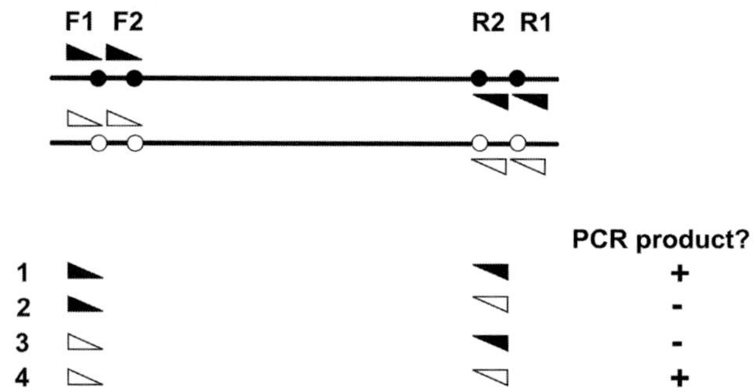

Fig. 20.2. Principle of linkage phasing. Heterozygous selector SNPs are depicted as black and white circles, forward allele-specific primers (F1 and F2) and allele-specific reverse primers (R1 and R2) as black or white arrowheads. To determine the linkage phase of selector SNPs, PCR is performed using all four possible combinations of each allele-specific forward and reverse primer, in this example using F1-R2 combinations (rows 1–4).

3. Retain the positive PCR products from the phasing assay and store at –20°C. These separated haplotypes will be needed later for haplotyping internal SNPs during crossover breakpoint mapping (*see* **Section 3.3.1**).

3.2.4. Amplification of Crossover Molecules from Sperm DNA

Sperm crossover analysis is feasible once suitable pairs of allele-specific primers for primary PCR and for nested secondary PCR have been developed. Modification of primers (*see* **Section 3.2.1**) may prove necessary to develop primer pairs that operate with good efficiency and specificity at the *same* annealing temperature.

One further problem is recombination frequency. Human crossover hotspots vary widely in sperm crossover frequency, from 0.0004% to 1% at currently characterized hotspots, and the crossover frequency over the test interval will not be known a priori. An initial pilot crossover assay over widely varying inputs of sperm DNA is therefore necessary. Given that a haploid human genome contains 3 pg DNA, then 6 pg sperm DNA will contain on average one molecule of each haplotype. Not all DNA molecules will amplify during PCR due to primer inefficiency and DNA damage; our experience indicates that typically 50% of molecules will generate PCR products. Thus 12 pg genomic DNA will contain one amplifiable molecule of each haplotype, and a PCR seeded with 1,000 molecules should contain an input of 12 ng DNA.

Crossover assays can also generate spurious positive PCRs, usually because of incomplete specificity of an allele-specific primer leading to bleed-through of one or other progenitor haplotype in the assay. Jumping PCR artifacts arising from annealing between progenitor haplotypes can also occur, though these are generally rare (*see also* **Section 3.2.4.2**). Both can be controlled in the pilot assay by including blood DNA from the semen donor. We have never seen an authentic crossover in blood DNA, which therefore provides an ideal negative control.

It is also advisable to include positive controls in the pilot assay by spiking some PCRs with genomic DNA from an individual carrying haplotypes showing the opposite linkage phase between 5′ and 3′ selector sites; such individuals should have been identified in the linkage phasing assay (*see* **Section 3.2.3**). Obviously, such genomic DNA should be handled with extreme caution during PCR setup, given its ability to cause major contamination of the assay with "crossover" molecules.

3.2.4.1. First Round of PCR

To detect crossover molecules, amplify multiple aliquots of sperm DNA using the outside primer pair F1-R1 in recombinant phase (**Fig. 20.2**). A typical 96-well plate setup would contain six rows of sperm DNA with 100, 200, 400, 800, 1,600, and 3,200 molecules per well (1.2–38.4 ng DNA per PCR), plus two rows of blood DNA (one row at 800 and one at 3,200 molecules per well). Positive controls are obtained by spiking a few wells with 0.5 μL 200 ng/mL genomic DNA (eight amplifiable molecules) from a control individual (*see* **Section 3.2.4**). This setup will screen 76,000 molecules for crossovers and will detect crossover molecules even at recombination frequencies as low as 0.0013%.

1. Set up PCRs as in **Section 3.2.2**, **Step** 1, but in total volumes of 8 μL. Start PCRs with an initial denaturation step at 96°C for 1 min, followed by cycling conditions determined in the optimization experiments. Generally, 26 cycles of amplification should be used in the primary PCR.
2. Immediately dilute PCR products with 35 μL 5 mM Tris-HCl pH 7.5 plus 2 μg/mL high molecular weight herring or salmon DNA carrier, or if necessary proceed to S1 nuclease digestion (*see* **Section 3.2.4.2** and **Note 5**).

3.2.4.2. S1 Nuclease Digestion (Optional, *see* **Note 5**)

1. Set up the same number of fresh 0.2 mL PCR tubes as you have in the first round of PCR, or a fresh plate. To each fresh tube, or each well in the fresh plate, add 5 μL S1 digestion mix containing 0.7 U/μL S1 nuclease.
2. Take 0.5 μL of each primary PCR product and add to S1 digestion mix. Mix by pipetting up and down three times.

3. Leave samples to digest at room temperature for 20 min.
4. Add 45 μL of S1 dilution buffer to each digest to bring the pH above 7. Keep on ice. Store S1 digests at –20°C. They are stable for at least 6 months.

3.2.4.3. Second Round of PCR

1. Set up each of the secondary (nested) PCRs in a 10 μL total volume. Seed each PCR with 0.5 μL of diluted primary PCR product or 1.6 μL of S1-digested primary PCR product. Use allele-specific primers targeted towards inner SNP selector sites in recombinant phase. As in **Section 3.2.4.1**., start all PCRs with an initial denaturation step at 96°C for 1 min, followed by optimized cycling conditions. Amplify the secondary PCRs for 30–35 cycles.
2. Add 2 μL all-purpose loading dye to each tube, and run out 5 μL aliquots on an agarose gel in the presence of 0.5 μg/mL ethidium bromide in both the gel and the buffer. A successful assay will show as positive secondary PCRs increasing in frequency with increasing sperm DNA input, a lack of PCR products in blood DNA reactions, and strongly positive signals in the positive control PCRs.

3.2.4.4. Calculating Crossover Frequencies

Estimate the crossover frequency from the numbers of positive and negative PCR reactions at each input of sperm DNA. To illustrate the different steps of crossover scoring, the hypothetical crossover assay shown in **Fig. 20.3** (eight input pools of 2,000, 600, and 200 molecules each, with 7, 3, and 1 crossover positive reactions recovered, respectively) is used as an example throughout the sections below.

1. Estimate the frequency of crossovers (m) in input DNA for each input pool by Poisson approximation:

   m = – ln (number of negative reactions/total number of reactions).
2. Maximum-likelihood methods (software written by A.J.J., available upon request) can be used to estimate crossover frequency more accurately by combining data from all input pool sizes.

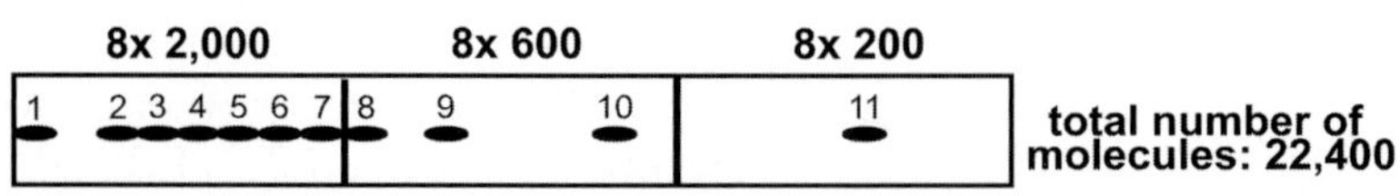

Fig. 20.3. Schematic representation of crossover-positive secondary PCRs at input pool sizes of 2,000, 600, and 200 DNA molecules, with eight PCRs per pool size. Here, 11 reactions (labeled 1–11) are positive for a crossover (e.g. 7/8 reactions at 2,000 molecule input). The input pools shown would contain 2.08, 0.47, and 0.13 crossovers per reaction at 2,000, 600, and 200 molecule input, respectively (*see* **Section 3.2.4.4, Step 1**), giving approximate crossover frequencies of 1.0×10^{-3}, 0.78×10^{-3}, and 0.65×10^{-3} respectively.

The mean crossover frequency calculated for the example data set in **Fig. 20.3** is 0.89×10^{-3}, with lower and upper 95% confidence intervals of 0.41×10^{-3} and 1.72×10^{-3}.

3.2.4.5. Bulk Recovery of Crossover Molecules

1. Use the crossover frequency estimates from **Section 3.2.4.4** to estimate optimal DNA pool sizes for efficient recovery of crossover molecules. Accurate definition of crossover profiles in the test interval requires the recovery and mapping of about 100 crossovers.
2. Recover crossovers from a single 96-well plate by using varying inputs of sperm DNA containing from 0.6 to 1.5 crossover molecules per well. Set up PCRs such that at the lowest inputs 45% of PCR reactions will be positive (and consequently 74% of positive reactions will be derived from a single input crossover molecule). The corresponding figures at the highest inputs are 88% and 43%, indicating that the majority of positive PCRs will contain DNA from more than one crossover, though not all molecules will necessarily show different exchange points. Mixed products derived from two or more crossovers with different exchange points can be readily identified during crossover breakpoint mapping (***see*** **Section 3.3.4**).

3.2.4.6. Reciprocal Crossover Assays

The crossover assay shown in **Fig. 20.1** recovers black-to-white crossover molecules. Reciprocal crossover products can be recovered by changing all allele-specific primers to their alternative allele, to recover white-to-black crossovers. The assay must take into account the different annealing specificities of the alternative primers. Assays performed in each orientation are independent of each other and provide a useful test of reciprocity; equal crossover frequencies should be seen in both assay orientations. Also, reciprocal crossover data provide a powerful probe for detecting differences in recombination initiation frequencies on the two haplotypes in an individual *(31)*.

3.3. Mapping Locations of Crossovers

3.3.1. Third Round of PCR

1. Re-amplify secondary PCRs using universal PCR primers located just internal to the secondary PCR primer SNP sites. Set up PCRs in total volumes of 10–20 μL, seeded with 0.5 μL of secondary PCR product. Reserve two wells on the PCR plate and seed with one or other separated progenitor haplotype prepared in **Section 3.2.3, Step 3**. Amplify for 35 cycles to ensure that all reactions now amplify, even those seeded with blood DNA and crossover negative sperm reactions.
2. Add 2.5–5 μL all-purpose loading dye, and check that PCR amplification has occurred to a uniform high level by electrophoresis of 0.6 μL of PCR product/loading dye mix on an agarose gel. Only a few reactions need be tested.

*3.3.2. Transferring Reactions onto Dotblots (See **Note 6**)*

1. Cut two pieces of 3MM paper, plus as many pieces of nylon membrane as you need replica filters, to the size of your 96-well dotblot manifold (we use model DHM-96, Scie-PLAS, Southam, UK). Soak paper and membranes in ddH_2O for a few minutes.
2. Place the two pieces of 3MM paper on the dotblot manifold. Overlay with one piece of nylon membrane. Assemble the dotblot manifold, making sure that the securing nuts are uniformly fastened but not over-tightened. Apply a vacuum to the assembled manifold.
3. Add 30 μL of denaturing solution (0.5 M NaOH, 2 M NaCl, 25 mM EDTA) per replica filter (e.g. 180 μL denaturing mix if you wish to make six replica filters) directly to PCRs from **Section 3.3.1** and mix well by pipetting up and down, or by sealing the plate, mixing and centrifuging.
4. Using a multichannel pipette, load 35 μL denatured DNA into each well of the manifold. Once all samples are loaded and the liquid has been pulled through the membrane, add 150 μL 2x SSC to each well to neutralize DNA. Bake dotblots for 10 min at +80°C and covalently link the DNA to the membrane by UV exposure in a UV crosslinker.

3.3.3. ASO Hybridization

1. Label each ASO by mixing 1 μL 10x kinase mix, 0.2 μL [γ-^{32}P]-ATP, 7.8 μL ddH_2O, and 0.35 μL T4 polynucleotide kinase (10 U/μL) in a screw-top microcentrifuge tube at room temperature (*see* **Note 7**). Add 1 μL of 8 μg/mL ASO. Incubate at 37°C in a heat block for 1–3 h. Add 20 μL kinase stop solution. The probe can be used immediately for hybridization, or stored at +4°C or –20°C for a few days.
2. Soak dotblots in 3x SSC. Transfer each to a small hybridization bottle with the DNA side facing inwards.
3. Pre-hybridize dotblots at 49°C for 5 min in 2 mL TMAC hybridization solution. Discard solution. Add 2.5 mL fresh hybridization solution. Incubate in a rotator oven at 49°C for 5 min.
4. Add ^{32}P-labeled ASO and 40 μL unlabeled 8 μg/mL ASO of the opposite allele as competitor. Hybridize in a rotator oven at 49°C for 1–2 h.
5. Wash membranes inside bottles in two changes of 2–3 mL TMAC wash solution at 53°C in a rotator oven for approximately 5 min per wash, followed by one 15 min wash with 4 mL TMAC wash solution.
6. Rinse membranes twice in ~5–10 mL of 3x SSC in the bottles at room temperature, then transfer the membranes to a Pyrex tray with 3x SSC and wash at room temperature a few times.

7. Blot excess liquid off membranes with 3MM paper, seal them in Saran wrap and autoradiograph at room temperature or at –80°C with an intensifier screen. Do not allow the blot to dry out (*see* **Note 8**).
8. Combine all membranes in a Pyrex dish and remove the probe with at least ten changes of 100 mL boiling 0.1% SDS. Monitor probe removal using a Geiger-Muller meter; >99% of label should be removed. Re-probe the same membrane with the ASO corresponding to the alternative allele of same SNP site. Dotblots can be re-probed at least 10 times without loss of signal.

3.3.4. Mapping Crossover Breakpoints

ASO typing reveals the status of each internal SNP site in each crossover-positive PCR. The presence of progenitor haplotypes on the same dotblot allows internal alleles to be assigned to one or other progenitor haplotype, essential for the determination of crossover breakpoints. Including crossover negative and blood PCRs can provide useful information on bleed-through PCR amplification of progenitor haplotypes and help identify potentially problematic allele-specific primers used in crossover recovery. Hypothetical examples of dotblot data are shown in **Fig. 20.4**.

Crossover-positive reactions can either be "simple" (see e.g. dot number 11 in **Fig. 20.4B**, **C**), or "mixed", i.e. giving a positive signal upon hybridization with both alleles of some SNP sites (see e.g. dot number 1 in **Fig. 20.4B**, **C**). "Simple" ones contain only one class of crossover; they might contain more than one crossover molecule, but all crossovers have their breakpoint in the same inter-SNP interval. "Mixed" reactions contain two or more different crossovers. We score them as two crossovers, one with a breakpoint 5′ to where a succession of one or more mixed SNP sites starts, and the other 3′ to where the mixture finishes (**Table 20.3B**). The estimate of the number of crossovers will always be conservative - simple reactions may contain more than one crossover, but only one will be scored; mixed reactions may contain more than two crossovers, but only two will be scored. The true number of crossovers will therefore be higher, and can be estimated for each inter-marker interval by Poisson correction. This involves combining data (number of PCRs positive and negative for a given type of crossover) over all input pool sizes. Note that some PCRs will not be scorable for certain types of crossover; specifically, if a PCR contains a succession of two or more mixed sites, then each interval between *mixed* sites will not be scorable for a crossover in that interval, since such crossovers will be masked by the two definite exchanges that create the succession of mixed sites. Such PCRs should be ignored in the Poisson correction.

Crossovers may occasionally be observed in the end intervals (furthest 5′ or 3′ SNP interval assayed by ASO hybridization, see dot 7 in **Fig. 20.4C**). Crossovers of this type could either have

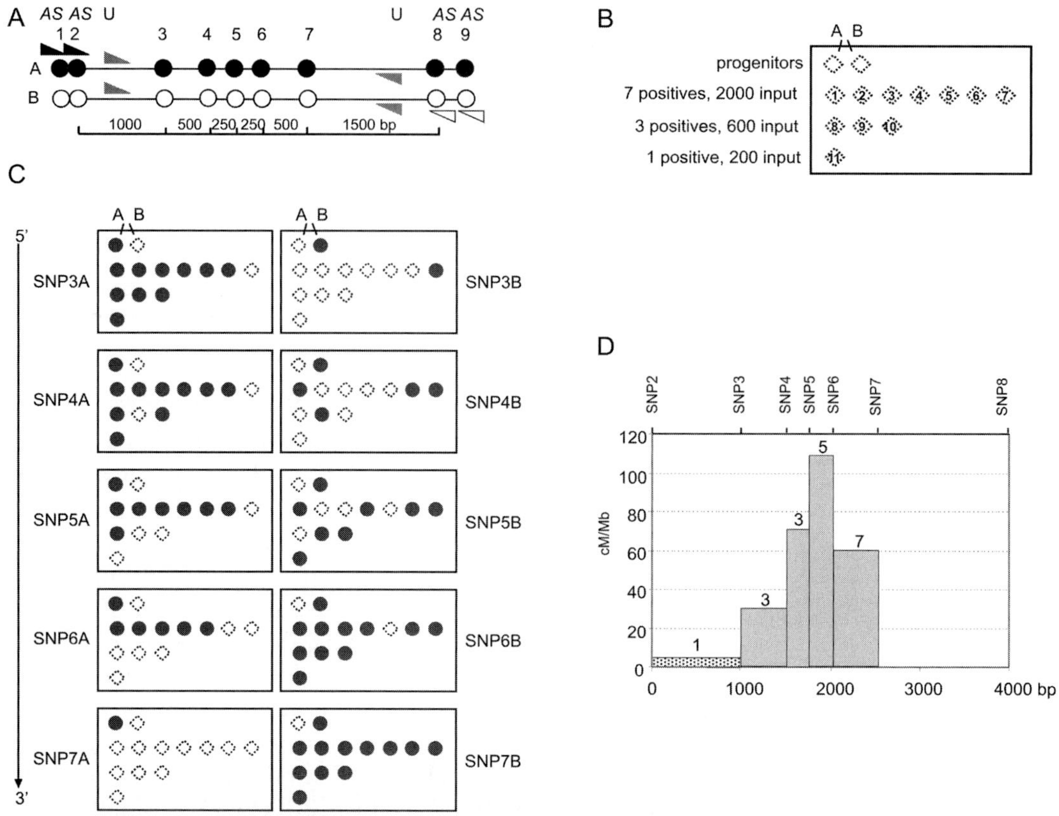

Fig. 20.4. Principle of crossover breakpoint mapping. (**A**) In the crossover assay, if forward primers are allele-specific for the A haplotype (black), and reverse primers for the B haplotype (white; allele-specific primers *AS*), "A to B" crossovers are amplified. Crossover-positive reactions and progenitor haplotypes A and B are re-amplified with universal primers (U). Five internal heterozygous SNP sites (numbered 3–7) and their inter-marker distances are shown. SNPs 1, 2, 8, and 9 are selector sites for allele-specific PCR. (**B**) Dotblot positions of progenitor haplotypes A and B, plus the 11 hypothetical crossover positives from the different input pools (1–11) from **Fig. 20.3** are shown. (**C**) Dotblots on the left are probed with ASOs targeted towards haplotype A alleles at heterozygous SNP sites (5′ to 3′, top to bottom: SNPs 3–7), while dotblots on the right are probed with ASOs specific for haplotype B. The haplotypic origin of each allele is identified from progenitor haplotype dots. Filled circles are positive for that particular haplotype, open dashed circles are negative (*see* **Table 20.3A** for crossover scoring). (**D**) *Gray bars* indicate recombination activity for each SNP interval (*see* **Table 20.4A, B** and **Section 3.3.4. Step 2** for calculations). The Poisson-corrected number of crossovers (rounded to nearest full number) observed in each interval is shown above each bar.

their breakpoint in the interval between the inner selector site and the nearest internal heterozygous SNP site, or they could be PCR artifacts arising from bleed-through amplification of one or other haplotype. Analysis of crossover-negative PCRs and of blood DNA can show whether such bleed-through amplification is a reasonable explanation. If not, then score these end-interval crossovers but interpret with caution. If bleed-through is a significant problem, then ignore the end interval when determining the crossover profile over the test interval.

Table 20.3
Example of crossover scoring. (A) In the crossover assay example given in Fig. 20.3, eight pools of 2,000, 600, and 200 amplifiable molecules each were screened (24 PCRs, 22,400 molecules in total). Eleven of the 24 PCRs were positive for crossovers (dots 1–11 on blot in Fig. 20.4B, and rows 1–11 in this table). Progenitor haplotypes A and B (black and white) are indicated at top. The status of heterozygous SNPs is shown for each crossover-positive PCR. Five (dots 1–4 and 6 in Fig. 20.4C) of the 11 positive reactions contain two or more different crossovers as shown by mixed SNP sites (M). (B) Resolution of crossover mixtures. For example, reaction 1 is resolved as 1a and 1b (*see* Section 3.3.4). Note that dot 1 could contain an undetectable crossover between SNP4 and SNP5 and/or between SNP5 and SNP6. (C) Crossovers ordered according to breakpoint location, 5′ to 3′, and numbers of crossovers per SNP interval. The single crossover observed in the 5′ end interval is shown in parentheses

	Number on blot	input	(SNP2)	SNP3	SNP4	SNP5	SNP6	SNP7	(SNP8)
			heterozygous markers						
		Progenitor A		A	A	A	A	A	
		Progenitor B		B	B	B	B	B	
A	1	2000		A	M	M	M	B	
	2	2000		A	A	A	M	B	
	3	2000		A	A	A	M	B	
	4	2000		A	A	M	M	B	
	5	2000		A	A	A	A	B	
	6	2000		A	M	M	B	B	
	7	2000		B	B	B	B	B	
	8	600		A	A	A	B	B	
	9	600		A	B	B	B	B	
	10	600		A	A	B	B	B	
	11	200		A	A	B	B	B	
B	1a	2000		A	A	A	A	B	
	1b	2000		A	B	B	B	B	
	2a	2000		A	A	A	A	B	
	2b	2000		A	A	A	B	B	
	3a	2000		A	A	A	A	B	
	3b	2000		A	A	A	B	B	
	4a	2000		A	A	A	A	B	
	4b	2000		A	A	B	B	B	
	5	2000		A	A	A	A	B	
	6a	2000		A	A	A	B	B	
	6b	2000		A	B	B	B	B	
	7	2000		B	B	B	B	B	
	8	600		A	A	A	B	B	
	9	600		A	B	B	B	B	
	10	600		A	A	B	B	B	
	11	200		A	A	B	B	B	

(continued)

Table 20.3 (continued)

C	7	2000	B	B	B	B	B
	1b	2000	A	B	B	B	B
	6b	2000	A	B	B	B	B
	9	600	A	B	B	B	B
	4b	2000	A	A	B	B	B
	10	600	A	A	B	B	B
	11	200	A	A	B	B	B
	2b	2000	A	A	A	B	B
	3b	2000	A	A	A	B	B
	6a	2000	A	A	A	B	B
	8	600	A	A	A	B	B
	2a	2000	A	A	A	A	B
	1a	2000	A	A	A	A	B
	3a	2000	A	A	A	A	B
	4a	2000	A	A	A	A	B
	5	2000	A	A	A	A	B

1. Score ASO signals on dotblots. Order crossovers in a table according to the location of their breakpoints, 5′ to 3′. Count the number of crossovers for each SNP interval (**Table 20.3C**). Maximum-likelihood methods (software written by A.J. Jeffreys, available upon request) can then be used to estimate the mean frequency of crossovers in each interval, taking into account in which input pools the crossovers were observed (**Table 20.4A**).
2. Adjust the number of molecules screened for each SNP interval where mixtures are observed by excluding pools that contain sequential mixed sites spanning the SNP interval (see intervals SNP4-SNP5 and SNP5-SNP6 in **Table 20.4A**). For each SNP interval, multiply the maximum-likelihood Poisson estimate of recombination frequency by the total number of molecules screened, to yield the Poisson-corrected number of crossovers adjusted for any missing data from PCRs showing sequential mixed sites (**Table 20.4B**). Calculate the recombination activity in cM/Mb for each SNP interval as follows:
(Poisson-corrected number of crossovers/total number of molecules screened) $\times$ 100 = recombination fraction in cM,
then:
(recombination fraction in cM/length of interval in bp) $\times$ 10^6 = recombination activity per unit physical length in cM/Mb.

Recombination activity across the example interval in **Fig. 20.4A** is shown in **Fig. 20.4D** and **Table 20.4B**.

Table 20.4
Calculating crossover frequencies for each SNP interval. (A) Total number of positive plus negative reactions for each input pool, per SNP interval. Crossover-positive pools that contain crossover mixtures spanning more than one interval (in gray italics) are excluded for all SNP sites that lie between the two SNP intervals where crossover breakpoints were scored. For example, crossover 6 is excluded for the SNP4-SNP5 interval. The total number of molecules screened is lower for the SNP4-SNP5 and SNP5-SNP6 intervals (open box), as follows: 22,400 – (2 × 2,000)=18,400, because two of the 2,000-molecule input pools are excluded for these intervals. (B) The Poisson-corrected number of crossovers in each interval is determined, and if necessary adjusted upwards to take into account any PCRs in which mixed sites make crossover detection impossible. Thus the Poisson-corrected number of crossovers in interval SNP4-SNP5 is adjusted by the factor 22,400/18,400. Recombination activity per interval then is calculated from the corrected number of crossovers and inter-marker distance. End interval crossovers (in parentheses) are scored as normal but we exclude them from further analyses if they are likely to result from parental haplotype bleed-through amplification (*see* Section 3.3.4)

A

	number of crossover positive + negative reactions for each SNP interval, in each input pool					
input pool	SNP2 to SNP3	SNP3 to SNP4	SNP4 to SNP5	SNP5 to SNP6	SNP6 to SNP7	SNP7 to SNP8
2000	1+7	2+6	1+5	3+3	5+3	0+8
600	0+8	1+7	1+7	1+7	0+8	0+8
200	0+8	0+8	1+7	0+8	0+8	0+8
molecules screened	22400	22400	18400	18400	22400	22400
			exclude crossover positives 1 and 6	*exclude crossover positives 1 and 4*		
Poisson-est. crossover frequency	-	0.00015	0.00018	0.00027	0.00030	-

B

	SNP2 to SNP3	SNP3 to SNP4	SNP4 to SNP5	SNP5 to SNP6	SNP6 to SNP7	SNP7 to SNP8	
uncorrected number of crossovers	(1)	3	3	4	5	0	**total: 16**
Poisson-corr. number of crossovers	0	3.4	3.3	5.0	6.7	0	**total: 19.4**
inter-marker distance (bp)	1000	500	250	250	500	1500	
cM/Mb	(4.5)	30	72	108	60	0	

3.4. Detecting Gene Conversions and Crossovers Using the Half-Crossover Assay

Recombination can generate both crossovers and nonexchange gene conversions. The latter often involves the transfer of just a single SNP marker between haplotypes and cannot be detected in crossover assays. An alternative approach that is capable of detecting such events is the half-crossover assay (**Fig. 20.5**) in which nested allele-specific PCR on sperm DNA is directed to markers only at one end of the test interval, with universal primers being used at the other end. In the absence of recombination, PCR products will only be generated from the selected haplotype. Recombinants can be detected by the presence of markers from the unselected haplotype, with crossovers and conversions each giving distinct patterns of positive markers. PCR amplifications are essentially as described in **Sections 3.2.4.1** and **3.2.4.3**, with recombinant detection as described in **Sections 3.3.2** and **3.3.3** (*see* **Note 9**).

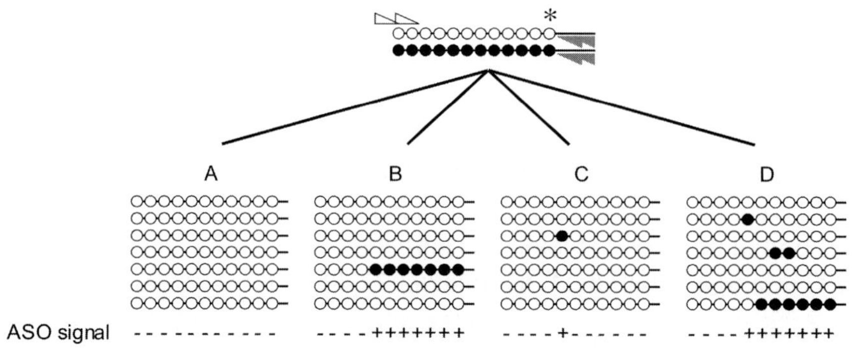

Fig. 20.5. Detecting crossovers and nonexchange gene conversions by the half-crossover assay. Multiple small pools (labeled A–D) of sperm DNA from a man carrying multiple SNP heterozygosities (white and black circles) are amplified by nested long PCR using allele-specific primers (white arrowheads) directed to white alleles plus universal primers (gray arrowheads). PCR products are then probed using ASOs specific to alleles from the unselected black haplotype. Various outcomes are possible in each pool. A, PCR products are derived only from white progenitor molecules. B, the pool contains a crossover in addition to progenitors, yielding a continuous block of positive markers distal to the selector sites. Note that any additional crossovers in this pool with exchange points mapping 3′ to the observed crossover would not be detectable. C, the pool contains a nonexchange gene conversion molecule, in this case affecting just one SNP. D, a complex pool containing two conversions and one crossover. The double-site conversion is masked by the crossover, and the single site conversion plus crossover appear as a single crossover with the exchange point mapping to the wrong interval. Pools B and D can be distinguished by further allele-specific PCR directed for example to the terminal SNP (marked with an *asterisk*), which allows crossovers and conversions to be separately amplified from nested PCR products prior to mapping.

1. For primary PCR, set up a long PCR master mix (800 μL) containing the outermost allele-specific primer, the distal universal primer plus typically 48 ng sperm DNA. Array 8 μL aliquots into a 96-well plate to give 0.48 ng sperm DNA per PCR (40 amplifiable molecules of each haplotype for this input). Amplify for 26 cycles using annealing temperature conditions optimal for the allele-specific primer. Note that optimization is simpler than for the full crossover assay since only one allele-specific primer is being used.
2. Dilute the primary PCR products with 35 μL 5 mM Tris-HCl pH 7.5 plus 2 μg/mL high molecular weight herring or salmon DNA as carrier. Seed 15 μL secondary PCRs containing the nested allele-specific primer and nested universal primer with 0.6 μL diluted primary PCR product and amplify under optimized conditions for 36 cycles.
3. Test 0.6 μL of a selection of secondary PCRs by agarose gel electrophoresis and staining with ethidium bromide. Check that the correct-sized PCR product (typically 5–10 kb in these assays) has been generated and that PCR product yields are high and uniform over the reactions.
4. Before dotblotting, add the nonselected haplotype, amplified separately, to five wells on the secondary PCR plate such that the ratio of nonselected to selected haplotype PCR products

is 1:10, 1:30, 1:100, 1:300, and 1:1,000. These wells will serve as positive controls to determine the level of allele-specificity of hybridization for each ASO, as well as estimating the proportion of recombinant DNA in pools showing a putative crossover or conversion.

5. Denature the secondary PCRs and prepare as many replica dotblots as required (*see* **Section 3.3.2**).
6. Hybridize dotblots with each ASO to detect alleles from the nonselected haplotype (*see* **Section 3.3.3**).
7. Score autoradiographs for the presence of a crossover or nonexchange conversion in each secondary PCR. Crossovers will be clearly evident from the presence of a succession of positive markers extending continuously towards the distal end of the test interval (**Fig. 20.5**). Conversions involving more than one SNP site will also be clear from adjacent markers being positive. Single-site conversions will show as single markers with a positive signal, and with a signal level corresponding to at least 1% of PCR product being derived from the recombinant. Crossovers (which are always genuine recombinant molecules) provide a useful internal control to demonstrate proper signal intensities. Very faint (<<1%) single-site signals should be rejected as PCR mis-incorporations. Blood DNA, which is free from recombinants, can be processed in parallel to give more information on the likely incidence and signal intensities of mis-incorporation artifacts.
8. The gene conversion frequency for each SNP site can be calculated by excluding all crossover-positive PCRs in which conversions might be masked (**Fig. 20.5**), and counting the number of conversion-positive (p) and conversion-negative (n) PCRs for the remaining $p+n$ reactions. The Poisson-corrected total number of conversions (c) per PCR is estimated by $-\ln[n/(p+n)]$, and the conversion frequency is thus given by c/t, where t is the mean number of amplifiable molecules of each haplotype present per PCR (in this example, 40).
9. If desired, crossover-positive PCRs can be further analyzed by allele-specific PCR directed to a distal SNP site (**Fig. 20.5**). This allows crossover molecules and noncrossovers (including progenitors) to be amplified separately prior to mapping.

3.5. DNA Enrichment by Allele-Specific Hybridization (DEASH)

DEASH is a pre-PCR DNA enrichment strategy that both facilitates the detection of de novo noncrossover gene conversion events and serves to verify their authenticity *(24)*. By hybridizing a biotinylated ASO to denatured sperm DNA, molecules containing a known single-nucleotide variant can be captured on streptavidin-coated paramagnetic beads whilst those bearing the alternative variant can be removed by washing. On subsequent release from the beads, this specifically enriched single-stranded DNA can be analyzed by allele-specific PCR (*see* **Section 3.2**). Gene conversion and crossover events are thus detected by enriching for a variant specific to one haplotype then amplifying pools of this enriched DNA using primers specific to the opposite haplotype (*see* **Fig. 20.6**).

The enrichment procedure reduces the complexity of the DNA and therefore improves the signal-to-noise ratio in the PCR such that pools of enriched DNA derived from large amounts of genomic DNA can be screened for recombinants. Since gene conversion tracts are very short *(25)*, the ability to detect such events depends on having SNP markers very close to the center of the hotspot. If there

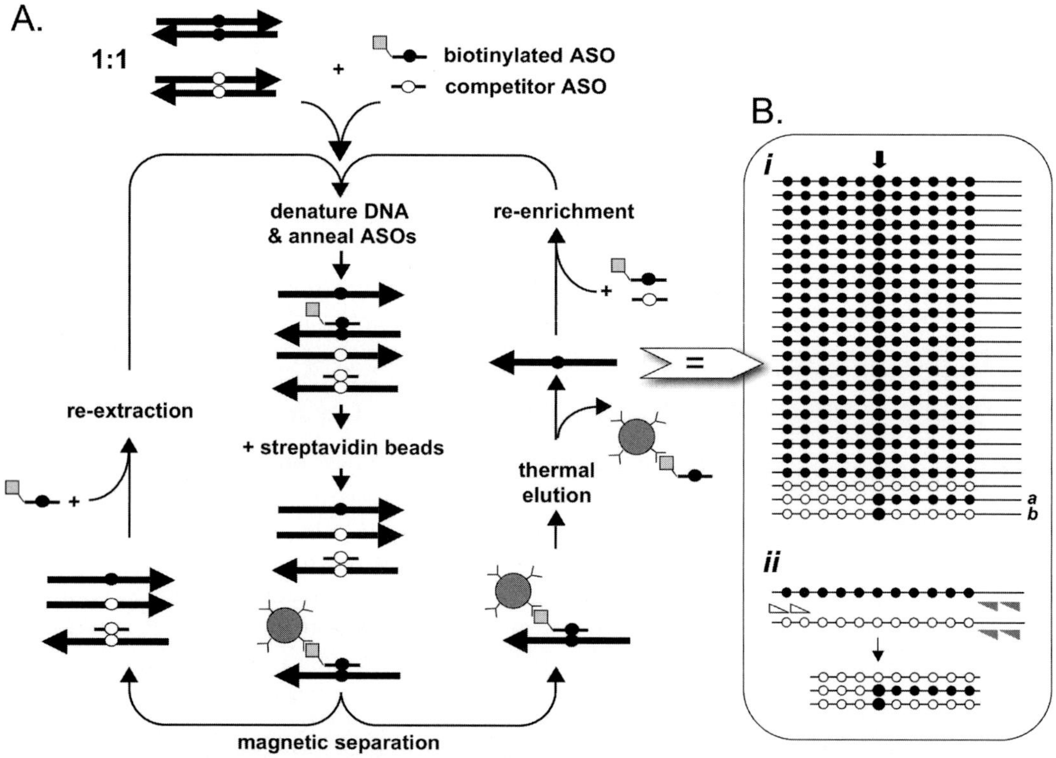

Fig. 20.6. Principle of DEASH and subsequent detection of recombinants. (**A**) Digested double-stranded genomic DNA containing equal numbers of molecules differing by the chosen base substitution (*circle*) is mixed with a biotinylated ASO (bio-ASO, in black with biotin shown by a gray square) and the corresponding competitor ASO (*white*). The DNA is thermally denatured and the ASOs anneal as the temperature is subsequently reduced. Streptavidin-coated paramagnetic beads are then added, allowing the bio-ASO/target hybrid to be separated from the rest of the mixture with a magnet. The single-stranded target is released by thermal elution. The unbound fraction can be re-extracted by adding more bio-ASO, thus increasing the yield. Recovered fractions can be further enriched if additional bio-ASO and competitor ASO are added and the procedure repeated. (**B**) The recovered single-stranded DNA is greatly enriched for the selected haplotype (arrow indicates selector site, additional circles represent other SNP heterozygosities carried by this individual) (i). This facilitates detection of genuine recombinants exchanged at the selector site (a, crossover; b, gene conversion) using pools of the enriched DNA in a half-crossover assay (ii). Two-rounds of nested PCR are performed using allele-specific primers directed to the nonselected haplotype (white arrowheads) together with universal, i.e. nonallele specific, primers (gray arrowheads). Recombinants are detected by the presence of markers from the selected parental haplotype (black markers) as determined by ASO typing.

is more than one SNP requiring analysis, then separate enrichment of each SNP will need to be carried out. The level of enrichment achievable can be 300-fold or higher. Unlike the half-crossover assay alone (**Section 3.4**), pre-processing DNA with DEASH allows the detection of very rare events, with a frequency of 10^{-6} or lower, even if they only affect a single SNP site. These single-site conversions should only affect the selected SNP and not a nonselected SNP. This provides validation for such events and helps establish that they are not PCR mis-incorporations (*see also* **Section 3.4.7**).

3.5.1. The DEASH Procedure

To maximize the yield of recovered DNA, cleave sperm genomic DNA with a restriction enzyme that cuts just outside the region of interest; undigested DNA gives very poor yields. We have successfully used the DEASH protocol to enrich for genomic restriction fragments up to 12.5 kb in size. Digest according to the manufacturer's instructions, supplemented if necessary with spermidine trichloride to a final concentration of 4 mM to improve digestion. After digestion, ethanol precipitate the genomic DNA, wash with 80% ethanol, dry and dissolve in 5 mM Tris-HCl (pH 7.5) to a final concentration of 0.2–1 mg/mL (*see also* **Section 3.1.1**, **Steps 8–9**).

It is necessary to determine empirically the annealing temperature (A°) of each biotinylated ASO before bulk genomic DNA is processed in order to maximize both yield and the degree of allele separation. As a guide, the relationship between optimal annealing temperature and GC content of a bio-ASO approximates to 29.6°C + 0.318(%GC)°C in the buffer used for hybridization *(24)*. However, the efficiency of allele separation is very much ASO-dependent (*see* **Table 20.5**).

Table 20.5
Examples of biotinylated ASOs used in DEASH experiments to recover gene conversion events

SNP type (variants)	Bio-ASO (5′ to 3′)	%GC	A°	target (kb)	%yield	purification
Indel (-/AGA)	Bio-GAAAAAAA***GA***GGAGTGGA	39	49	6.6	60	7×
Ti (C/T)	Bio-AAAAGGAGTG**C**GTGTGTG	50	50	6.6	60	13×
Tv (G/C)	Bio-AATAGGGAGA**G**GATAAAG	39	42	10.2	60	100×
Ti (A/G)	Bio-GGTGAGAGGA**A**AGAAGTG	50	45	6.6	65	120×
Tv (C/G)	Bio-AATAGGGAGA**C**GATAAAG	39	42	12.5	20	200×
Tv (G/C)	Bio-GGACGGGCTG**G**GGTTCCT	72	55	9.4	30	>200×
Ti (C/T)	Bio-ACTAATGATT**C**ACTAACA	28	40	7.0	45	300×
Tv (T/G)	Bio-GTTTGTTGTTT**T**TTGTTT	22	35	7.0	45	>300×

Indel, insertion/deletion; Ti, transition; Tv, transversion. SNP variants are shown in bold, and the insertion site for the + variant of the indel is shown in bold italics.

3.5.2. Annealing Temperature Titration

Throughout the procedure use dedicated "pre-PCR" pipettes for manipulations. Typically, we test annealing temperatures over the range of 30–60°C in 5°C increments using 0.1 μg of digested genomic DNA in a 20 μL volume. The following should be scaled according to the number of annealing temperatures to be tested.

1. Set "annealing" waterbath to A°C and "elution" waterbath to 60°C.
2. Prepare 210 μL of 1x DHB for bead preparation, keep at room temperature; 21 μL 10x DHB plus 189 μL ddH_2O.
3. Prepare 100 μL of DHB/BSA wash solution and place in "annealing" waterbath; 10 μL 10x DHB, 88 μL ddH_2O plus 2 μL 500 μg/mL BSA.
4. Aliquot 50 μL of ED solution, keep at room temperature.
5. Aliquot a further 100 μL of ED solution and place in "elution" waterbath.
6. Gently agitate to re-suspend the M-280 Streptavidin Dynabeads, then aliquot 21 μL into a siliconized 1.5 mL Eppendorf tube. Pre-wash the beads twice in 1x DHB buffer: place the tube in the Dynal magnetic particle concentrator, leave for 10 s, then twist each way until the beads have formed a small mass, aspirate off liquid, then remove tube from concentrator and re-suspend bead mass in 100 μL 1x DHB by gentle pipetting, repeat procedure. Finally, re-pellet beads and re-suspend in 1.5 μL of 1x DHB (i.e. 1/14th the original volume of the beads). Keep on ice until needed.
7. In a 0.2 mL PCR tube add 2 μL 10x DHB, 1.5 μL 5 μM biotinylated ASO, 1 μL 30 μM nonbiotinylated competitor ASO, 0.1 μg digested genomic DNA, and make up to 20 μL with ddH_2O.
8. Place tube in a PCR machine and denature the DNA at 96°C for 75 s, then step down the temperature to allow the bio-ASO to anneal to the target DNA starting at A+9°C and going down to A+1°C in 1°C steps for 20 s per step. Anneal at A°C for 2 min.
9. During the annealing step, pre-warm a siliconized 1.5 mL Eppendorf tube containing 1.3 μL of washed beads in a waterbath set at A°C.
10. Transfer the contents of the PCR tube to the pre-warmed tube containing the beads, mix *extremely gently*, and incubate for 10 min at A°C. Ensure the beads remain suspended during this time by *very carefully* flicking the contents of the tube every couple of minutes. Binding of the bio-ASO to the streptavidin-coated beads appears to be complete within about 5 min, even for large DNA targets.

11. Use the magnetic concentrator to pellet the beads. Remove the supernatant containing the unbound DNA, but keep this fraction if you wish to re-extract DNA from it (*see* **Section 3.5.4, Step 2**)
12. Re-suspend the beads and captured DNA hybrids in 50 μL of the DHB/BSA wash solution. Pipette very gently and transfer to a fresh siliconized 1.5 mL tube and incubate at A°C for 1 min.
13. Repeat **Steps 11–12**, to remove the last traces of unbound DNA.
14. Repeat **Steps 11–12**, this time re-suspending the beads in the 50 μL of room temperature elution buffer ED and transferring to a fresh siliconized tube at the same temperature.
15. Repeat **Step 14**, using the 60°C elution buffer.
16. Incubate for 2 min at 60°C to release the target DNA from the biotinylated ASO/bead complex. Use the magnetic concentrator to remove the beads and recover the eluted DNA (supernatant) into a siliconized 1.5 mL tube.
17. Turn up "elution" waterbath containing the remaining elution buffer to 70°C.
18. Repeat **Steps 15–16** to release any remaining DNA.

3.5.3. Determining Yield and Allele Separation

Yields of each haplotype are estimated by comparing allele-specific PCR amplification (*see* **Section 3.2**) of a range of concentrations of input DNA with recovered DNA on ethidium bromide-stained gels. This also allows assessment of the degree of allele separation (**Fig. 20.7**). The PCR is limited to the exponential phase of amplification and yield estimation must take into account the fact that unfractionated double-stranded DNA results in twice as much PCR product as an equal number of recovered single-stranded molecules. Typically, 10 μL PCRs are seeded with 4% of each eluate (2 μL) and the control input series corresponds to that expected for 100% down to 0.5% yields. In addition to evaluating the optimal annealing temperature, the best elution temperature can also be determined. Note that the elution buffer ED is 0.14x PCR buffer less dNTPs and BSA; PCR setup should take this buffer into account.

3.5.4. Processing Bulk Genomic DNA by DEASH

1. Up to 2 μg of digested genomic DNA can be processed per 20-μL volume as described in **Section 3.5.1**. If larger quantities are required then set up multiple 40 μL volumes with 4 μg of DNA each and double the volumes described in the protocol. In our experience, it is difficult to keep the beads suspended in larger volumes than this.
2. Yields can be increased by re-extracting the remaining target from the unbound fraction obtained in **Step 11 of Section 3.5.1**. Place the unbound fraction in a 0.2 mL PCR

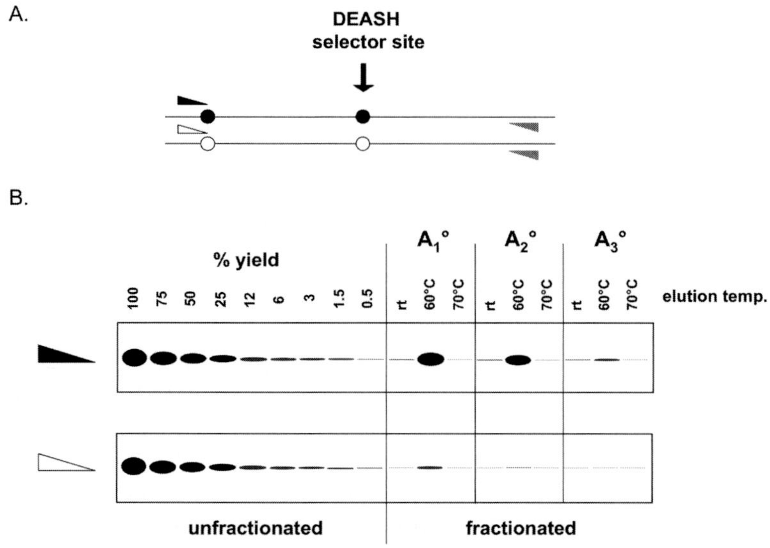

Fig. 20.7. Determining optimal yield and degree of purification post-DEASH. (**A**) Restricted DNA is enriched for molecules carrying the black allele (*circle*) using a biotinylated ASO as indicated by the *arrow*. Yield and purification are established by setting up two parallel sets of PCRs, one using an allele-specific primer directed against the selected haplotype (*black triangle*), the other using an allele-specific primer directed against the other haplotype (*white triangle*). The same reverse strand primer is used for both (*gray triangle*). Comparisons are made with the signals obtained from unfractionated DNA. (**B**) Hypothetical results from the DEASH eluates obtained using three different annealing temperatures (A_1°, A_2°, A_3°) and three sequential elution temperatures (room temperature [rt], 60°C, 70°C). Regardless of annealing temperature, little target is released either at the rt or 70°C elution steps; the majority of DNA can be recovered by simply raising the elution temperature to 60°C. The yield of selected haplotype is determined as ~75%, ~50%, and ~3% for annealing temperatures A_1°, A_2°, A_3° respectively. The degree of purification is determined from the nonselected haplotype PCRs; for example, for A_2°, the signal is approximately equivalent to that from the 0.5% input control, such that the fraction is enriched 100 fold (= 50/0.5) for the selected haplotype. This degree of purification appears better than that obtained at A_1° (12.5 fold [75/6]), so bulk fractionation would best be carried out at A_2°.

tube and add 5 μM biotinylated ASO to a final concentration of 0.375 μM. Proceed from **Step 8 of Section 3.5.1** using the optimal annealing and elution temperatures. This re-extraction can be set up immediately on separating the first unbound DNA; re-extraction thus proceeds through the denaturation, annealing, and binding steps whilst washing and elution is carried out on the initially bound DNA. More than 95% of the recoverable DNA will be obtained from three rounds of re-extraction. However, the degree of purification can decrease several fold with each re-extraction.

3. Recovered DNA can also be subjected to a second round of purification using the same biotinylated ASO. The same degree of purification is achieved at each cycle such that a 20-fold purification obtained with one round equates to

a 400-fold purification after the second. However, the percentage recovery per round remains roughly constant such that the yield declines proportionally. To re-purify the eluted DNA in ED buffer, add DHB to 1x (note that ED is already 0.14x DHB), biotinylated ASO to 0.375 μM and nonbiotinylated competitor ASO to 1.5 μM. Proceed as for the first purification, starting at **Step 8 of Section 3.5.1**.

4. When target fragments contain Alu elements, yields can sometimes be improved by adding Alu competitor DNA to suppress networking of the genomic DNA via Alu-Alu hybridization during the annealing step of the DEASH procedure. The competitor can be made by PCR using consensus sequence-derived primers Alu-F and Alu-R. Initially seed a 10 μL PCR containing 0.9 μL of 11.1x PCR buffer (*see* **Section 2.2.1**) along with 0.5 μM of each primer and 50 ng of human genomic DNA. Cycle for [96°C *20* s, 52°C *30* s, 70°C for *1* min] for two cycles, followed by a further two cycles annealing at 55°C and then finally 12 cycles annealing at 58°C. The initial low annealing temperature should allow amplification of diverged Alu elements whilst the later higher annealing temperature is optimal for this primer pair. Re-amplify by seeding 400 μL of PCR mix with 4 μL of the PCR product and splitting mix into eight 50-μL reactions. Use 17 cycles at the higher annealing temperature, gel-purify the 270 bp product, ethanol precipitate and dissolve in 100 μL 5 mM Tris-HCl pH 7.5. Estimate yield by gel electrophoresis. Irradiate on a UV transilluminator for 5 min to render the DNA recalcitrant to further PCR amplification. Add the Alu competitor to a final concentration of 2.5 μg/mL at **Step 7 of Section 3.5.2**.
5. Screen enriched DNA for recombinants as described in **Section 3.5.5**.

3.5.5. Detecting Gene Conversions and Crossovers in DNA Enriched by DEASH

Recombinant detection in the half-crossover assay is limited by the progenitor haplotype molecules present in each pool. As described in **Section 3.5**, DEASH allows genomic DNA to be fractionated to enrich for molecules containing a specific allele at a given SNP, thus depleting fractionated DNA in one of the two progenitor haplotypes. As shown in **Fig. 20.6B**, a half-crossover assay directed to the depleted haplotype will remove the selected progenitor haplotype, greatly facilitating the detection of crossover and conversion events. By way of example, suppose that DEASH has been performed on 12 μg sperm DNA (10^6 amplifiable molecules of each haplotype) to give a 200-fold enrichment of the selected haplotype and a 50% yield. The enriched single-stranded DNA will therefore contain 500,000 molecules of the selected haplotype and 2,500 remaining molecules of the unselected haplotype. The DNA can thus be analyzed by the half-crossover method, amplifying the unselected haplotype exactly as in **Section 3.4**, in a single plate with just 26 (2,500/96) progenitor molecules per PCR, allowing the detection of recombinants (both crossovers and conversions) as rare as 0.0002% (1/500,000). Recombinant detection and interpretation, and the Poisson correction of data, are performed exactly as in the half-crossover assay (*see* **Section 3.4**).

4. Notes

1. At this step, it is not necessary to add salt, as there is enough SSC and SDS in the sample to allow the DNA to precipitate.
2. Allele-specific primers can sometimes behave slightly differently in each batch of 11.1x PCR buffer. It is best to make a large enough batch of 11.1x buffer to last through optimizations and the desired crossover assays. If you run out of 11.1x buffer in the middle of a crossover assay, you may need to re-optimize the allele-specific primers in the new batch of buffer.
3. We find that using this relatively low extension temperature of 65°C increases the efficiency of amplification, especially for AT-rich amplicons, and reduces the incidence of jumping-PCR artifacts arising by annealing of incompletely extended PCR products. This reduction is especially important in the final crossover assay in which jumping PCR could generate false recombinants.
4. An ideal allele-specific primer will show, at one or more annealing temperatures, both good efficiency with PCR product yields from the correct homozygote comparable to those obtained with a universal primer, and good specificity with little if any detectable PCR product being generated from DNA from someone homozygous for the alternative allele. Low-efficiency primers can be modified by adding synthetic 5′ extensions (*see* **Section 3.2.1**) and low-specificity primers can be shortened by one or more bases; these modifications can sometimes, but not always, substantially improve primer performance.
5. This step removes single-stranded DNA from PCR products prior to re-amplification with nested primers. S1 digestion improves the specificity of nested allele-specific primers and removes PCR artifacts such as panhandle loop structures that can interfere with crossover detection *(18)*. However, most crossover assays can be performed successfully without this step.

6. Tertiary PCR products can be typed for internal SNP markers in several ways, for example sequencing. Here, we describe a straightforward dotblot hybridization protocol that requires no specialist equipment. We type SNPs by hybridization of ASOs to dotblots of PCR products using a modification of the TMAC hybridization protocol *(32)*. TMAC allows oligonucleotides of differing GC content to be hybridized at the same temperature, greatly facilitating typing. Note that TMAC is toxic!
7. If you are labeling several ASOs, make a master mix of the kinase mix plus [γ-^{32}P]-ATP plus T4 kinase, then aliquot 9 μL per ASO and add the ASO to each aliquot.
8. If the ASO shows any lack of specificity, try re-washing for 20 min in TMAC wash solution at a higher temperature (e.g. 56°C or 60°C) as in **Section 3.3.3**, **Steps 5–7**.
9. Most ASOs are very specific, showing >100:1 discrimination between alleles. However, recombinant detection is limited by PCR mis-incorporations, which can arise in small DNA pools early during PCR to give artifact signals that can appear to be single-site conversions. Different SNPs show different artifact levels, with transition SNPs (C/T, A/G) being most prone to switching by mis-incorporation. In practice, both mis-incorporation and ASO specificity mean that pool sizes have to be limited to 50 molecules at most, and preferably less. One 96-well plate will therefore allow at most only 4,800 molecules to be screened for recombinants, limiting analysis to only the most active hotspots with recombination frequencies above 0.1%.

Acknowledgments

Development of these methods was carried out at the University of Leicester with generous support by grants to A.J.J. from The Medical Research Council, The Louis-Jeantet Foundation and The Royal Society. L.K. was supported by the Osk. Huttunen Foundation, the Finnish Cultural Foundation, the Instrumentarium Science Foundation, and the Helsingin Sanomat Centennial Foundation.

References

1. Cullen, M., Erlich, H., Klitz, W., and Carrington, M. (1995) Molecular mapping of a recombination hotspot located in the second intron of the human *TAP2* locus. *Am. J. Hum. Genet.* **56,** 1350–1358.
2. Cullen, M., Noble, J., Erlich, H., Thorpe, K., Beck, S., Klitz, W., et al. (1997) Characterization of recombination in the *HLA* class II region. *Am. J. Hum. Genet.* **60,** 397–407.

3. Cruciani, F., Bernardini, L., Santolamazza, P., Modiano, D., Torroni, A., and Scozzari, R. (2003) Linkage disequilibrium analysis of the human adenosine deaminase (*ada*) gene provides evidence for a lack of correlation between hot spots of equal and unequal homologous recombination. *Genomics* **82,** 20–33.
4. Smith, R. A., Ho, P. J., Clegg, J. B., Kidd, J. R., and Thein, S. L. (1998) Recombination breakpoints in the human beta-globin gene cluster. *Blood* **92,** 4415–4421.
5. Twells, R. C., Mein, C. A., Phillips, M. S., Hess, J. F., Veijola, R., Gilbey, M., et al. (2003) Haplotype structure, LD blocks, and uneven recombination within the *LRP5* gene. *Genome Res.* **13,** 845–855.
6. Hedrick, P. W. (1987) Gametic disequilibrium measures: proceed with caution. *Genetics* **117,** 331–341.
7. Ardlie, K. G., Kruglyak, L., and Seielstad, M. (2002) Patterns of linkage disequilibrium in the human genome. *Nat. Rev. Genet.* **3,** 299–309.
8. Consortium, T. I. H. (2005) A haplotype map of the human genome. *Nature* **437,** 1299–1320.
9. Myers, S., Bottolo, L., Freeman, C., McVean, G., and Donnelly, P. (2005) A fine-scale map of recombination rates and hotspots across the human genome. *Science* **310,** 321–324.
10. Zavattari, P., Deidda, E., Whalen, M., Lampis, R., Mulargia, A., Loddo, M., et al. (2000) Major factors influencing linkage disequilibrium by analysis of different chromosome regions in distinct populations: demography, chromosome recombination frequency and selection. *Hum. Mol. Genet.* **9,** 2947–2957.
11. Laan, M., and Paabo, S. (1997) Demographic history and linkage disequilibrium in human populations. *Nat. Genet.* **17,** 435–438.
12. Tapper, W., Collins, A., Gibson, J., Maniatis, N., Ennis, S., and Morton, N. E. (2005) A map of the human genome in linkage disequilibrium units. *Proc. Natl. Acad. Sci. USA* **102,** 11835–11839.
13. Crawford, D. C., Bhangale, T., Li, N., Hellenthal, G., Rieder, M. J., Nickerson, D. A., et al. (2004) Evidence for substantial fine-scale variation in recombination rates across the human genome. *Nat. Genet.* **36,** 700–706.
14. Fearnhead, P., Harding, R. M., Schneider, J. A., Myers, S., and Donnelly, P. (2004) Application of coalescent methods to reveal fine-scale rate variation and recombination hotspots. *Genetics* **167,** 2067–2081.
15. Jeffreys, A. J., Neumann, R., Panayi, M., Myers, S., and Donnelly, P. (2005) Human recombination hot spots hidden in regions of strong marker association. *Nat. Genet.* **37,** 601–606.
16. McVean, G. A., Myers, S. R., Hunt, S., Deloukas, P., Bentley, D. R., and Donnelly, P. (2004) The fine-scale structure of recombination rate variation in the human genome. *Science* **304,** 581–584.
17. Jeffreys, A. J., Kauppi, L., and Neumann, R. (2001) Intensely punctate meiotic recombination in the class II region of the major histocompatibility complex. *Nat. Genet.* **29,** 217–222.
18. Jeffreys, A. J., Murray, J., and Neumann, R. (1998) High-resolution mapping of crossovers in human sperm defines a minisatellite-associated recombination hotspot. *Mol. Cell* **2,** 267–273.
19. Jeffreys, A. J., Ritchie, A., and Neumann, R. (2000) High resolution analysis of haplotype diversity and meiotic crossover in the human TAP2 recombination hotspot. *Hum. Mol. Genet.* **9,** 725–733.
20. Kauppi, L., Stumpf, M. P., and Jeffreys, A. J. (2005) Localized breakdown in linkage disequilibrium does not always predict sperm crossover hot spots in the human MHC class II region. *Genomics* **86,** 13–24.
21. May, C. A., Shone, A. C., Kalaydjieva, L., Sajantila, A., and Jeffreys, A. J. (2002) Crossover clustering and rapid decay of linkage disequilibrium in the Xp/Yp pseudoautosomal gene SHOX. *Nat. Genet.* **31,** 272–275.
22. Tiemann-Boege, I., Calabrese, P., Cochran, D. M., Sokol, R., and Arnheim, N. (2006) High-resolution recombination patterns in a region of human chromosome 21 measured by sperm typing. *PLoS Genet.* **2,** e70.
23. Jeffreys, A. J., Kauppi, L., and Neumann, R. (2001) Intensely punctate meiotic recombination in the class II region of the major histocompatibility complex. *Nat. Genet.* **29,** 217–222.
24. Jeffreys, A. J., and May, C. A. (2003) DNA enrichment by allele-specific hybridization (DEASH): a novel method for haplotyping and for detecting low-frequency base substitutional variants and recombinant DNA molecules. *Genome Res.* **13,** 2316–2324.
25. Jeffreys, A. J., and May, C. A. (2004) Intense and highly localized gene conversion activity in human meiotic crossover hot spots. *Nat. Genet.* **36,** 151–156.

26. Holloway, K., Lawson, V. E., and Jeffreys, A. J. (2006) Allelic recombination and de novo deletions in sperm in the human beta-globin gene region. *Hum. Mol. Genet.* **15,** 1099–1111.

27. Jeffreys, A. J., and Neumann, R. (2005) Factors influencing recombination frequency and distribution in a human meiotic crossover hotspot. *Hum. Mol. Genet.* **14,** 2277–2287.

28. Jeffreys, A. J., Tamaki, K., MacLeod, A., Monckton, D. G., Neil, D. L., and Armour, J. A. (1994) Complex gene conversion events in germline mutation at human minisatellites. *Nat. Genet.* **6,** 136–145.

29. Cheng, S., Chang, S. Y., Gravitt, P., and Respess, R. (1994) Long PCR. *Nature* **369,** 684–685.

30. Dean, F. B., Hosono, S., Fang, L., Wu, X., Faruqi, A. F., Bray-Ward, P., et al. (2002) Comprehensive human genome amplification using multiple displacement amplification. *Proc. Natl. Acad. Sci. USA* **99,** 5261–5266.

31. Jeffreys, A. J., and Neumann, R. (2002) Reciprocal crossover asymmetry and meiotic drive in a human recombination hot spot. *Nat. Genet.* **31,** 267–271.

32. Wood, W. I., Gitschier, J., Lasky, L. A., and Lawn, R. M. (1985) Base composition-independent hybridization in tetramethylammonium chloride: a method for oligonucleotide screening of highly complex gene libraries. *Proc. Natl. Acad. Sci. USA* **82,** 1585–1588.

Index

Note: The letters '*t*', '*f*' and 'n' following the locators refer to tables, figures and note numbers respectively.

D

H

I

J

L

M

N

O

P

Q

R

S

T